HVAC
Control System
Design Diagrams

Other HVAC Titles of Interest

CHEN & DEMSTER, *Variable Air Volume Systems for Environmental Quality*

COHEN, *Ventilation for Indoor Air Quality*

GEBHART, *Heat Conduction and Mass Diffusion*

GLADSTONE & HUMPHREYS, *Mechanical Estimating Guidebook for HVAC*

GLADSTONE, ET AL., *HVAC Testing, Adjusting, and Balancing Field Manual*

GRIMM & ROSALER, *HVAC Systems and Components Handbook*

HAINES & WILSON, *HVAC Systems Design Handbook*

HARTMAN, *Direct Digital Controls for HVAC Systems*

LEVENHAGEN & SPETHMANN, *HVAC Controls and Systems*

MOORE, *Environmental Control Systems: Heating, Cooling, and Lighting*

MULL, *HVAC Principles and Applications Manual*

PARMLEY, *HVAC Design Data Sourcebook*

RISHEL, *HVAC Pump Handbook*

ROSALER, *HVAC Maintenance and Operations Handbook*

SUN, *Air Handling Systems*

WANG, ET AL, *Handbook of Air Conditioning and Refrigeration*

WATSON & CHAPMAN, *Radiant Heating and Cooling Handbook*

To order or receive additional information on these or any other McGraw-Hill titles, in the United States please call 1-800-722-4726. In other countries, contact your local McGraw-Hill representative.

HVAC
Control System
Design Diagrams

John I. Levenhagen

McGraw-Hill

New York San Francisco Washington, D.C. Auckland Bogotá Caracas Lisbon London
Madrid Mexico City Milan Montreal New Delhi San Juan Singapore Sydney Tokyo Toronto

Library of Congress Cataloging-in-Publication Data

Levenhagen, John I.
    HVAC control system diagrams / John I. Levenhagen.
        p. cm.
    Includes index.
    ISBN 0-07-038129-1 (alk. paper)
        1. Heating—Control—Charts, diagrams, etc.   2. Ventilation—
    Control—Charts, diagrams, etc.   3. Air conditioning—Control—
    Charts, diagrams, etc.   I. Title
TH7466.5L4797   1998
697—dc21   98-34733
                                                       CIP

McGraw-Hill

A Division of The McGraw-Hill Companies

1 2 3 4 5 6 7 8 9 0   ACM/AGM   9 0 3 2 1 0 9 8

ISBN 0-07-038129-1

The sponsoring editor for this book was Linda Ludewig, the editing supervisor was Sally Glover, and the production supervisor was Pamela Pelton. It was set in Melior per the CON2 Specs by Patricia Caruso and Michele Zito of McGraw-Hill's Professional Book Group compositions unit, Hightstown, N.J.

Printed and bound by Quebecor/Martinsburg.

McGraw-Hill books are available at special quantity discounts to use as premiums and sales promotions, or for use in corporate training programs. For more information, please write to the Director of Special Sales, McGraw-Hill, 11 West 19th Street, New York, NY 10011. Or contact your local bookstore.

This book is printed on recycled, acid-free paper containing a minimum of 50 percent recycled, de-inked fiber.

Contents

SECTION III *Air-Handling Units*

Preface

This handbook is intended to be a design manual for the practicing design engineer who is interested in historical or typical ways to control commercial HVAC systems. The book presents generic systems sketches with diagram descriptions and suggestions for further research and other ways to control a particular system.

The text is broken down into three basic divisions that classify, as much as possible, the equipment used in commercial HVAC systems. The divisions are:

1. Primary equipment
2. Terminal equipment
3. Air-handling systems

The table of contents is arranged with section headings and chapters so that a designer can refer to an item in question easily without paging through the entire book.

An attempt was made in the creation of this handbook to include most all of the control systems for each and every HVAC system (and equipment) used in commercial HVAC designs. It must be remembered, however, that designers and engineers are basically innovative, and hence many control systems have been designed by individual engineers to be different and even unique for particular projects. Therefore, this book *does not contain all possible control systems*.

Additions can be made to any design in the book—for example, in some systems, adding isolation valves when switching from heating to cooling, or summer to winter. Switching can also be performed by clocks, computers, or other devices.

In all cases in this handbook, the hardware used for the control systems will be a matter of designer choice. The designer will be able to choose whether or not he or she wishes to use pneumatic, electric, or digital electronic controls. All the systems shown in this handbook can be converted from one type of controls to another.

Acknowledgments

First of all I wish to recognize the encouragement that I received in putting together this reference book from my wife Theresa. Her help and assistance over the years it took to gather the data was invaluable.

I also wish to recognize the assistance of the many HVAC and controls engineers from Johnson Controls, Honeywell, Landis Gyr & Staffea (formerly Powers Regulator Co.), as well as Seibe Co (Formerly Barber Coleman Co.).

Michael Prentice a skilled CAD operator was a great part of this book, as he created all the excellent graphics shown in the book.

That skill was invaluable, in as much as the book is 90% graphics and very little text. The reader will appreciate that fact when looking for drawings that are useful for a particular application.

SECTION I

Primary Systems

CHAPTER 1

Hot Water Boilers

Hot Water Boiler Controls

This diagram shows typical controls on a low-pressure hot water boiler. There are two pressure controls—an operating one and a high-limit safety control. There is an "altitude gage" and a low-water cutoff. The boiler, unlike steam boilers, must be filled with water at all times.

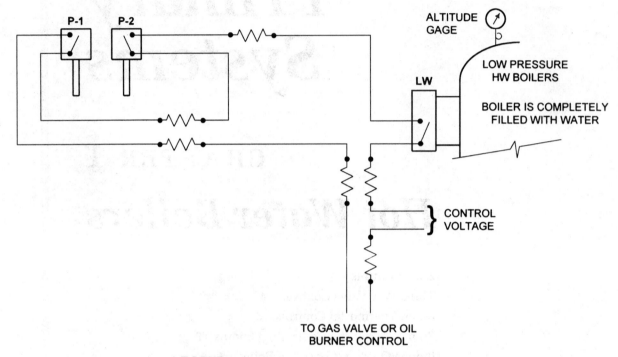

LEGEND:

P-1 = OPERATING TEMPERATURE CONTROL
P-2 = HI LIMIT TEMPERATURE CONTROL
LW = LOW WATER CUTOFF

Zone Controls

This diagram shows typical controls on a low-pressure hot water boiler being used with room zone controls. The room thermostats T-1 and T-2 operate zone valves V-1 and V-2 with the circulator controlled so that if any one valve opens, the circulator runs. The boiler is fired at a constant temperature by its aquastat.

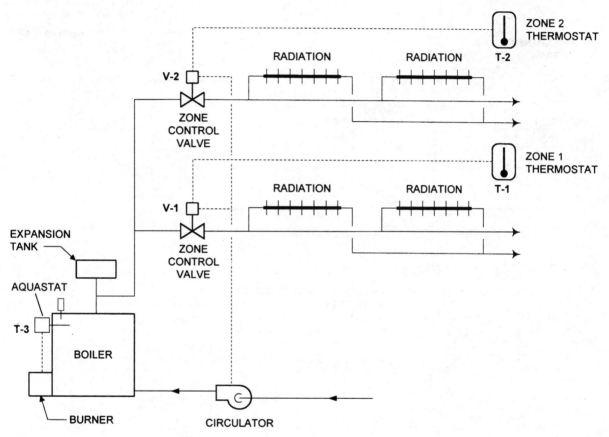

LEGEND:

T-1   = MODULATING ROOM THERMOSTAT
T-2   = MODULATING ROOM THERMOSTAT
T-3   = BOILER THERMOSTAT
V-1   = ZONE VALVE
V-2   = ZONE VALVE

Three-Way Valve Controls

This diagram shows typical controls on a low-pressure hot water boiler being used with room zone controls. The room thermostat T-1 controls a 3-way valve V-1 and the circulator runs all the time. The boiler is fired at a constant temperature by its aquastat.

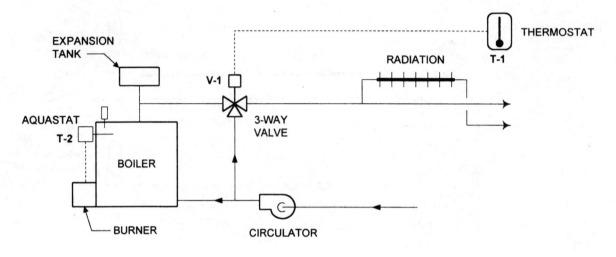

LEGEND:

T-1 = MODULATING ROOM THERMOSTAT
T-2 = BOILER AQUASTAT
V-1 = 3-WAY VALVE

Room Thermostat Controls

This diagram shows typical controls on a low-pressure hot water boiler being used with room controls. The room thermostat T-1 controls the boiler through its aquastat directly. The circulator runs as long as there is hot water sensed by a strapon on the boiler discharge line.

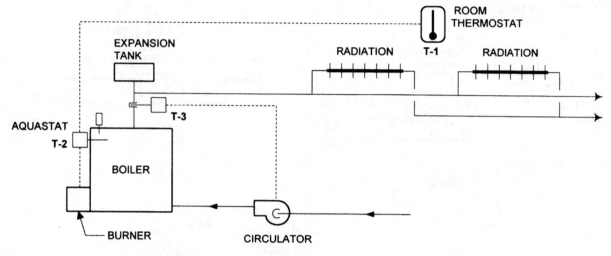

LEGEND:

T-1   = TWO POSITION ROOM THERMOSTAT
T-2   = BOILER AQUASTAT
T-3   = STRAP-ON THERMOSTAT

Room Thermostat Controlling Pumps

This diagram shows typical controls on a low-pressure hot water boiler being used with a indoor/outdoor controllers T-2 and T-3 to give a variable hot water temperature in accordance with outdoor air temperature. The room thermostat T-1 controls the circulator directly.

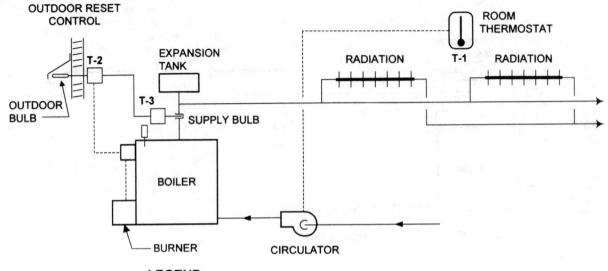

LEGEND:

T-1 = TWO POSITION ROOM THERMOSTAT
T-2 = INDOOR/OUTDOOR THERMOSTAT
T-3 = BOILER AQUASTAT

Indoor/Outdoor Control of Boiler

This diagram shows typical controls on a low pressure hot water boiler being used with an indoor/out-door controller T-2 operating through relay R-1 the 3-way valve V-1. The room thermostat T-1 modulates the 3-way valve.

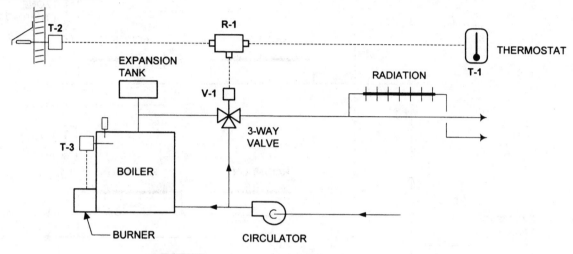

LEGEND:

T-1 = MODULATING ROOM THERMOSTAT
T-2 = OUTDOOR THERMOSTAT
T-3 = BOILER AQUASTAT
V-1 = 3-WAY MIXING VALVE
R-1 = SIGNAL SELECTOR RELAY

Zone Control by Circulator Control

This diagram shows typical controls on a low-pressure hot water boiler being used with the zone room thermostats T-1 and T-2 controlling the two circulators directly. The aquastat T-3 fires the boiler at a constant temperature.

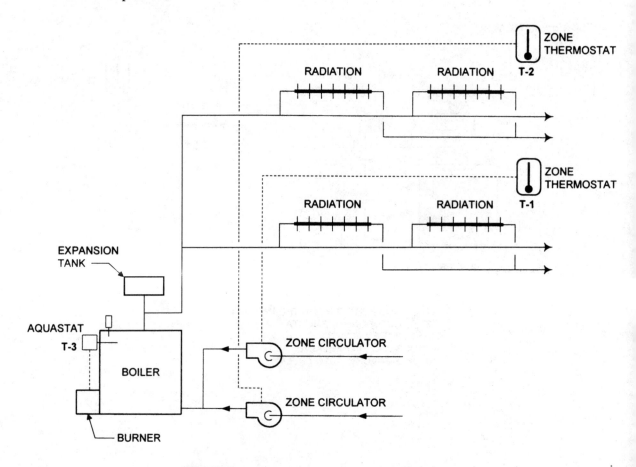

LEGEND:

T-1   = TWO POSITION ZONE ROOM THERMOSTAT
T-2   = TWO POSITION ZONE ROOM THERMOSTAT
T-3   = BOILER AQUASTAT

Three-way Valve Control through Indoor/Outdoor Stat

This diagram shows typical controls on a low-pressure hot water boiler being used with a master-sub-master control system operating a 3-way valve V-1 on the hot water system. For clarity, room thermostats are not shown.

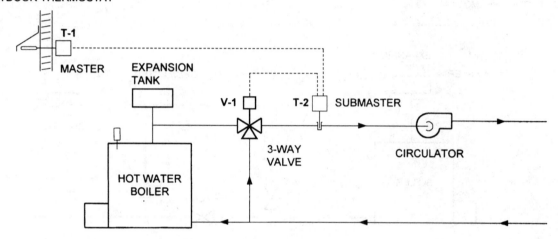

LEGEND:

T-1	= OUTDOOR MASTER THERMOSTAT
T-2	= HOT WATER SUPPLY SUBMASTER THERMOSTAT
V-1	= 3-WAY HW SUPPLY VALVE

Control through Return Water

This diagram shows typical controls on a low-pressure hot water boiler being used with a master-sub-master control system T-1 and T-2 operating two valves, V-1 and V-2, to bypass hot water when not needed. There is a differential pressure controller DP-1 operating a bypass valve V-3.

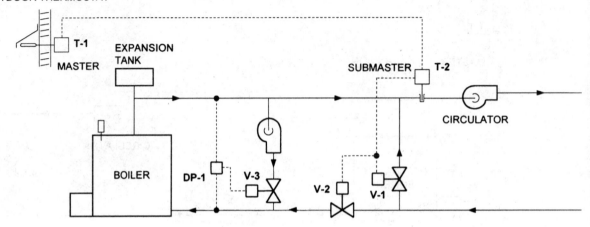

LEGEND:

T-1 = OUTDOOR MASTER THERMOSTAT
T-2 = INDOOR SUBMASTER THERMOSTAT
V-1 = MODULATING BYPASS VALVE
V-2 = MODULATING RETURN LINE VALVE
V-3 = BOILER BYPASS VALVE
DP-1 = DIFFERENTIAL PRESSURE CONTROL

Indoor/Outdoor Controls

This diagram shows typical controls on a low-pressure hot water boiler being used with a master-sub-master control system T-1 and T-2 operating two valves, V-1 and V-2, to bypass hot water when not needed. There are two separate circulators, and one of them near the boiler automatically bypasses the boiler when needed.

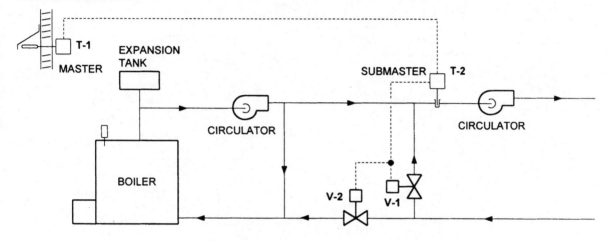

LEGEND:

T-1 = OUTDOOR MASTER THERMOSTAT
T-2 = INDOOR SUBMASTER THERMOSTAT
V-1 = MODULATING BYPASS VALVE
V-2 = MODULATING RETURN LINE VALVE

CHAPTER 2

Steam Boilers

Indoor/Outdoor Controls

This diagram shows typical controls on a low-pressure steam boiler with feed pump and condensate tank. An outdoor thermostat, T-2, resets the control point of a room thermostat, T-1, which controls a zone valve, V-1, to feed the radiators. The boiler is operated under a constant pressure by the pressure stat on the boiler.

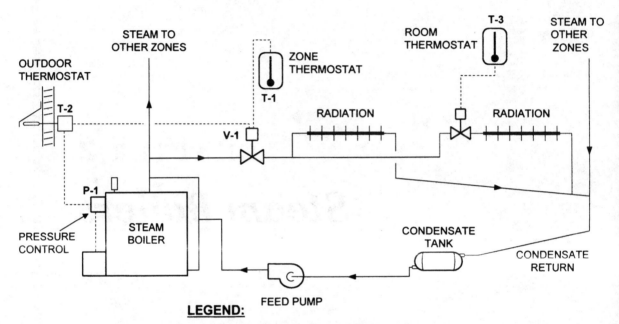

LEGEND:

T-1 = TWO POSITION ROOM THERMOSTAT
T-2 = OUTDOOR THERMOSTAT
T-3 = ROOM THERMOSTAT
V-1 = ZONE VALVE
P-1 = BOILER PRESSURE CONTROL

CHAPTER 3

Heat Exchangers

Hot-Water-to-Hot-Water Exchanger

This diagram shows typical controls of a water-to-water heat exchanger. Thermostat T-1 modulates valve V-1, supplying hot water to the exchanger.

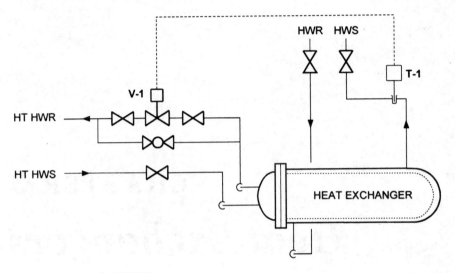

LEGEND:

T-1   = MODULATING CAPILLARY THERMOSTAT
V-1   = HT HOT WATER VALVE

Steam-to-Hot-Water Exchanger

This diagram shows typical controls of a steam-to-water heat exchanger. Thermostat T-1 senses the OA temperature and resets the control point of the submaster thermostat T-2 to give a variable hot water temperature in accordance with the OA temperature by controlling valve V-1, supplying the steam to the converter. The flow switch F-1 also enters the picture by making sure that the water is flowing before allowing the control system to operate.

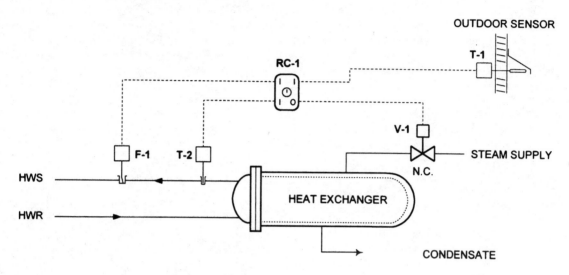

LEGEND:

T-1	=	OUTDOOR SENSOR
T-2	=	HOT WATER SUPPLY SENSOR
RC-1	=	RECEIVER CONTROLLER
F-1	=	FLOW SENSOR

CHAPTER 4

Dual CW/HW Piping Systems

Common Pipe Switchover Systems

This diagram shows typical controls of a hot water, chilled water common pipe system with valves on the various loads (heating cooling coils as an example) and a switch over valve on the return to the chiller(s) and heater(s) when switching from heating to cooling and visa versa. Care needs to be taken when switching over from one to the to the other, that the return lines are not too hot or too cold to damage the boiler or the chiller.

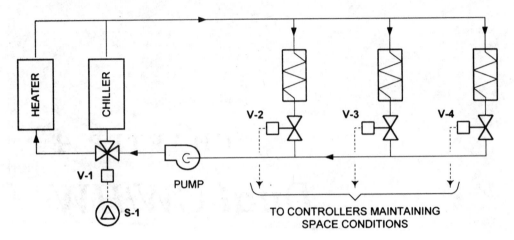

LEGEND:

S-1 = SUMMER/WINTER (HEAT/COOL) SWITCH
V-1 = 3-WAY SWITCHOVER VALVE ON RETURN LINES TO CHILLER & BOILER (HEATER)
V-2 = TWO WAY VALVE ON UNIT COILS
V-3 = TWO WAY VALVE ON UNIT COILS
V-4 = TWO WAY VALVE ON UNIT COILS

Three-Pipe Systems

This diagram shows typical controls of a hot water, chilled water three pipe system with common return lines but separate chilled water and hot water supply lines. The valves V-1 and V-2 on the coil loads are sequences so that they are never both open at the same time. However, care must be taken to insure that the return lines supply either the chiller or the boiler when the temperatures are correct. That is the purpose of T-1 T-2 and T-3 as well as 3-way valve V-3. T-1 senses the return water temperature and operates through the safety thermostats T-2 and T-3 to switch valve V-3.

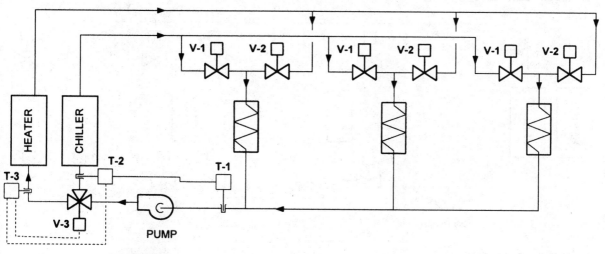

LEGEND:

T-1 = MODULATING PIPE THERMOSTAT
T-2 = HI LIMIT SAFETY THERMOSTAT
T-3 = LOW LIMIT SAFETY THERMOSTAT
V-1 = SPECIAL CW VALVE ON UNIT COIL WITH HESITATION SPRING
V-2 = SPECIAL HW VALVE ON UNIT COIL WITH HESITATION SPRING
V-3 = 3-WAY BYPASS VALVE ON RETURN LINE

Four-Pipe Systems

This diagram shows typical controls of a hot water, chilled water four pipe system with 4 valves on each coil. The valves have hesitation springs in them and the supply line valves and the return line valves operate as a team. There is a dead-band in the operation of both the supply and return line valves so that hot water and chilled water are never mixed. The hot water and the chilled water are modulated to the unit coil by the room controllers.

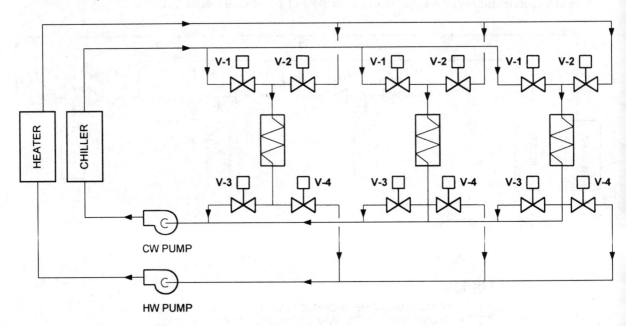

LEGEND:

V-1   = CW SUPPLY TWO WAY VALVE
V-2   = HW SUPPLY TWO WAY VALVE
V-3   = CW RETURN TWO WAY VALVE
V-4   = HW RETURN TWO WAY VALVE

CHAPTER 5

Constant Volume CW/HW Piping Systems

Secondary Circuits with Return DP Control

This diagram shows typical controls of a hot water or chilled water secondary piping system from the primary systems. The secondary pump takes water from the primary piping, and as the building loads change, the DP-1 (differential pressure controller) senses the change through the orifice plate and modulates valve V-1 accordingly.

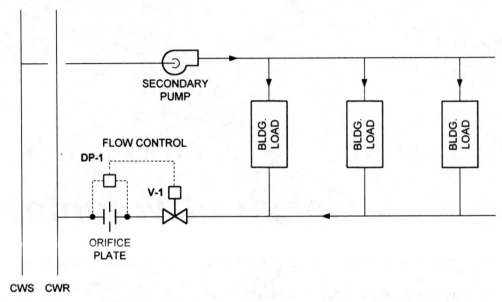

LEGEND:

V-1   = MODULATING SECONDARY RETURN LINE VALVE
DP-1 = DIFFERENTIAL PRESSURE CONTROLLER

Secondary Circuits with Return DP Control

This diagram shows typical controls of a hot water or chilled water secondary piping system from the primary systems. The secondary pump takes water from the primary piping, and as the building loads change, the DP-1 (differential pressure controller) senses the change by sensing the differential pressure between the secondary supply and return lines and modulates valve V-1 accordingly.

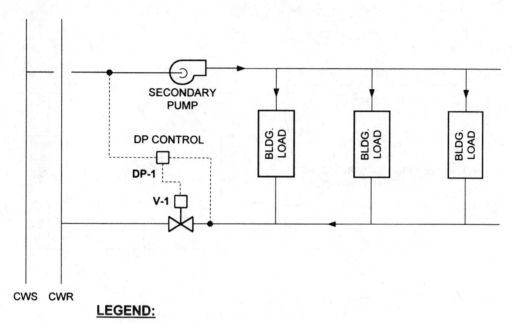

CWS   CWR

LEGEND:

V-1 = SECONDARY SYSTEM RETURN LINE VALVE
DP-1 = DIFFERENTIAL PRESSURE CONTROLLER

Secondary Circuits with DP Control on Primary Supply and Return Piping

This diagram shows typical controls of a hot water or chilled water secondary piping system from the primary systems. The secondary pump takes water from the primary piping, and as the building loads change, the DP-1 (differential pressure controller) senses the change by sensing the differential pressure between the primary supply and return lines. A VFD, as an example, varies the speed of the secondary pump.

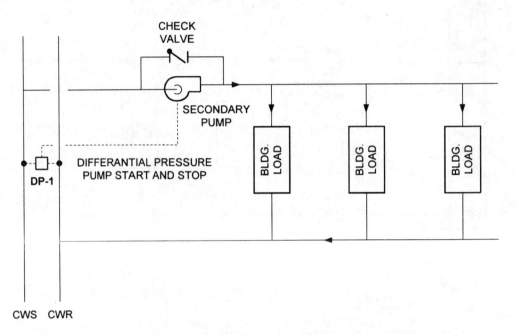

CHAPTER 6

Variable Volume CW/HW Piping Systems

VVF with Primary Secondary Systems

Valve V-1 is set to maintain a fixed temperature to the secondary system by the operator. Valve V-3 is controlled by the room controller to maintain space conditions.

A thermostat T-1 in the return to the primary system modulates the valve V-2 to restrict the flow on chilled water back to the system as required.

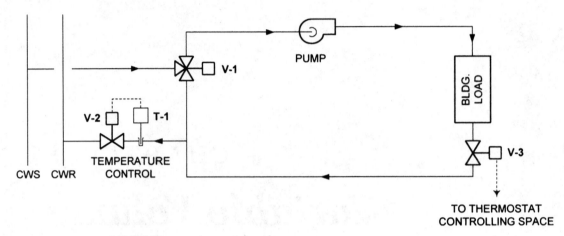

LEGEND:

T-1 = MODULATING PIPE THERMOSTAT
V-1 = MODULATING 3-WAY VALVES
V-2 = CW SECONDARY RETURN VALVE
V-3 = TWO WAY BUILDING LOAD RETURN LINE VALVE

VVF Through Chillers

This diagram shows typical controls of a chiller with a differential pressure sensor in a line between the supply and return system to vary the flow as the loads change.

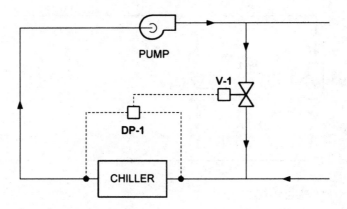

LEGEND:

V-1   = CW BYPASS VALVE
DP-1 = DIFFERENTIAL PRESSURE CONTROLLER

VVF Through Bypass

This diagram shows typical controls of a chilled water hot water system using a differential pressure controller operating valve V-1 between the supply and return lines.

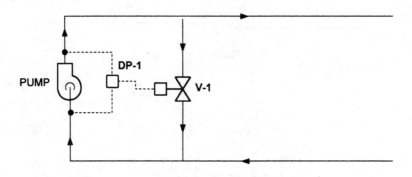

LEGEND:

V-1   = BYPASS VALVE
DP-1 = DIFFERENTIAL PRESSURE CONTROLLER

VVF Control of Direct Return Systems

This diagram shows typical controls of a chilled water hot water system using a differential pressure controllers on a direct return system and unit coils with control valves V-1.

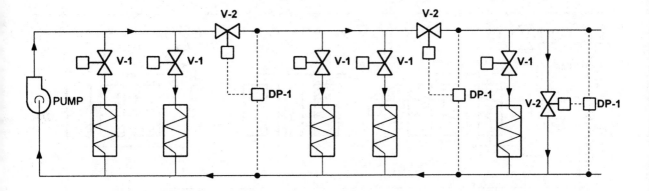

LEGEND:

V-1 = UNIT LOAD VALVES
V-2 = SYSTEM BYPASS VALVES
DP-1 = DIFFERENTIAL PRESSURE SENSOR/CONTROLLER

VVF of Some Subcircuits

This diagram shows typical controls of a chilled water hot water system using a differential pressure controllers controlling bypass valves V-1 on a direct return sub systems.

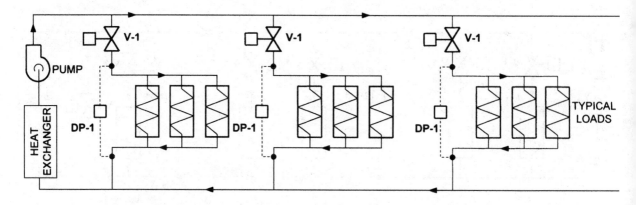

LEGEND:

V-1 = MODULATING LOAD VALVES
DP-1 = DIFFERENTIAL PRESSURE SENSOR/CONTROLLER

VVF Control of Reverse Return Systems

This diagram shows typical controls of a chilled water hot water system using a differential pressure controller sensing the pressures between the supply and return lines and controlling bypass valve V-1 on the supply line. The suggested use of valves V-2 and V-3 is to keep a supply of water going to the end of the circuit so that the water does not get stagnant at that point.

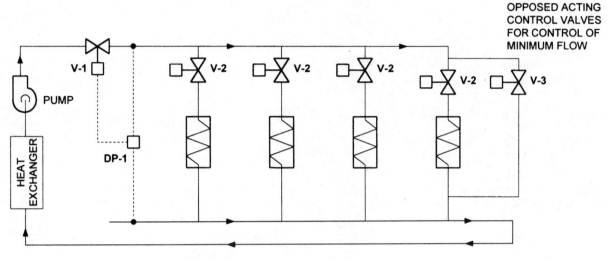

LEGEND:

V-1 = PRESSURE CONTROL VALVE
V-2 = LOAD VALVES
V-3 = BYPASS VALVE
DP-1 = DIFFERENTIAL PRESSURE SENSOR/CONTROLLER

CHAPTER 7

Chiller Controls

Control of Chillers through Valve Monitoring

This diagram shows typical controls of a chiller with multiple loads by monitoring the valve V-1 positions on the chilled water coils. The thermostat T-2 acts as a low limit to protect the chiller.

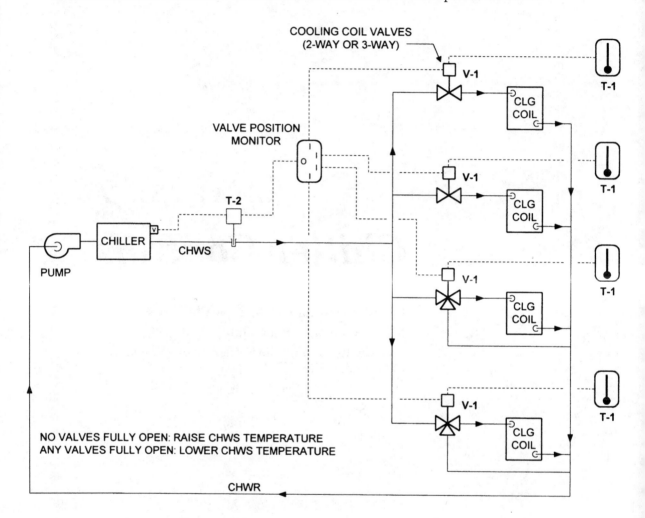

NO VALVES FULLY OPEN: RAISE CHWS TEMPERATURE
ANY VALVES FULLY OPEN: LOWER CHWS TEMPERATURE

LEGEND:

T-1   =   UNIT ROOM THERMOSTAT
T-2   =   MODULATING LOW LIMIT CWS THERMOSTAT
V-1   =   UNIT COIL VALVES(TWO WAY OR 3-WAY)

Constant Flow through Multiple Chillers

This diagram shows typical controls of multiple chillers with a constant flow system. The thermostats T-2 on the return lines to the chillers operate the chillers with low limit thermostats T-1 limiting the temperature of the chilled water. This system requires 3-way valves on the loads, as there is no variable flow system.

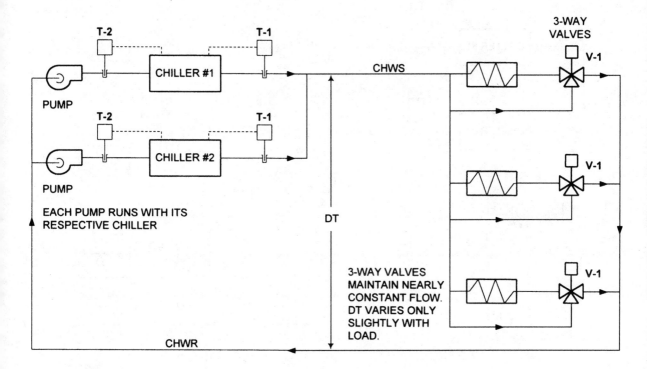

LEGEND:

T-1 = CHWS MODULATING USED WITH CENTRIFUGAL CHILLERS
T-2 = CHWR MODULATING USED WITH RECIPROCATING CHILLERS
V-1 = UNIT COIL 3-WAY VALVES

VVF Control of Multiple Chillers

This diagram shows typical controls of multiple chillers within a variable volume flow system. Two-way valves are used on the loads and a differential pressure controller DP-1 senses the pressures across the supply and return lines to control a bypass valve V-1 as the loads (valves V2) change.

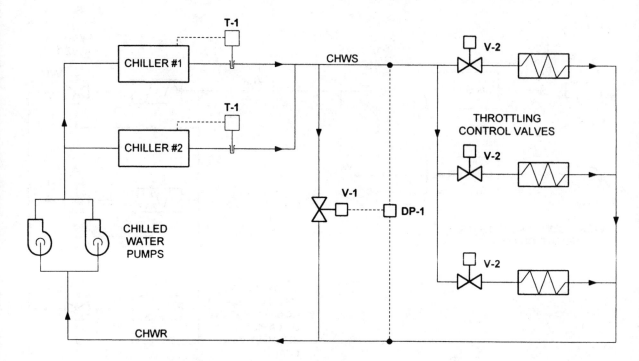

LEGEND:

T-1 = LOW LIMIT CHILLER THERMOSTATS
V-1 = DIFFERENTIAL BYPASS VALVE
V-2 = UNIT COIL 2-WAY VALVES
DP-1 = DIFFERENTIAL PRESSURE CONTROLLER

VVF Control of Multiple Chillers with VVF Control of Secondary Pumps

This diagram shows typical controls of multiple chillers within a variable volume secondary flow system and constant flow primary system. This keeps the flow through the chillers at a fixed level and allows for variable flow in the secondary systems. The differential pressure controller DP-1 controls the variable speed secondary pumps and loads have two way valves on them. There is a "hydraulic coupling" between the points A and B on the diagram which permits the water to flow either direction as required by the secondary pumps.

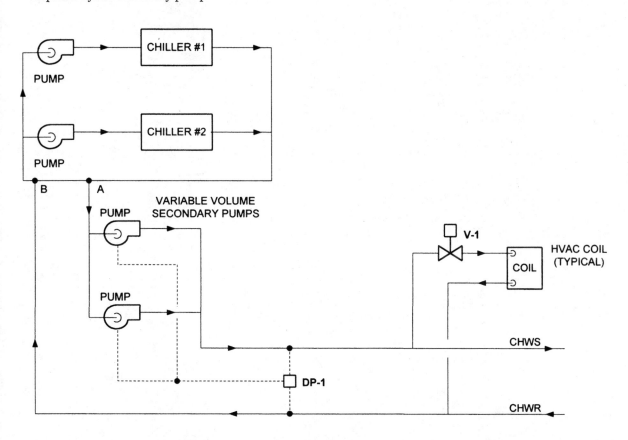

LEGEND:

DP-1 = DIFFERENTIAL PRESSURE SENSOR/CONTROLLER
V-1   = UNIT COIL TWO WAY VALVES

CHAPTER 8

Miscellaneous Controls

Radiant Systems Controls

This diagram shows typical controls of a hot water floor radiant system as might be used in classrooms at a school. An outdoor thermostat, T-1, resets the control point of a submaster thermostat, T-2, which operates a 3-way valve on the radiant system, mixing hot water and return water to the floor coils. It is possible to also "zone" this system with zone valves in each area.

OUTDOOR
THERMOSTAT

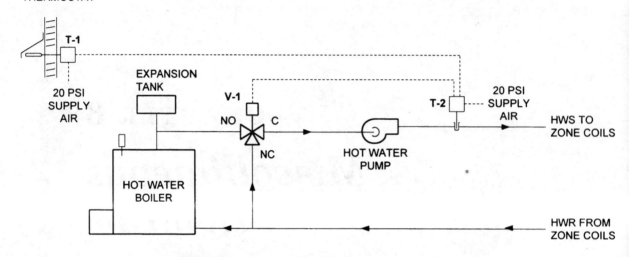

LEGEND:

T-1 = MASTER OUTDOOR THERMOSTAT
T-2 = SUBMASTER HOT WATER SUPPLY THERMOSTAT
V-1 = 3-WAY MIXING VALVE

Run-around Coil Controls

This diagram shows typical "run-around" controls where the building exhaust air (typically warm) can be used to heat water to be fed back into the system. Thermostat T-1 controls valve V-1 to heat the hot water to the reheat coil that can heat the in-coming ventilation air.

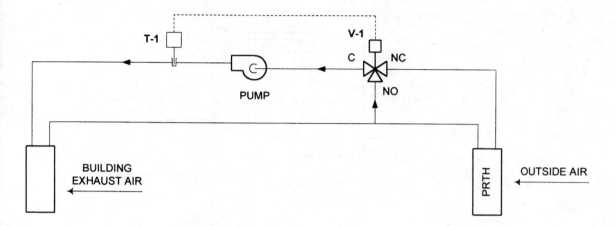

LEGEND:

T-1   = MODULATING THERMOSTAT IN HW LINE
V-1   = MODULATING 3-WAY VALVE

Evaporative Condenser Controls

This diagram shows a typical method (although there are others) of controlling the head pressure in an evaporative condenser by sensing the pressure in the system and controlling the evap sump pump. The fan motor and the inlet dampers to the evap can also be sequenced.

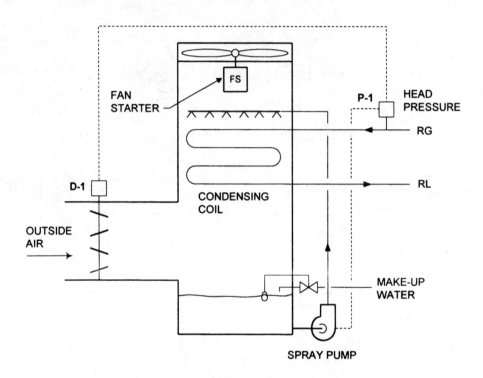

LEGEND:

D-1 = OUTSIDE AIR DAMPER MOTOR
P-1 = COMPRESSOR HEAD PRESSURE CONTROL

SECTION II

Terminal Systems

CHAPTER 9

Radiation

Steam Radiation—Pneumatic Controls

This diagram shows a simple system where room thermostats modulate valves V-1 on the radiation in the space. The thermostats and valves are pneumatic.

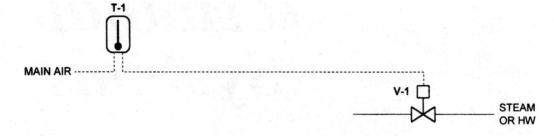

LEGEND:

T-1   = MODULATING PNEUMATIC ROOM THERMOSTAT
V-1   = PNEUMATIC STEAM OR HW VALVE ON RADIATION

Steam Radiation—Electric Controls

This diagram shows a simple system where room thermostats modulate valves V-1 on the radiation in the space. The thermostats and valves are electric.

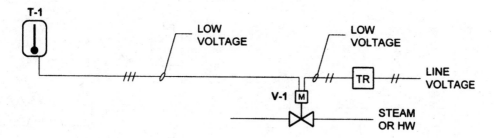

<u>**LEGEND:**</u>

T-1 = PROPORTIONAL MODULATING ROOM THERMOSTAT
V-1 = MODULATING 3-WIRE STEAM OR HW VALVE
TR = LINE/LOW VOLTAGE TRANSFORMER

Electric Baseboard Controls

This diagram shows a simple system where room thermostats pressure electric switches to cycle the electric coils on the baseboard heating. The thermostat can also operate a SCR S-1 to modulate the electric coils.

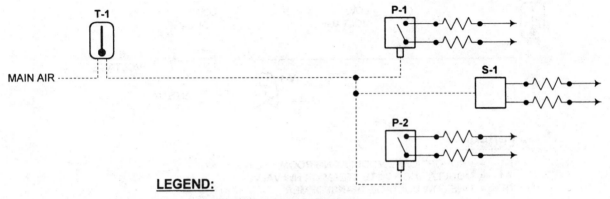

LEGEND:

T-1 = PNEUMATIC ROOM THERMOSTAT
P-1 = PNEUMATIC/ELECTRIC SWITCH
P-2 = OPTIONAL PNEUMATIC/ELECTRIC SWITCH
S-1 = OPTIONAL SCR CONTROLLER

Fan Coil Units

Heating Only—Electric Controls

This diagram shows a simple fan coil system that is supplied with hot water in the heating season. There is no cooling with this unit. The room thermostat T-1 cycles the unit on and off on demand and the strap-on T-2 makes sure there is hot water available before the unit cycles on.

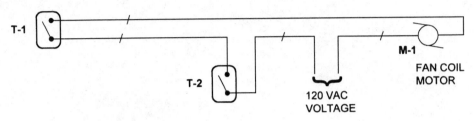

LEGEND:

T-1 = LINE VOLTAGE ROOM THERMOSTAT
T-2 = STRAP-ON THERMOSTAT
M-1 = LINE VOLTAGE FAN COIL MOTOR

Heating Only—Pneumatic Controls

This diagram shows a simple fan coil system that is supplied with hot water in the heating season. There is no cooling with this unit. The room thermostat T-1 cycles the unit on and off on demand through the PE switch PE-1 and the strap-on T-2 makes sure there is hot water available before the unit cycles on. The controls are pneumatic/electric in this case.

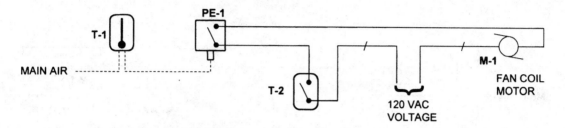

LEGEND:

T-1 = LINE VOLTAGE ROOM THERMOSTAT
T-2 = STRAP-ON THERMOSTAT
M-1 = LINE VOLTAGE FAN COIL MOTOR
PE-1 = PRESSURE/ELECTRIC SWITCH

Heating Only—Pneumatic Return Air Controls

This diagram shows a simple fan coil system that is supplied with hot water in the heating season.

There is no cooling with this unit. The return air thermostat T-1 cycles the unit on and off on demand through the PE switch PE-1 and the strap-on T-2 makes sure there is hot water available before the unit cycles on. The controls are pneumatic/electric in this case.

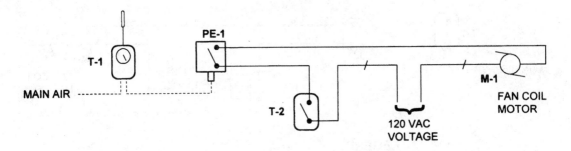

LEGEND:

T-1   = PNEUMATIC RETURN AIR THERMOSTAT
T-2   = STRAP-ON THERMOSTAT
M-1   = LINE VOLTAGE FAN COIL MOTOR
PE-1 = PRESSURE/ELECTRIC SWITCH

Heating Only—Electric Return Air Controls

This diagram shows a simple fan coil system that is supplied with hot water in the heating season. There is no cooling with this unit. The return air thermostat T-1 cycles the unit on and off on demand. The strap-on T-2 makes sure there is hot water available before the unit cycles on. The controls are electric in this case.

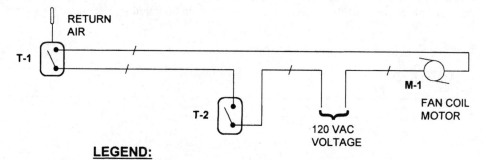

RETURN AIR

T-1

T-2

120 VAC VOLTAGE

M-1

FAN COIL MOTOR

LEGEND:

T-1 = ELECTRIC RETURN AIR THERMOSTAT
T-2 = STRAP-ON THERMOSTAT
M-1 = LINE VOLTAGE FAN COIL MOTOR

Heating/Cooling, No Valves Electric Controls

This diagram shows a simple fan coil system that is supplied with hot water in the heating season, and chilled water in the cooling system. The room air thermostat T-1 has built into it a switch that allows the thermostat to in effect reverse its action when the switch is operated from heating to cooling and visa versa. The thermostat is a single-pole double-throw type because on a rise in temperature in the heating season the stat must stop the fan. On a rise in temperature in the cooling season, the stat must operate the fan. This means that the switch on the stat must be operated by the occupant summer and winter.

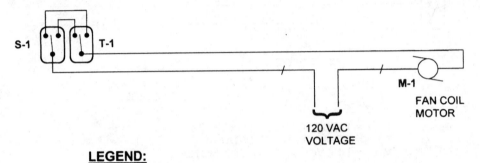

LEGEND:

T-1 = S.P.D.T. ELECTRIC LINE VOLTAGE ROOM THERMOSTAT
S-1 = S.P.D.T. ELECTRIC SWITCH BUILT INTO ROOM THERMOSTAT
M-1 = LINE VOLTAGE FAN COIL MOTOR

Heating/Cooling Electric Strap-on Controls

This diagram shows a simple fan coil system that is supplied with hot water in the heating season, and chilled water in the cooling system. The room air thermostat T-1 has a single-pole double-throw switch in it and with the strap-on T-2 that is also SPDT the system is switched from heating to cooling and visa versa by the strap-on. When hot water is in the pipes, the strap-on switches the system to heating and visa versa.

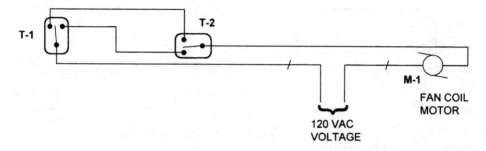

LEGEND:

T-1   = S.P.D.T. ELECTRIC LINE VOLTAGE ROOM THERMOSTAT
T-2   = S.P.D.T. ELECTRIC STRAP-ON THERMOSTAT
M-1   = LINE VOLTAGE FAN COIL MOTOR

Heating/Cooling with R.A. Electric with Strap-on

This diagram shows a simple fan coil system that is supplied with hot water in the heating season, and chilled water in the cooling system. The return air thermostat T-1 has a single-pole double-throw switch in it, and with the strap-on T-2 that is also SPDT the system is switched from heating to cooling and visa versa by the strap-on. When hot water is in the pipes, the strap-on switches the system to heating and visa versa.

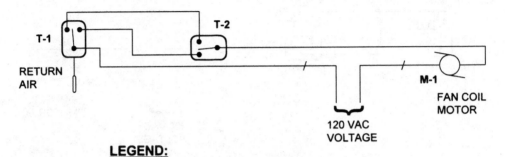

LEGEND:

T-1 = S.P.D.T. ELECTRIC LINE VOLTAGE R.A. THERMOSTAT
S-1 = S.P.D.T. STRAP-ON THERMOSTAT
M-1 = LINE VOLTAGE FAN COIL MOTOR

Heating/Cooling, Valves, Dual Air

This diagram shows a simple fan coil system that is supplied with hot water in the heating season and chilled water in the cooling season. The room thermostat is a special pneumatic thermostat that switches itself from direct acting to reverse acting when the supply air is changed from 15 psia to 20 psia. The valve can be one action only (normally open as an example).

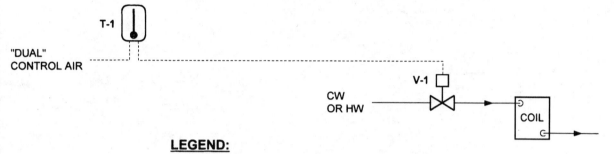

LEGEND:

T-1 = HEATING/COOLING ROOM THERMOSTAT
V-1 = MODULATING PNEUMATIC VALVE

Heating/Cooling, Valves, R.A. Dual Air

This diagram shows a simple fan coil system that is supplied with hot water in the heating season, and chilled water in the cooling season. The return air thermostat T-1 is a special pneumatic thermostat that switches itself from direct acting to reverse acting when the supply air is changed from 15 psia to 20 psia. The valve can be one action only (normally open as an example).

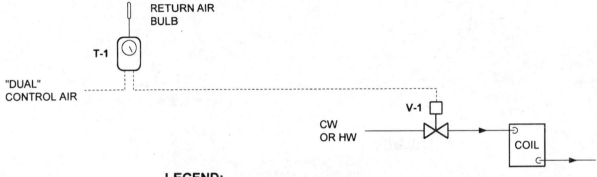

LEGEND:

T-1 = HEATING/COOLING UNIT RETURN AIR THERMOSTAT
V-1 = MODULATING PNEUMATIC VALVE

Heating/Cooling, Three-Pipe System, Room Stat

This diagram shows a fan coil system that is supplied with both hot water and chilled water all the time. The room thermostat T-1 controls a 3-way valve V-1 on the supply to the coil of the unit. The valve has a special "hesitation spring" in it so that at no time does the valve mix the hot water and chilled water.

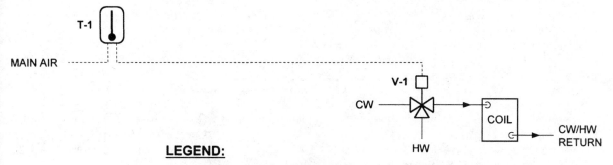

LEGEND:

T-1   = MODULATING ROOM THERMOSTAT
V-1   = MODULATING "SPECIAL" HW/CW VALVE

Heating/Cooling, Three-Pipe System, R.A. Stat

This diagram shows a fan coil system that is supplied with both hot water and chilled water all the time. The R.A. thermostat T-1 controls a 3-way valve V-1 on the supply to the coil of the unit. The valve has a special "hesitation spring" in it so that at no time does the valve mix the hot water and chilled water.

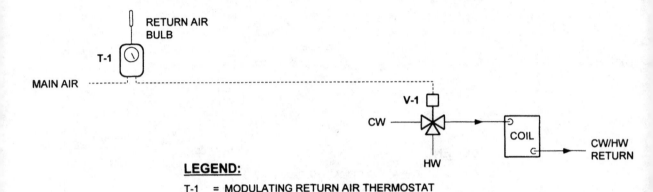

LEGEND:

T-1   = MODULATING RETURN AIR THERMOSTAT
V-1   = MODULATING "SPECIAL" HW/CW VALVE

Heating/Cooling, Four-Pipe System, R.A Stat

This diagram shows a fan coil system that is supplied with both hot water and chilled water all the time. The R.A. thermostat T-1 controls a 3-way valve V-1 on the supply line and a 3-way valve V-2 on the return line. The valves are sequenced so that as the unit is fed hot water through the supply valve the return line valve insures that the hot water is sent back to the hot water return line. The valve on the return line is a 3-way bypass.

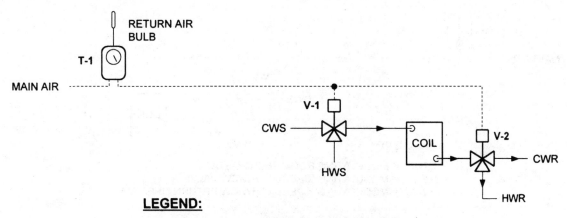

LEGEND:

T-1   = MODULATING RETURN AIR THERMOSTAT
V-1   = MODULATING "SPECIAL" HW/CW VALVE
V-2   = MODULATING "SPECIAL" HW/CW RETURN LINE VALVE

Heating/Cooling, Four-Pipe System, Room Stat

This diagram shows a fan coil system that is supplied with both hot water and chilled water all the time. The room thermostat T-1 controls a 3-way valve V-1 on the supply line and a 3-way valve V-2 on the return line. The valves are sequenced so that as the unit is fed hot water through the supply valve the return line valve insures that the hot water is sent back to the hot water return line. The valve on the return line is a 3-way bypass.

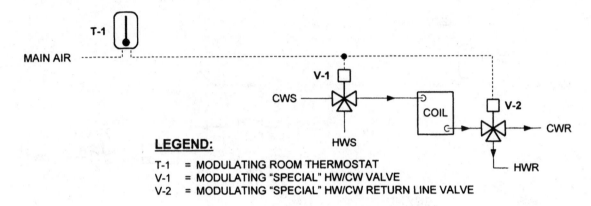

LEGEND:

T-1 = MODULATING ROOM THERMOSTAT
V-1 = MODULATING "SPECIAL" HW/CW VALVE
V-2 = MODULATING "SPECIAL" HW/CW RETURN LINE VALVE

Heating/Cooling, Two-Coil System, Room Stat

This diagram shows a fan coil system that consists of two coils, one for hot water and one for chilled water. The room thermostat T-1 controls a valve V-1 on the hot water supply line and a valve V-2 on the chilled water line. The valves are sequenced so that heating and cooling are not on at the same time.

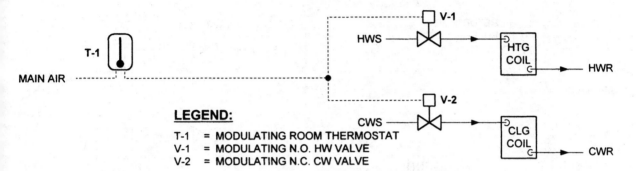

LEGEND:

T-1 = MODULATING ROOM THERMOSTAT
V-1 = MODULATING N.O. HW VALVE
V-2 = MODULATING N.C. CW VALVE

Heating/Cooling, Two-Coil System, R.A. Stat

This diagram shows a fan coil system that consists of two coils, one for hot water and one for chilled water. The return air thermostat T-1 controls a valve V-1 on the hot water supply line and a valve V-2 on the chilled water line. The valves are sequenced so that heating and cooling are not on at the same time.

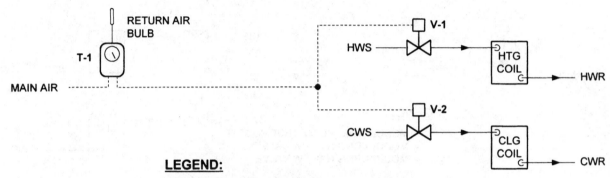

LEGEND:

T-1 = MODULATING RETURN AIR THERMOSTAT
V-1 = MODULATING N.O. HW VALVE
V-2 = MODULATING N.C. CW VALVE

CHAPTER 11

Induction Units

Heating/Cooling, Room Control, with Valves

This diagram shows an induction unit with a room thermostat T-1 controlling the valve V-1 on the hot-water/chilled-water coil. The room thermostat must be the type that can switch from direct acting to reverse acting with a change in supply pressure. The induction unit has no fan and delivers air to the space by inducing room air through a jet of primary air.

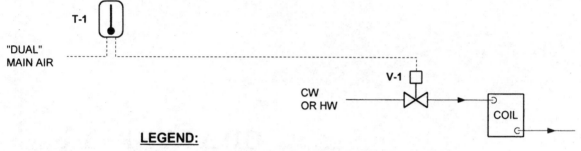

LEGEND:

T-1   = HEATING/COOLING MODULATING ROOM THERMOSTAT
V-1   = CHILLED WATER/HOT WATER MODULATING VALVE

Heating/Cooling, R.A. Control

This diagram shows an induction unit with a return air thermostat T-1 controlling the valve V-1 on the hot-water/chilled-water coil. The thermostat must be the type that can switch from direct acting to reverse acting with a change in supply pressure. The induction unit has no fan and delivers air to the space by inducing room air through a jet of primary air.

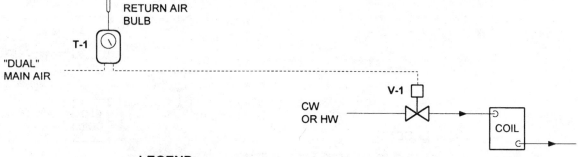

LEGEND:

T-1 = HEATING/COOLING MODULATING RETURN AIR THERMOSTAT
V-1 = CHILLED WATER/HOT WATER MODULATING VALVE

Heating/Cooling, Two Coils, Room Control

This diagram shows an induction unit with a room thermostat T-1 controlling the two valves V-1 and V-2 on the two coils in the induction unit. The thermostat modulates the valves in sequence so that they do not heat and cool at the same time. The induction unit has no fan and delivers air to the space by inducing room air through a jet of primary air.

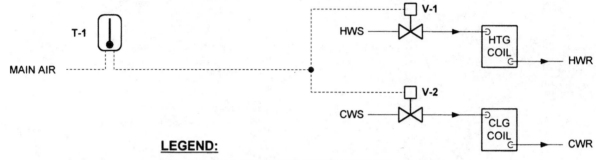

LEGEND:

T-1	=	MODULATING ROOM THERMOSTAT
V-1	=	MODULATING N.O. HW VALVE
V-2	=	MODULATING N.C. CW VALVE

Heating/Cooling, Two Coils, R.A. Control

This diagram shows an induction unit with a return air thermostat T-1 controlling the two valves V-1 and V-2 on the two coils in the induction unit. The thermostat modulates the valves in sequence so that they do not heat and cool at the same time. The induction unit has no fan and delivers air to the space by inducing room air through a jet of primary air.

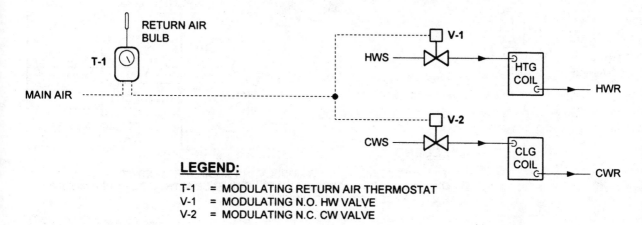

LEGEND:

T-1 = MODULATING RETURN AIR THERMOSTAT
V-1 = MODULATING N.O. HW VALVE
V-2 = MODULATING N.C. CW VALVE

Heating/Cooling, Face and Bypass, Room Stat

This diagram shows an induction unit with a room thermostat T-1 controlling the face and bypass dampers around the induction unit coil. The thermostat must be the type that changes action from direct to reverse when it is supplied with different air pressures. The induction unit has no fan and delivers air to the space by inducing room air through a jet of primary air. There are no valves on the coil.

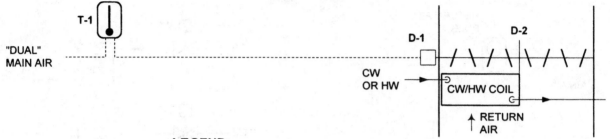

LEGEND:

T-1 = HEATING/COOLING MODULATING ROOM THERMOSTAT
D-1 = CHILLED WATER/HOT WATER COIL FACE & BYPASS DAMPER MOTOR
D-2 = FACE & BYPASS DAMPERS

Heating/Cooling, Face and Bypass, R.A. Stat

This diagram shows an induction unit with a return air thermostat T-1 controlling the face & bypass dampers around the induction unit coil. The thermostat must be the type that changes action from direct to reverse when it is supplied with different air pressures. The induction unit has no fan and delivers air to the space by inducing room air through a jet of primary air. There are no valves on the coil.

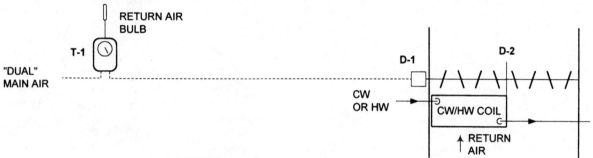

LEGEND:

T-1 = HEATING/COOLING MODULATING RETURN AIR THERMOSTAT
D-1 = CHILLED WATER/HOT WATER COIL FACE & BYPASS DAMPER MOTOR
D-2 = FACE & BYPASS DAMPERS

CHAPTER 12

Unit Ventilators

Steam/HW UV ASHRAE Cycle I

This diagram shows a classroom unit ventilator operating with ASHRAE cycle I. The unit operates as follows:

Whenever the fan runs the EP E-1 is energized allowing control air to be supplied to the valve V-1 and OA and RA damper motor D-1. When the room thermostat T-1 calls for heat the valve is open and the OA dampers are closed with the unit taking all return air. When the room thermostat reaches the set point it operates the valve and damper to reduce the heat and cool off the room. Low-limit thermostat T-2 exhausts its line to prevent too low a temperature from entering the space.

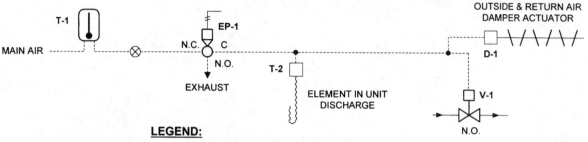

LEGEND:

T-1 = MODULATING ROOM THERMOSTAT
T-2 = LOW LIMIT UNIT THERMOSTAT
V-1 = N.O. COIL WATER VALVE
EP-1 = SOLENOID AIR VALVE
D-1 = OUTSIDE AND RETURN AIR DAMPER MOTOR

Steam/HW UV ASHRAE Cycle I, Optional Electric Coil, Face and Bypass Controls

This diagram shows a classroom unit ventilator operating with ASHRAE cycle I. The unit operates as follows:

Whenever the fan runs the EP E-1 is energized allowing control air to be supplied to the damper motor D-2 which controls the OA and RA dampers. When the room thermostat T-1 calls for heat the valve V-1 is open (or the optional electric heat controller calls for heat). If the face & bypass dampers are used instead of a coil valve, the face damper is also wide open. The OA dampers are closed with the unit taking all return air. When the room thermostat reaches the set point it operates the valve (the face & bypass damper or electric heat controller) to reduce the heat and cool off the room. Low-limit thermostat T-2 exhausts its line to prevent too low a temperature from entering the space.

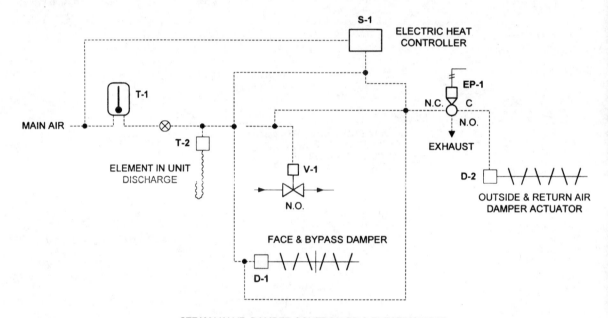

STEAM VALVE, DAMPER CONTROLLED & ELECTRIC HEAT

LEGEND:

T-1 = MODULATING ROOM THERMOSTAT
T-2 = LOW LIMIT UNIT THERMOSTAT
V-1 = N.O. COIL VALVE STEAM OR HW
EP-1 = SOLENOID AIR VALVE
D-1 = OPTIONAL FACE & BYPASS DAMPER MOTOR
D-2 = OUTSIDE & RETURN AIR DAMPER MOTOR
S-1 = ELECTRIC STEP CONTROLLER FOR ELECTRIC HEAT COIL

NOTE:
    S-1, D-1 & V-1 ARE NOT USED TOGETHER, ONLY ONE IS USED

Steam/HW UV Cycle I, with Sampling Chamber, Optional Electric Coil

This diagram shows a classroom unit ventilator operating with ASHRAE cycle I. The unit operates as follows:

Whenever the fan runs the EP E-1 is energized allowing control air to be supplied to the damper motor D-2 which controls the OA and RA dampers. When the return air thermostat T-1 calls for heat the valve V-1 is open (or the optional electric heat controller calls for heat). If the face & bypass dampers are used instead of a coil valve, the face damper is also wide open. The OA dampers are closed with the unit taking all return air. When the room thermostat reaches the set point it operates the valve (the face and bypass damper or electric heat controller) to reduce the heat and cool off the room. Low-limit thermostat T-2 exhausts its line to prevent too low a temperature from entering the space.

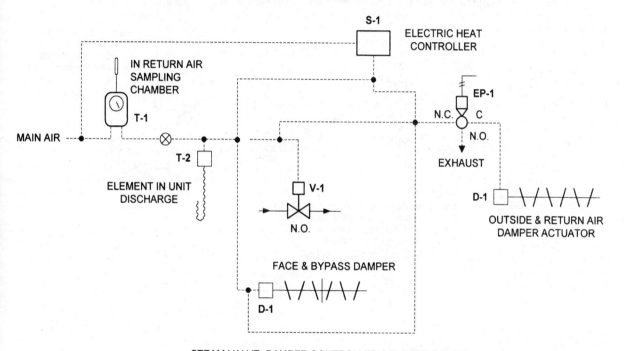

STEAM VALVE, DAMPER CONTROLLED & ELECTRIC HEAT

LEGEND:

T-1 = MODULATING CAPILLARY THERMOSTAT WITH BULB IN R.A.
T-2 = LOW LIMIT UNIT THERMOSTAT
V-1 = N.O. COIL VALVE STEAM OR HW
EP-1 = SOLENOID AIR VALVE
D-1 = OPTIONAL FACE & BYPASS DAMPER MOTOR
D-2 = OUTSIDE & RETURN AIR DAMPER MOTOR
S-1 = ELECTRIC STEP CONTROLLER FOR ELECTRIC HEAT COIL

NOTE:
    S-1, D-1 & V-1 ARE NOT USED TOGETHER, ONLY ONE IS USED

Steam/HW UV ASHRAE Cycle II

This diagram shows a classroom unit ventilator operating with ASHRAE cycle II. The unit operates as follows:

Whenever the fan runs the EP E-1 is energized allowing control air to be supplied to the damper motor D-2 which has a double spring in the motor. As the air pressure increases the damper motor opens the OA damper to a minimum position. A further increase in air pressure above certain level opens the OA damper 100%. The RA damper moves accordingly. When the room thermostat T-1 calls for heat, the valve V-1 is open. The OA dampers are fully closed with the unit taking all return air. When the room thermostat reaches the set point it operates damper motor to allow minimum ventilation air to enter the space. It also operates the valve V-1 to reduce the heat being supplied. Low-limit thermostat T-2 exhausts its line to prevent too low a temperature from entering the space.

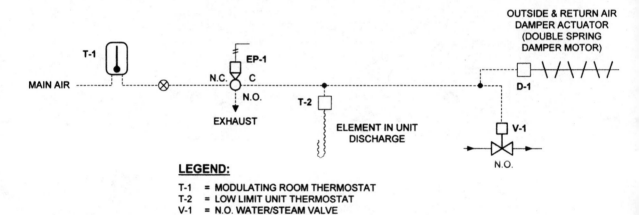

LEGEND:

T-1 = MODULATING ROOM THERMOSTAT
T-2 = LOW LIMIT UNIT THERMOSTAT
V-1 = N.O. WATER/STEAM VALVE
EP-1 = SOLENOID AIR VALVE
D-1 = OUTSIDE AND RETURN AIR (MIN. - MAX.) DAMPER MOTOR

Steam/HW UV ASHRAE Cycle II, R.A. Stat

This diagram shows a classroom unit ventilator operating with ASHRAE cycle II. The unit operates as follows:

Whenever the fan runs the EP E-1 is energized allowing control air to be supplied to the damper motor D-2 which has a double spring in the motor. As the air pressure increases the damper motor opens the OA damper to a minimum position. A further increase in air pressure above certain level opens the OA damper 100%. The RA damper moves accordingly. When the return air thermostat T-1 calls for heat, the valve V-1 is open. The OA dampers are fully closed with the unit taking all return air. When the room thermostat reaches the set point it operates damper motor to allow minimum ventilation air to enter the space. It also operates the valve V-1 to reduce the heat being supplied. Low limit thermostat T-2 exhausts its line to prevent too low a temperature from entering the space.

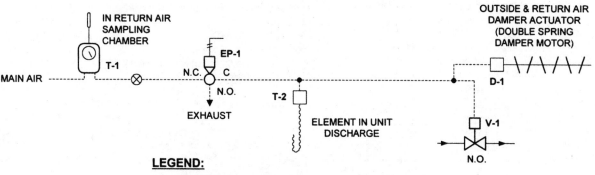

LEGEND:

T-1 = MODULATING CAPILLARY THERMOSTAT WITH BULB IN R.A.
T-2 = LOW LIMIT UNIT THERMOSTAT
V-1 = N.O. WATER/STEAM VALVE
EP-1 = SOLENOID AIR VALVE
D-1 = OUTSIDE AND RETURN AIR (MIN. - MAX.) DAMPER MOTOR

Steam/HW UV ASHRAE Cycle II, Optional Electric Coil, Room Stat

This diagram shows a classroom unit ventilator operating with ASHRAE cycle II. The unit operates as follows:

Whenever the fan runs the EP E-1 is energized allowing control air to be supplied to the damper motor D-2 which has a double spring in the motor. As the air pressure increases the damper motor opens the OA damper to a minimum position. A further increase in air pressure above certain level opens the OA damper 100%. The RA damper moves accordingly. When the room thermostat T-1 calls for heat the valve V-1 is open, (or the optional heat controller calls for heat through the electric coil) (the unit can also have an optional face & bypass damper across the coil, in which case the face damper would be wide open). The OA dampers are fully closed with the unit taking all return air. When the room thermostat reaches the set point it operates damper motor to allow minimum ventilation air to enter the space. It also operates the valve V-1 (electric coil/ face & bypass damper) to reduce the heat being supplied. Low-limit thermostat T-2 exhausts its line to prevent too low a temperature from entering the space.

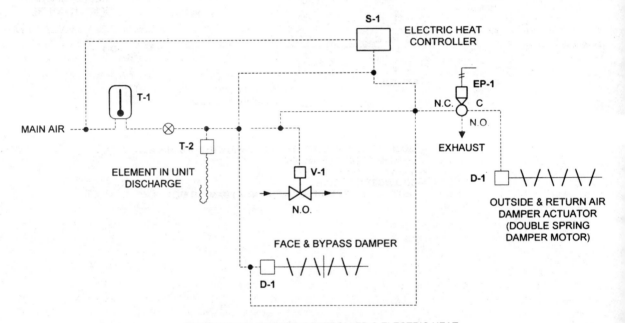

STEAM VALVE, DAMPER CONTROLLED & ELECTRIC HEAT

LEGEND:

T-1 = MODULATING ROOM THERMOSTAT
T-2 = LOW LIMIT UNIT THERMOSTAT
V-1 = N.O. COIL VALVE STEAM OR HW
EP-1 = SOLENOID AIR VALVE
D-1 = OPTIONAL FACE & BYPASS DAMPER MOTOR
D-2 = OUTSIDE & RETURN AIR (MIN. - MAX.) DAMPER MOTOR
S-1 = ELECTRIC STEP CONTROLLER FOR ELECTRIC HEAT COIL

NOTE:
    S-1, D-1 & V-1 ARE NOT USED TOGETHER, ONLY ONE IS USED

Steam/HW UV ASHRAE Cycle II, Optional Electric Coil, R.A. Stat

This diagram shows a classroom unit ventilator operating with ASHRAE cycle II. The unit operates as follows:

Whenever the fan runs the EP E-1 is energized allowing control air to be supplied to the damper motor D-2 which has a double spring in the motor. As the air pressure increases the damper motor opens the OA damper to a minimum position. A further increase in air pressure above certain level opens the OA damper 100%. The RA damper moves accordingly. When the return air thermostat T-1 calls for heat the valve V-1 is open, (or the optional heat controller calls for heat through the electric coil) (the unit can also have an optional face & bypass damper across the coil, in which case the face damper would be wide open). The OA dampers are fully closed with the unit taking all return air. When the return air thermostat reaches the set point it operates damper motor to allow minimum ventilation air to enter the space. It also operates the valve V-1 (electric coil/ face & bypass damper) to reduce the heat being supplied. Low-limit thermostat T-2 exhausts its line to prevent too low a temperature from entering the space.

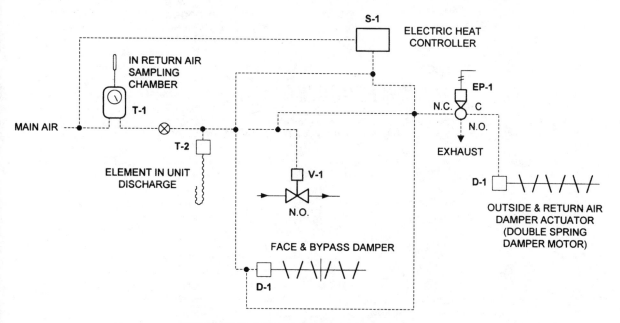

STEAM VALVE, DAMPER CONTROLLED & ELECTRIC HEAT

LEGEND:

T-1   = MODULATING CAPILLARY THERMOSTAT WITH BULB IN R.A.
T-2   = LOW LIMIT UNIT THERMOSTAT
V-1   = N.O. COIL VALVE STEAM OR HW
EP-1 = SOLENOID AIR VALVE
D-1   = OPTIONAL FACE & BYPASS DAMPER MOTOR
D-2   = OUTSIDE & RETURN AIR (MIN. - MAX.) DAMPER MOTOR
S-1   = ELECTRIC STEP CONTROLLER FOR ELECTRIC HEAT COIL

NOTE:
   S-1, D-1 & V-1 ARE NOT USED TOGETHER, ONLY ONE IS USED

Steam/HW UV ASHRAE Cycle III

This diagram shows a classroom unit ventilator operating with ASHRAE cycle III. The unit operates as follows:

Whenever the fan runs the EP E-1 is energized allowing control air to be supplied to the damper motor D-2 and valve V-1 as well as low limit thermostat T-2. When room thermostat T-1 calls for heat the valve V-1 is open. The OA dampers are fully closed with the unit taking all return air. When the room thermostat reaches the set point it passes air to the mixed air controller T-2 which controls the OA and RA damper motor. The valve V-1 is under control of the room thermostat.

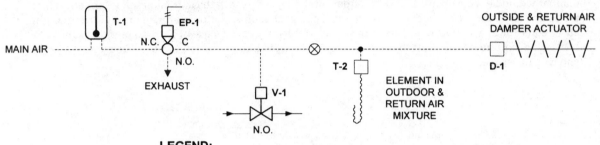

LEGEND:

T-1   = MODULATING ROOM THERMOSTAT
T-2   = MIXED AIR THERMOSTAT
V-1   = N.O. COIL VALVE
EP-1 = SOLENOID AIR VALVE
D-1   = OUTSIDE AND RETURN AIR DAMPER MOTOR

Steam/HW UV ASHRAE Cycle III, R.A. Stat

This diagram shows a classroom unit ventilator operating with ASHRAE cycle III. The unit operates as follows:

Whenever the fan runs the EP E-1 is energized allowing control air to be supplied to the damper motor D-2 and valve V-1 as well as low limit thermostat T-2. When return air thermostat T-1 calls for heat the valve V-1 is open. The OA dampers are fully closed with the unit taking all return air. When the room thermostat reaches the set point it passes air to the mixed air controller T-2 which controls the OA and RA damper motor. The valve V-1 is under control of the return air thermostat.

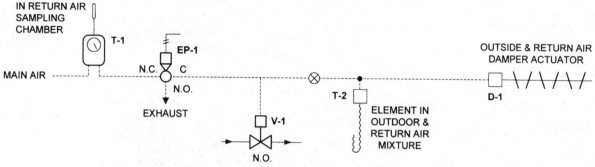

LEGEND:

T-1 = MODULATING CAPILLARY THERMOSTAT WITH BULB IN R.A.
T-2 = MIXED AIR THERMOSTAT
V-1 = N.O. COIL VALVE
EP-1 = SOLENOID AIR VALVE
D-1 = OUTSIDE AND RETURN AIR DAMPER MOTOR

Steam/HW UV ASHRAE Cycle III, Optional Electric Coil

This diagram shows a classroom unit ventilator operating with ASHRAE cycle III. The unit operates as follows:

Whenever the fan runs the EP E-1 is energized allowing control air to be supplied to the damper motor D-2. When room thermostat T-1 calls for heat the valve V-1 (optional electric heat controller or face and bypass dampers are positioned to give full heat) is open. The OA dampers are fully closed with the unit taking all return air. When the room thermostat reaches the set point it passes air to the mixed air controller T-2 which controls the OA and RA damper motor. The valve V-1 (electric heat controller or face & bypass dampers) is under control of the room thermostat.

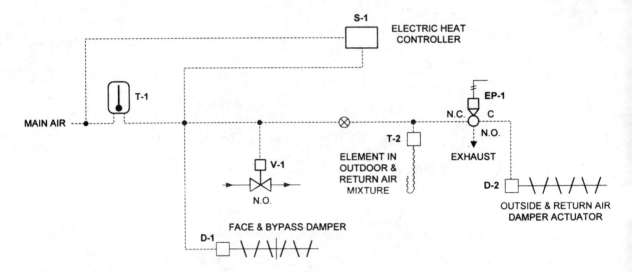

STEAM VALVE, DAMPER CONTROLLED & ELECTRIC HEAT

LEGEND:

T-1 = MODULATING ROOM THERMOSTAT
T-2 = MIXED AIR THERMOSTAT
V-1 = N.O. COIL VALVE STEAM OR HW
EP-1 = SOLENOID AIR VALVE
D-1 = FACE & BYPASS DAMPER MOTOR
D-2 = OUTSIDE & RETURN AIR DAMPER MOTOR
S-1 = ELECTRIC STEP CONTROLLER FOR ELECTRIC HEAT COIL

NOTE:
    S-1, D-1 & V-1 ARE NOT USED TOGETHER, ONLY ONE IS USED

Steam/HW UV ASHRAE Cycle W

This diagram shows a classroom unit ventilator operating with ASHRAE cycle W. The unit operates as follows:

Whenever the fan runs the EP E-1 is energized allowing control air to be supplied to the damper motor D-1 and valve as well as low limit thermostat T-2 in the fan discharge. When the room thermostat calls for heat the OA damper is closed and the return air damper is open and the valve V-1 is also open. When the space becomes satisfied the room thermostat passes air and closes the valve with the thermostat T-2 controlling the damper motor D-1 to maintain a fixed fan discharge temperature.

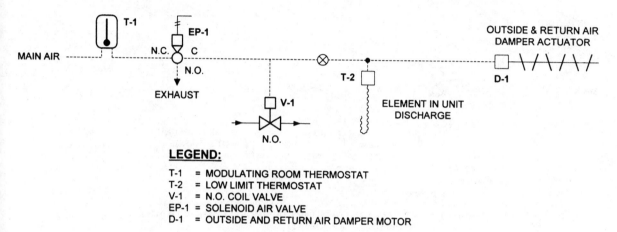

LEGEND:

T-1 = MODULATING ROOM THERMOSTAT
T-2 = LOW LIMIT THERMOSTAT
V-1 = N.O. COIL VALVE
EP-1 = SOLENOID AIR VALVE
D-1 = OUTSIDE AND RETURN AIR DAMPER MOTOR

Steam/HW UV ASHRAE Cycle W, R.A. Stat Control

This diagram shows a classroom unit ventilator operating with ASHRAE cycle W. The unit operates as follows:

Whenever the fan runs the EP E-1 is energized allowing control air to be supplied to the damper motor D-1 and valve as well as low limit thermostat T-2 in the fan discharge. When the return air thermostat calls for heat the OA damper is closed and the return air damper is open and the valve V-1 is also open. When the space becomes satisfied the return air thermostat passes air and closes the valve with the thermostat T-2 controlling the damper motor D-1 to maintain a fixed fan discharge temperature.

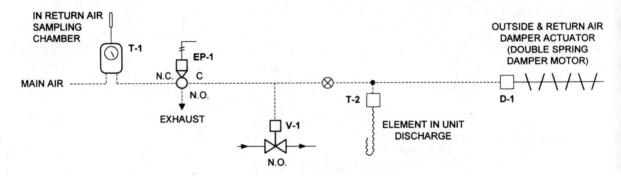

LEGEND:

T-1	= MODULATING CAPILLARY THERMOSTAT WITH BULB IN R.A.
T-2	= LOW LIMIT THERMOSTAT
V-1	= N.O. COIL VALVE
EP-1	= SOLENOID AIR VALVE
D-1	= OUTSIDE AND RETURN AIR (MIN. - MAX.) DAMPER MOTOR

Steam/HW UV ASHRAE Cycle W, Room Stat

This diagram shows a classroom unit ventilator operating with ASHRAE cycle W. The unit operates as follows:

Whenever the fan runs, the EP E-1 is energized, allowing control air to be supplied to the damper motor D-2. When the room thermostat T-1 calls for heat, the OA damper is closed and the return air damper is open. The valve V-1 (or optional electric heat controller) as well as optional face and bypass dampers are in full heating mode. When the space becomes satisfied the room thermostat passes air and closes the valve (or optional other sources of heat), with the thermostat T-2 controlling the damper motor D-1 to maintain a fixed fan discharge temperature.

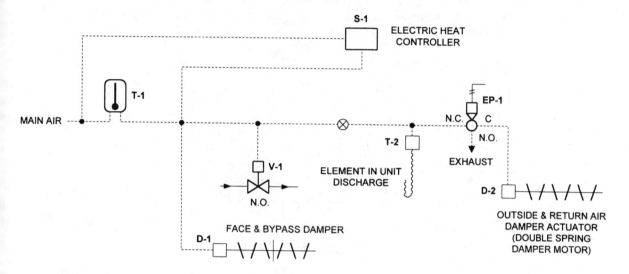

LEGEND:

T-1 = MODULATING ROOM THERMOSTAT
T-2 = LOW LIMIT THERMOSTAT
V-1 = N.O. COIL VALVE STEAM OR HW
EP-1 = SOLENOID AIR VALVE
D-1 = FACE & BYPASS DAMPER MOTOR
D-2 = OUTSIDE & RETURN AIR (MIN. - MAX.) DAMPER MOTOR
S-1 = ELECTRIC STEP CONTROLLER FOR ELECTRIC HEAT COIL

NOTE:
    D-1 & V-1 ARE NOT USED TOGETHER, ONLY ONE IS USED

Heating/Cooling UV (CW/HW) ASHRAE Cycle I

This diagram shows a classroom heating/cooling classroom unit ventilator operating with ASHRAE cycle I. This system requires "dual" (15-psi cooling and 20-psi heating air pressure). When the system is switched to heating (20-psi supply pressure), the operation is as follows:

With the fan running on a call for heat, the room thermostat lowers the air pressure and opens the valve V-1 on the coil. It also, through the 3-way air valve (NC to Common ports), lowers the air pressure to the damper motor D-1, reducing the amount of OA being taken in. The unit thermostat T-2 exhausts its air as the discharge temperature drops, to prevent a low discharge temperature.

When the supply air is switched to 15 psi, the thermostat is switched to reverse acting and the 3-way air valve is switched to pass only 5 psi (from the PRV P-1) to the damper motor, positioning it to a minimum amount of ventilation air and no more. The valve action is reversed so that the room thermostat can cool the space.

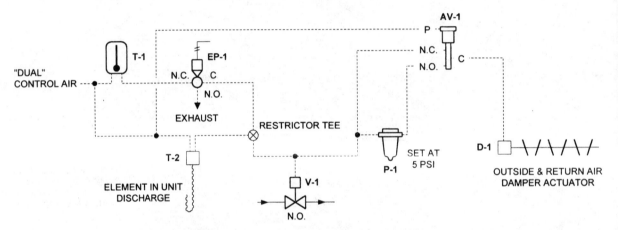

LEGEND:

T-1 = MODULATING HEATING/COOLING ROOM THERMOSTAT
T-2 = LOW LIMIT THERMOSTAT
V-1 = N.O. COIL VALVE (HW/CW)
EP-1 = SOLENOID AIR VALVE
D-1 = OUTSIDE & RETURN AIR DAMPER MOTOR
P-1 = PRESSURE REDUCING VALVE SET @ 5 PSI
AV-1 = 3-WAY SWITCHING AIR VALVE SET SWITCH ABOVE 15 PSI

NOTE:
DUAL AIR SUPPLY REQUIRED TO SWITCH SYSTEM FROM HEATING TO COOLING
AND VISA/VERSA. COIL MUST BE SUPPLIED WITH HW ON HEATING & CW ON COOLING

Heating/Cooling UV (CW/HW) ASHRAE Cycle I R.A. Stat

This diagram shows a classroom heating/cooling classroom unit ventilator operating with ASHRAE cycle I. This system requires "dual" (15-psi cooling and 20-psi heating air pressure). When the system is switched to heating (20-psi supply pressure), the operation is as follows:

With the fan running on a call for heat, the return air thermostat T-1 lowers the air pressure and opens the valve V-1 on the coil. It also, through the 3-way air valve (NC to Common ports), lowers the air pressure to the damper motor D-1, reducing the amount of OA being taken in. The unit thermostat T-2 exhausts its air as the discharge temperature drops, to prevent a low discharge temperature.

When the supply air is switched to 15 psi, the thermostat T-1 is switched to reverse acting and the 3-way air valve is switched to pass only 5 psi (from the PRV P-1) to the damper motor, positioning it to a minimum amount of ventilation air and no more. The valve action is reversed so that the thermostat can cool the space.

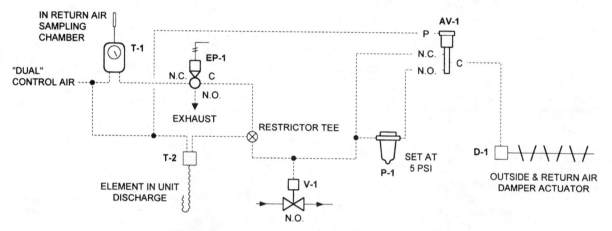

LEGEND:

T-1 = MODULATING HEATING/COOLING CAPILLARY R.A. THERMOSTAT
T-2 = LOW LIMIT THERMOSTAT
V-1 = N.O. COIL VALVE (HW/CW)
EP-1 = SOLENOID AIR VALVE
D-1 = OUTSIDE & RETURN AIR DAMPER MOTOR
P-1 = PRESSURE REDUCING VALVE SET @ 5 PSI
AV-1 = 3-WAY SWITCHING AIR VALVE SET SWITCH ABOVE 15 PSI

NOTE:
    DUAL AIR SUPPLY REQUIRED TO SWITCH SYSTEM FROM HEATING TO COOLING
    AND VISA/VERSA. COIL MUST BE SUPPLIED WITH HW ON HEATING & CW ON COOLING

Heating/Cooling UV (CW/HW) ASHRAE Cycle I Optional Face and Bypass Controls

This diagram shows a classroom heating/cooling classroom unit ventilator operating with ASHRAE cycle I. This system requires "dual" (15-psi cooling and 20-psi heating air pressure). When the system is switched to heating (20-psi supply pressure), the operation is as follows:

With the fan running on a call for heat, the room thermostat T-1 lowers the air pressure and opens the valve V-1 on the coil. It also modulates the face and bypass dampers D-1 in sequence and at the same time modulates the damper motor D-2 through the 3-way air valve (NC to Common ports) to take mostly return air. The unit thermostat T-2 exhausts its air as the discharge temperature drops, to prevent a low discharge temperature by operating the face & bypass dampers and the coil valve.

When the supply air is switched to 15 psi, the thermostat T-1 is switched to reverse acting and the 3-way air valve is switched to pass only 5 psi (from the PRV P-1) to the damper motor, positioning it to a minimum amount of ventilation air and no more. The valve action is reversed so that the room thermostat can cool the space. The EP switch E-1 is only connected to the damper motor, not the to the devices.

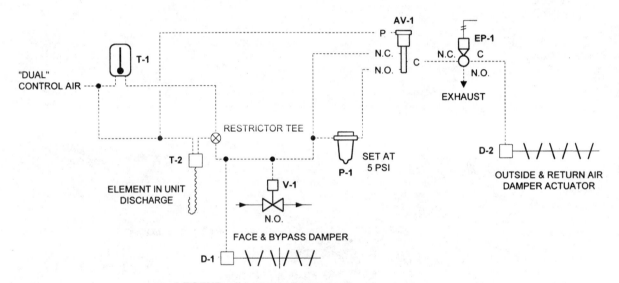

LEGEND:

T-1   = MODULATING HEATING/COOLING ROOM THERMOSTAT
T-2   = LOW LIMIT THERMOSTAT
V-1   = N.O. COIL VALVE (HW/CW)
EP-1 = SOLENOID AIR VALVE
D-1   = FACE & BYPASS DAMPER MOTOR
D-2   = OUTSIDE & RETURN AIR DAMPER MOTOR
P-1   = PRESSURE REDUCING VALVE SET @ 5 PSI
AV-1 = 3-WAY SWITCHING AIR VALVE SET SWITCH ABOVE 15 PSI

NOTES:
    DUAL AIR SUPPLY REQUIRED TO SWITCH SYSTEM FROM HEATING TO COOLING
    AND VISA/VERSA. COIL MUST BE SUPPLIED WITH HW ON HEATING & CW ON COOLING

    D-1 & V-1 ARE NOT USED TOGETHER, ONLY ONE IS USED

Heating/Cooling UV (CW/HW) ASHRAE Cycle I R.A Stat, Optional Face and Bypass Controls

This diagram shows a classroom heating/cooling classroom unit ventilator operating with ASHRAE cycle I. This system requires "dual" (15-psi cooling and 20-psi heating air pressure). When the system is switched to heating (20-psi supply pressure), the operation is as follows:

With the fan running on a call for heat, the return air thermostat T-1 lowers the air pressure and opens the valve V-1 on the coil. It also modulates the optional face and bypass dampers D-1 in sequence and at the same time modulates the damper motor D-2 through the 3-way air valve (NC to Common ports) to take mostly return air. The unit thermostat T-2 exhausts its air as the discharge temperature drops, to prevent a low discharge temperature by operating the face & bypass dampers and the coil valve. .

When the supply air is switched to 15 psi, the thermostat T-1 is switched to reverse acting and the 3-way air valve is switched to pass only 5 psi (from the PRV P-1) to the damper motor, positioning it to a minimum amount of ventilation air and no more. The valve action is reversed so that the room thermostat can cool the space. The EP switch E-1 is only connected to the damper motor, not the to the devices.

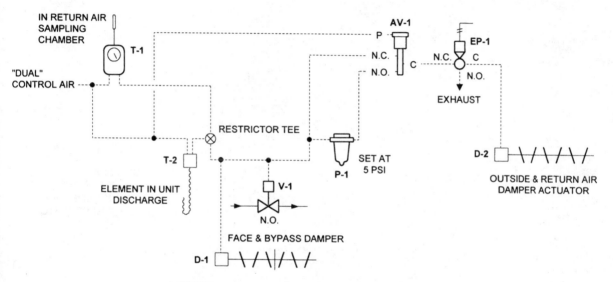

LEGEND:

T-1 = MODULATING HEATING/COOLING CAPILLARY R.A. THERMOSTAT
T-2 = LOW LIMIT THERMOSTAT
V-1 = N.O. COIL VALVE (HW/CW)
EP-1 = SOLENOID AIR VALVE
D-1 = FACE & BYPASS DAMPER MOTOR
D-2 = OUTSIDE & RETURN AIR DAMPER MOTOR
P-1 = PRESSURE REDUCING VALVE SET @ 5 PSI
AV-1 = 3-WAY SWITCHING AIR VALVE SET SWITCH ABOVE 15 PSI

NOTES:
DUAL AIR SUPPLY REQUIRED TO SWITCH SYSTEM FROM HEATING TO COOLING AND VISA/VERSA. COIL MUST BE SUPPLIED WITH HW ON HEATING & CW ON COOLING

D-1 & V-1 ARE NOT USED TOGETHER, ONLY ONE IS USED

Heating/Cooling UV (CW/HW) ASHRAE Cycle II

This diagram shows a classroom heating/cooling unit ventilator operating with ASHRAE cycle II. This system requires "dual" (15-psi cooling and 20-psi heating air pressure). When the system is switched to heating (20-psi supply pressure), the operation is as follows:

With the fan running on a call for heat the room thermostat T-1 lowers the air pressure and opens the valve V-1 on the coil. The OA and RA damper motor is also positioned to reduce the amount of OA and use mostly RA. When the space temperature becomes satisfied the room thermostat increases its pressure and begins to modulate the valve closed and the damper motor to the ventilation position.

The unit thermostat T-2 exhausts its air as the discharge temperature drops, to prevent a low discharge temperature by operating the valve and the damper motor.

When the supply air is switched to 15 psi, the thermostat T-1 is switched to reverse acting and the unit operates as a cooling unit. The room thermostat modulates the coil valve and the OA damper motor. The thermostat T-2 is switched by the change in supply pressure so that it can go to 57 degrees during the cooling cycle. The damper motor has a double spring in it so that as the pressure increases it goes to a minimum position and then hesitates until the pressure rises above a certain value whereupon it can go to maximum position.

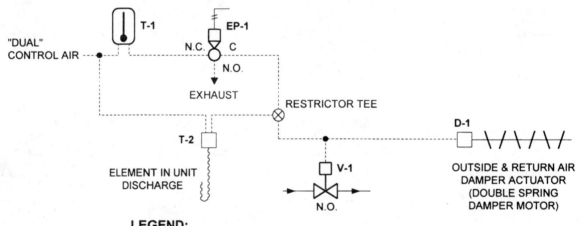

LEGEND:

T-1   = MODULATING HEATING/COOLING ROOM THERMOSTAT
T-2   = LOW LIMIT THERMOSTAT
V-1   = N.O. COIL VALVE (HW/CW)
EP-1 = SOLENOID AIR VALVE
D-1   = OUTSIDE & RETURN AIR (MIN. - MAX.) DAMPER MOTOR

NOTES:
    DUAL AIR SUPPLY REQUIRED TO SWITCH SYSTEM FROM HEATING TO COOLING
    AND VISA/VERSA. COIL MUST BE SUPPLIED WITH HW ON HEATING & CW ON COOLING

Heating/Cooling UV (CW/HW) ASHRAE Cycle II, R.A. Stat

This diagram shows a classroom heating/cooling unit ventilator operating with ASHRAE cycle II. This system requires "dual" (15-psi cooling and 20-psi heating air pressure). When the system is switched to heating (20psi supply pressure) the operation is as follows:

With the fan running on a call for heat the return air thermostat T-1 lowers the air pressure and opens the valve V-1 on the coil. The OA and RA damper motor is also positioned to reduce the amount of OA and use mostly RA. When the space temperature becomes satisfied the T-1 thermostat increases its pressure and begins to modulate the valve closed and the damper motor to the ventilation position.

The unit thermostat T-2 exhausts its air as the discharge temperature drops, to prevent a low discharge temperature by operating the valve and the damper motor.

When the supply air is switched to 15 psi, the thermostat T-1 is switched to reverse acting and the unit operates as a cooling unit. The T-1 thermostat modulates the coil valve and the OA damper motor. The thermostat T-2 is switched by the change in supply pressure so that it can go to 57 degrees during the cooling cycle. The damper motor has a double spring in it so that as the pressure increases it goes to a minimum position and then hesitates until the pressure rises above a certain value where-upon it can go to maximum position.

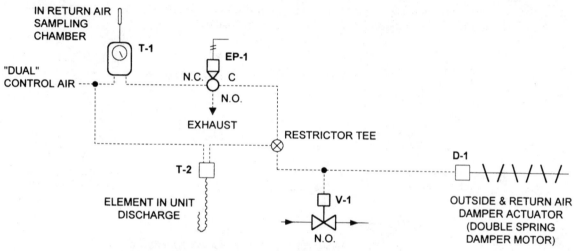

LEGEND:

T-1 = MODULATING HEATING/COOLING CAPILLARY R.A. THERMOSTAT
T-2 = LOW LIMIT THERMOSTAT
V-1 = N.O. COIL VALVE (HW/CW)
EP-1 = SOLENOID AIR VALVE
D-1 = OUTSIDE & RETURN AIR (MIN. - MAX.) DAMPER MOTOR

NOTES:
  DUAL AIR SUPPLY REQUIRED TO SWITCH SYSTEM FROM HEATING TO COOLING
  AND VISA/VERSA. COIL MUST BE SUPPLIED WITH HW ON HEATING & CW ON COOLING

Heating/Cooling UV (CW/HW) ASHRAE Cycle II, Room Stat with Optional Face and Bypass Dampers around the Coil

This diagram shows a classroom heating/cooling unit ventilator operating with ASHRAE cycle II. This system requires "dual" (15-psi cooling and 20-psi heating air pressure). When the system is switched to heating (20psi supply pressure) the operation is as follows:

With the fan running on a call for heat the room thermostat T-1 lowers the air pressure and opens the valve V-1 on the coil along with optional face & bypass dampers D-1. The OA and RA damper motor D-2 is also positioned to reduce the amount of OA and use mostly RA. When the space temperature becomes satisfied the T-1 thermostat increases its pressure and begins to modulate the valve closed and the face and bypass damper as well as the damper motor on the OA and RA dampers to the ventilation position.

The unit thermostat T-2 exhausts its air as the discharge temperature drops, to prevent a low discharge temperature by operating the valve and the damper motors.

When the supply air is switched to 15 psi, the thermostat T-1 is switched to reverse acting and the unit operates as a cooling unit. The T-1 thermostat modulates the coil valve the face & bypass damper motor as well as the OA damper motor. The thermostat T-2 is switched by the change in supply pressure so that it can go to 57 degrees during the cooling cycle. The damper motor has a double spring in it so that as the pressure increases it goes to a minimum position and then hesitates until the pressure rises above a certain value whereupon it can go to maximum position.

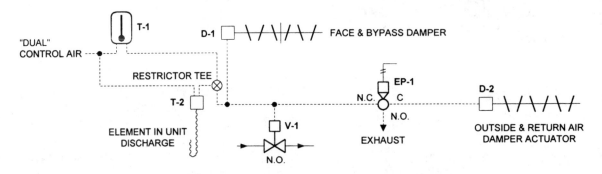

LEGEND:

T-1 = MODULATING HEATING/COOLING ROOM THERMOSTAT
T-2 = LOW LIMIT THERMOSTAT
V-1 = N.O. COIL VALVE (HW/CW)
EP-1 = SOLENOID AIR VALVE
D-1 = FACE & BYPASS DAMPER MOTOR
D-2 = OUTSIDE & RETURN AIR (MIN. - MAX.) DAMPER MOTOR

NOTES:
DUAL AIR SUPPLY REQUIRED TO SWITCH SYSTEM FROM HEATING TO COOLING
AND VISA/VERSA. COIL MUST BE SUPPLIED WITH HW ON HEATING & CW ON COOLING

Heating/Cooling UV (CW/HW) ASHRAE Cycle II, Optional Face and Bypass Control

This diagram shows a classroom heating/cooling unit ventilator operating with ASHRAE cycle II. This system requires "dual" (15-psi cooling and 20-psi heating air pressure). When the system is switched to heating (20-psi supply pressure) the operation is as follows:

With the fan running on a call for heat the return air thermostat T-1 lowers the air pressure and opens the valve V-1 on the coil along with optional face & bypass dampers D-1. The OA and RA damper motor D-2 is also positioned to reduce the amount of OA and use mostly RA. When the space temperature becomes satisfied the T-1 thermostat increases its pressure and begins to modulate the valve closed and the face and bypass damper as well as the damper motor on the OA and RA dampers to the ventilation position.

The unit thermostat T-2 exhausts its air as the discharge temperature drops, to prevent a low discharge temperature by operating the valve and the damper motors.

When the supply air is switched to 15 psi, the thermostat T-1 is switched to reverse acting and the unit operates as a cooling unit. The T-1 thermostat modulates the coil valve the face & bypass damper motor as well as the OA damper motor. The thermostat T-2 is switched by the change in supply pressure so that it can go to 57 degrees during the cooling cycle. The damper motor has a double spring in it so that as the pressure increases it goes to a minimum position and then hesitates until the pressure rises above a certain value where-upon it can go to maximum position.

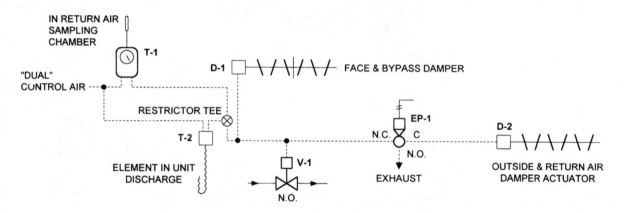

LEGEND:

T-1 = MODULATING HEATING/COOLING CAPILLARY R.A. THERMOSTAT
T-2 = LOW LIMIT THERMOSTAT
V-1 = N.O. COIL VALVE (HW/CW)
EP-1 = SOLENOID AIR VALVE
D-1 = FACE & BYPASS DAMPER MOTOR
D-2 = OUTSIDE & RETURN AIR (MIN. - MAX.) DAMPER MOTOR

NOTES:
   DUAL AIR SUPPLY REQUIRED TO SWITCH SYSTEM FROM HEATING TO COOLING
   AND VISA/VERSA. COIL MUST BE SUPPLIED WITH HW ON HEATING & CW ON COOLING

   D-1 & V-1 ARE NOT USED TOGETHER, ONLY ONE IS USED

Heating/Cooling UV (CW/HW) ASHRAE Cycle W

This diagram shows a classroom heating/cooling unit ventilator operating with ASHRAE cycle W. This system requires "dual" (15-psi cooling and 20-psi heating air pressure). When the system is switched to heating (20-psi supply pressure) the operation is as follows:

With the fan running on a call for heat the room thermostat T-1 lowers the air pressure and opens the valve V-1 on the coil. The OA and RA damper motor D-2 is also positioned to reduce the amount of OA and use mostly RA. When the space temperature becomes satisfied the T-1 thermostat increases its pressure and begins to modulate the valve closed and opens the OA damper through damper motor D-2.

The unit thermostat T-2 exhausts its air as the discharge temperature drops, to prevent a low discharge temperature by operating the valve and the damper motor.

When the supply air is switched to 15 psi, the thermostat T-1 is switched to reverse acting and the unit operates as a cooling unit. The T-1 thermostat modulates the coil valve and the OA damper motor. The thermostat T-2 is switched by the change in supply pressure so that it can go to 57 degrees during the cooling cycle. The thermostat T-2 only operates on the OA damper motor and not the coil valve.

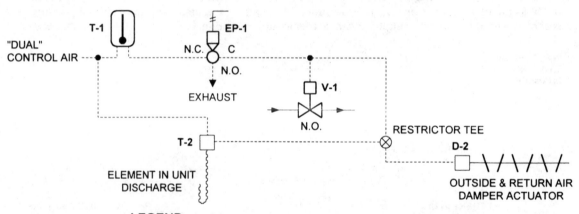

LEGEND:

T-1 = MODULATING HEATING/COOLING ROOM THERMOSTAT
T-2 = LOW LIMIT THERMOSTAT
V-1 = N.O. COIL VALVE (HW/CW)
EP-1 = SOLENOID AIR VALVE
D-1 = OUTSIDE & RETURN AIR DAMPER MOTOR

NOTES:
DUAL AIR SUPPLY REQUIRED TO SWITCH SYSTEM FROM HEATING TO COOLING
AND VISA/VERSA. COIL MUST BE SUPPLIED WITH HW ON HEATING & CW ON COOLING

Heating/Cooling UV (CW/HW) ASHRAE Cycle W R.A. Stat

This diagram shows a classroom heating/cooling unit ventilator operating with ASHRAE cycle W. This system requires "dual" (15-psi cooling and 20-psi heating air pressure). When the system is switched to heating (20-psi supply pressure), the operation is as follows:

With the fan running on a call for heat the return air thermostat T-1 lowers the air pressure and opens the valve V-1 on the coil. The OA and RA damper motor D-2 is also positioned to reduce the amount of OA and use mostly RA. When the space temperature becomes satisfied the T-1 thermostat increases its pressure and begins to modulate the valve closed and opens the OA damper through damper motor D-2.

The unit thermostat T-2 exhausts its air as the discharge temperature drops, to prevent a low discharge temperature by operating the valve and the damper motor.

When the supply air is switched to 15 psi, the thermostat T-1 is switched to reverse acting and the unit operates as a cooling unit. The T-1 thermostat modulates the coil valve and the OA damper motor. The thermostat T-2 is switched by the change in supply pressure so that it can go to 57 degrees during the cooling cycle. The thermostat T-2 only operates on the OA damper motor and not the coil valve.

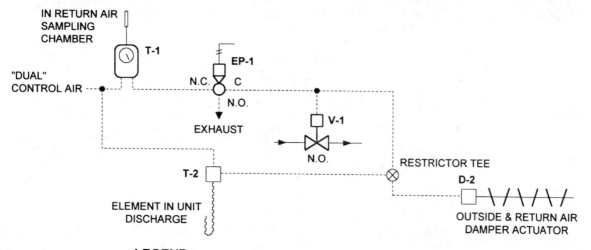

LEGEND:

T-1 = MODULATING HEATING/COOLING CAPILLARY R.A. THERMOSTAT
T-2 = LOW LIMIT THERMOSTAT
V-1 = N.O. COIL VALVE (HW/CW)
EP-1 = SOLENOID AIR VALVE
D-1 = OUTSIDE & RETURN AIR DAMPER MOTOR

NOTES:
    DUAL AIR SUPPLY REQUIRED TO SWITCH SYSTEM FROM HEATING TO COOLING
    AND VISA/VERSA. COIL MUST BE SUPPLIED WITH HW ON HEATING & CW ON COOLING

Heating/Cooling UV (CW/HW) ASHRAE Cycle W Optional Face and Bypass Damper or Valve Control

This diagram shows a classroom heating/cooling unit ventilator operating with ASHRAE cycle W. This system requires "dual" (15-psi cooling and 20-psi heating air pressure). When the system is switched to heating (20-psi supply pressure), the operation is as follows:

With the fan running on a call for heat the room thermostat T-1 lowers the air pressure and opens the V-1 and modulates the face & bypass dampers D-1 on the coil. The OA and RA damper motor D-2 is also positioned to reduce the amount of OA and use mostly RA. When the space temperature becomes satisfied the T-1 thermostat increases its pressure and begins to modulate the valve closed and the face damper closed and at the same time opens the OA damper through damper motor D-2. The face & bypass damper can be along with the coil valve or as an optional piece of equipment.

The unit thermostat T-2 exhausts its air as the discharge temperature drops, to prevent a low discharge temperature by operating the damper motor D-2 on the OA/RA dampers.

When the supply air is switched to 15 psi, the thermostat T-1 is switched to reverse acting and the unit operates as a cooling unit. The T-1 thermostat modulates the coil valve the face & bypass damper motor D-1 and the OA damper motor D-2. The thermostat T-2 is switched by the change in supply pressure so that it can go to 57 degrees during the cooling cycle. The thermostat T-2 only operates on the OA damper motor and not the coil valve.

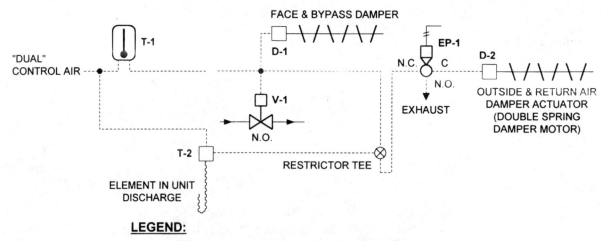

LEGEND:

T-1 = MODULATING HEATING/COOLING ROOM THERMOSTAT
T-2 = LOW LIMIT THERMOSTAT
V-1 = N.O. COIL VALVE (HW/CW)
EP-1 = SOLENOID AIR VALVE
D-1 = FACE & BYPASS DAMPER MOTOR
D-2 = OUTSIDE & RETURN AIR (MIN. - MAX.) DAMPER MOTOR

NOTES:
DUAL AIR SUPPLY REQUIRED TO SWITCH SYSTEM FROM HEATING TO COOLING
AND VISA/VERSA. COIL MUST BE SUPPLIED WITH HW ON HEATING & CW ON COOLING

Heating/Cooling UV (CW/HW) ASHRAE Cycle With Optional Face and Bypass or Valve Control with R.A. Control

This diagram shows a classroom heating/cooling unit ventilator operating with ASHRAE cycle W. This system requires "dual" (15-psi cooling and 20-psi heating air pressure). When the system is switched to heating (20-psi supply pressure) the operation is as follows:

With the fan running on a call for heat the return air thermostat T-1 lowers the air pressure and opens the valve V-1 and modulates the face & bypass dampers D-1 on the coil. The OA and RA damper motor D-2 is also positioned to reduce the amount of OA and use mostly RA. When the space temperature becomes satisfied the T-1 thermostat increases its pressure and begins to modulate the valve closed and the face damper closed and at the same time opens the OA damper through damper motor D-2. The face & bypass damper can be along with the coil valve or as an optional piece of equipment.

The unit thermostat T-2 exhausts its air as the discharge temperature drops, to prevent a low discharge temperature by operating the damper motor D-2 on the OA/RA dampers.

When the supply air is switched to 15 psi, the thermostat T-1 is switched to reverse acting and the unit operates as a cooling unit. The T-1 thermostat modulates the coil valve the face & bypass damper motor D-1 and the OA damper motor D-2. The thermostat T-2 is switched by the change in supply pressure so that it can go to 57 degrees during the cooling cycle. The thermostat T-2 only operates on the OA damper motor and not the coil valve.

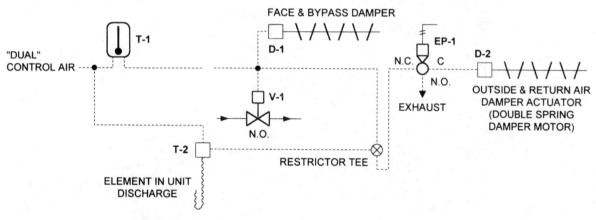

LEGEND:

T-1   = MODULATING HEATING/COOLING CAPILLARY R.A. THERMOSTAT
T-2   = LOW LIMIT THERMOSTAT
V-1   = N.O. COIL VALVE (HW/CW)
EP-1 = SOLENOID AIR VALVE
D-1   = FACE & BYPASS DAMPER MOTOR
D-2   = OUTSIDE & RETURN AIR (MIN. - MAX.) DAMPER MOTOR

NOTES:
    DUAL AIR SUPPLY REQUIRED TO SWITCH SYSTEM FROM HEATING TO COOLING
    AND VISA/VERSA. COIL MUST BE SUPPLIED WITH HW ON HEATING & CW ON COOLING

Heating/Cooling UV (DX Cooling HW Heating) ASHRAE Cycle I, with Optional Electric Heating Coil or Steam Coil

This diagram shows a classroom heating/cooling unit ventilator operating with ASHRAE cycle I. This system has a DX coil for the cooling and a steam or hot water coil for the heating.

On a call for heat the OA damper is closed and the RA damper is open. The valve V-1 is open or the electric heat controller has the electric coil on. The unit is operating at full heating capacity. As the room reaches set point the T-1 pressure increases to 7 psi and the OA damper admits 100% OA. The RA damper is closed. The next action is that the valve is closed. The electric heat controller opens its switches and the coil is off. The unit delivering full natural ventilation. The low limit T-2 prevents the discharge temperature from falling below its set point.

When the outside air is no longer providing cooling, the room thermostat passes more air and the PE switch P-1 closes one of its circuits which energizes E-2 which in turn operates AV-1 to return the OA damper to a minimum position through the PRV PR-1. That action also closes the coil valve. The electric heat controller is also de-energized. On a further rise in room temperature the T-1 operates a second circuit in P-1 to start the refrigeration system. As the set point is reached the T-1 opens the P-1 switches and the system returns to a standard ventilation cycle. P-1 has two circuits in it. One for the compressor circuit and one for the E-1 solenoid air valve.

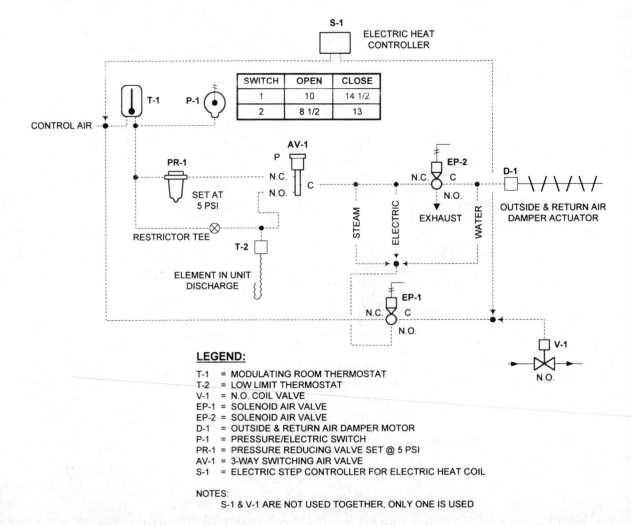

SWITCH	OPEN	CLOSE
1	10	14 1/2
2	8 1/2	13

LEGEND:

T-1 = MODULATING ROOM THERMOSTAT
T-2 = LOW LIMIT THERMOSTAT
V-1 = N.O. COIL VALVE
EP-1 = SOLENOID AIR VALVE
EP-2 = SOLENOID AIR VALVE
D-1 = OUTSIDE & RETURN AIR DAMPER MOTOR
P-1 = PRESSURE/ELECTRIC SWITCH
PR-1 = PRESSURE REDUCING VALVE SET @ 5 PSI
AV-1 = 3-WAY SWITCHING AIR VALVE
S-1 = ELECTRIC STEP CONTROLLER FOR ELECTRIC HEAT COIL

NOTES:
S-1 & V-1 ARE NOT USED TOGETHER, ONLY ONE IS USED

Heating/Cooling UV (DX Cooling/HW Heating) ASHRAE Cycle I with Optional Electric Heating Coil or Steam Coil with Return Air Thermostat

This diagram shows a classroom heating/cooling unit ventilator operating with ASHRAE cycle I. This system has a DX coil for the cooling and a steam or hot water coil for the heating.

On a call for heat the OA damper is closed and the RA damper is open. The valve V-1 is open or the electric heat controller has the electric coil on. The unit is operating at full heating capacity. As the return air temperature at T-1 reaches set point the T-1 pressure increases to 7 psi and the OA damper admits 100% OA. The RA damper is closed. The next action is that the valve is closed. The electric heat controller opens its switches and the coil is off. The unit is delivering full natural ventilation. The low limit T-2 prevents the discharge temperature from falling below its set point.

When the outside air is no longer providing cooling, the room thermostat passes more air and the PE switch P-1 closes one of its circuits which energizes E-2 which in turn operates AV-1 to return the OA damper to a minimum position through the PRV PR-1. That action also closes the coil valve. The electric heat controller is also de-energized. On a further rise in return air temperature, T-1 operates a second circuit in P-1 to start the refrigeration system. As the set point is reached the T-1 opens the P-1 switches and the system returns to a standard ventilation cycle. P-1 has two circuits in it. One for the compressor circuit and one for the E-1 solenoid air valve.

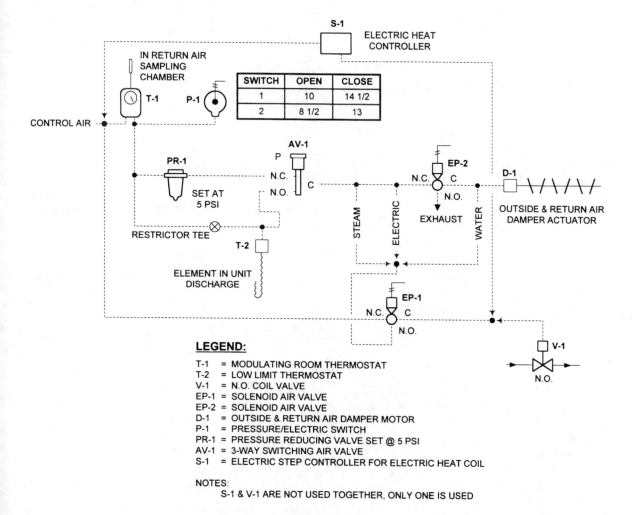

SWITCH	OPEN	CLOSE
1	10	14 1/2
2	8 1/2	13

LEGEND:

T-1	=	MODULATING ROOM THERMOSTAT
T-2	=	LOW LIMIT THERMOSTAT
V-1	=	N.O. COIL VALVE
EP-1	=	SOLENOID AIR VALVE
EP-2	=	SOLENOID AIR VALVE
D-1	=	OUTSIDE & RETURN AIR DAMPER MOTOR
P-1	=	PRESSURE/ELECTRIC SWITCH
PR-1	=	PRESSURE REDUCING VALVE SET @ 5 PSI
AV-1	=	3-WAY SWITCHING AIR VALVE
S-1	=	ELECTRIC STEP CONTROLLER FOR ELECTRIC HEAT COIL

NOTES:
   S-1 & V-1 ARE NOT USED TOGETHER, ONLY ONE IS USED

Heating/Cooling UV (Dx Cooling/HW Heating) ASHRAE Cycle II, with Optional Electric Coil or Steam Coil

This diagram shows a classroom heating/cooling unit ventilator operating with ASHRAE cycle II. This system has a DX coil for the cooling and a steam or hot water coil for the heating.

On a call for heat the OA damper is closed and the RA damper is open. The valve V-1 is open or the electric heat controller has the electric coil on. The unit is operating at full heating capacity. As the room reaches set point, the T-1 pressure increases and the OA damper admits minimum OA. The RA damper is closed proportionately. This admits the ventilation air. On a further increase in pressure from T-1 the OA damper begins to open further admitting up to 100% OA. The electric heat controller has the electric coil off. The RA damper at this point is closed. The unit is delivering its full natural cooling capacity. The low limit thermostat T-2 in the fan discharge limits the fan discharge temperature. E-2 closes the OA dampers and opens the valve when the fan shuts down.

When the unit can no longer cool the space the T-1 closes PE switch P-1 and this energizes E-1 which allows the PRV PR-1 to bring the OA damper system to a minimum OA. On a further rise in room temperature the PE switch P-1 starts the refrigeration system. P-1 has two circuits in it. One for the compressor circuit and one for the E-1 solenoid air valve.

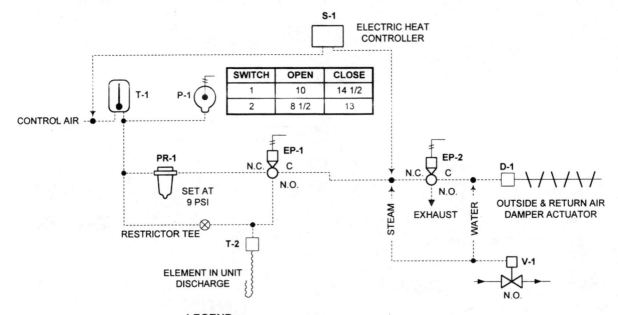

SWITCH	OPEN	CLOSE
1	10	14 1/2
2	8 1/2	13

LEGEND:

T-1	=	MODULATING ROOM THERMOSTAT
T-2	=	LOW LIMIT THERMOSTAT
V-1	=	N.O. COIL VALVE
EP-1	=	SOLENOID AIR VALVE
EP-2	=	SOLENOID AIR VALVE
D-1	=	OUTSIDE & RETURN AIR DAMPER MOTOR
P-1	=	PRESSURE/ELECTRIC SWITCH
PR-1	=	PRESSURE REDUCING VALVE SET @ 9 PSI
S-1	=	ELECTRIC STEP CONTROLLER FOR ELECTRIC HEAT COIL

NOTES:
  S-1 & V-1 ARE NOT USED TOGETHER, ONLY ONE IS USED

Heating/Cooling UV (Dx Cooling/HW Heating) ASHRAE Cycle II with Optional Electric Coil or Steam Coil and R.A. COntrol

This diagram shows a classroom heating/cooling unit ventilator operating with ASHRAE cycle II. This system has a DX coil for the cooling and a steam or hot water coil for the heating.

On a call for heat the OA damper is closed and the RA damper is open. The valve V-1 is open or the electric heat controller has the electric coil on. The unit is operating at full heating capacity. As the return air reaches set point, the T-1 pressure increases and the OA damper admits minimum OA. The RA damper is closed proportionately. This admits the ventilation air. On a further increase in pressure from T-1 the OA damper begins to open further admitting up to 100% OA. The electric heat controller has the electric coil off. The RA damper at this point is closed. The unit is delivering its full natural cooling capacity. The low limit thermostat T-2 in the fan discharge limits the fan discharge temperature. E-2 closes the OA dampers and opens the valve when the fan shuts down.

When the unit can no longer cool the space the T-1 closes PE switch P-1 and this energizes E-1 which allows the PRV PR-1 to bring the OA damper system to a minimum OA. On a further rise in room temperature the PE switch P-1 starts the refrigeration system. P-1 has two circuits in it. One for the compressor circuit and one for the E-1 solenoid air valve.

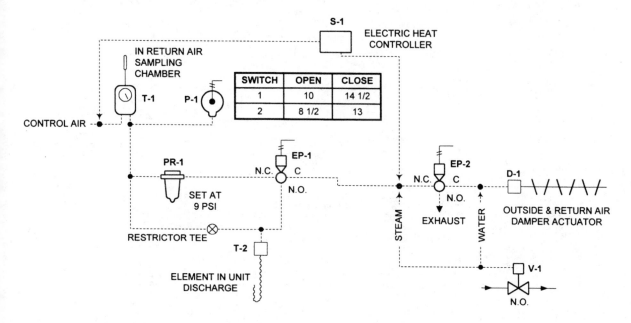

SWITCH	OPEN	CLOSE
1	10	14 1/2
2	8 1/2	13

LEGEND:

T-1 = MODULATING RETURN AIR CAPILLARY THERMOSTAT
T-2 = LOW LIMIT THERMOSTAT
V-1 = N.O. COIL VALVE
EP-1 = SOLENOID AIR VALVE
EP-2 = SOLENOID AIR VALVE
D-1 = OUTSIDE & RETURN AIR DAMPER MOTOR
P-1 = PRESSURE/ELECTRIC SWITCH
PR-1 = PRESSURE REDUCING VALVE SET @ 9 PSI
S-1 = ELECTRIC STEP CONTROLLER FOR ELECTRIC HEAT COIL

NOTES:
     S-1 & V-1 ARE NOT USED TOGETHER, ONLY ONE IS USED

Heating/Cooling UV (Dx Cooling/HW Heating) ASHRAE Cycle III with Optional Electric or Steam Coil

This diagram shows a classroom heating/cooling unit ventilator operating with ASHRAE cycle III. This system has a DX coil for the cooling and a steam, hot water or electric coil for the heating. The valves V-1 and V-2 as well as the heat controller S-1 are optional and not used together.

On a call for heat the OA damper is closed and the RA damper is open. The valve V-1 is open or the electric heat controller has the electric coil on. The unit is operating at full heating capacity. As the room temperature reaches set point the T-1 pressure increases and closes the valve V-1 to the coil. The low limit thermostat T-2 controls the OA and RA damper motor to maintain a minimum fan discharge temperature. The room thermostat controls the heat source without low limit interference. The EP switch E-1 closes the OA damper whenever the fan is not running.

When the OA can no longer provide cooling the room thermostat increases its signal and the second circuit of the PE switch P-1 closes and that energizes E-2 which in turn operates AV-1. That action with a regulated supply from PR-1 to the damper returns the OA damper to a minimum position. As the signal from the T-1 increases, a second switch in the P-1 closes to start the refrigeration system. If the system uses the steam valve, the action of E-2 places full air pressure on V-2 to be sure it is off during the cooling cycle.

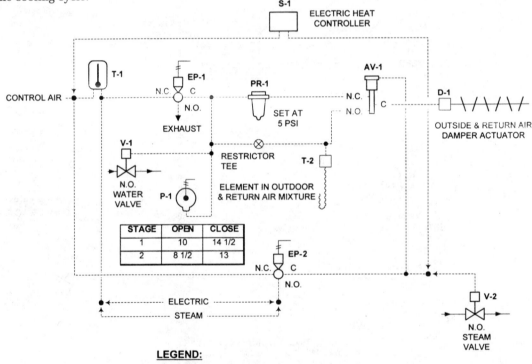

STAGE	OPEN	CLOSE
1	10	14 1/2
2	8 1/2	13

LEGEND:

T-1 = MODULATING ROOM THERMOSTAT
T-2 = LOW LIMIT THERMOSTAT
V-1 = N.O. WATER VALVE
V-2 = N.O. STEAM VALVE
EP-1 = SOLENOID AIR VALVE
EP-2 = SOLENOID AIR VALVE
D-1 = OUTSIDE & RETURN AIR DAMPER MOTOR
P-1 = PRESSURE/ELECTRIC SWITCH
PR-1 = PRESSURE REDUCING VALVE SET @ 5 PSI
AV-1 = 3-WAY SWITCHING AIR VALVE
S-1 = ELECTRIC STEP CONTROLLER FOR ELECTRIC HEAT COIL

NOTES:
S-1, V-1 & V-2 ARE NOT USED TOGETHER, ONLY ONE IS USED

Heating/Cooling UV (Dx Cooling/HW Heating) ASHRAE Cycle III with Optional Electric Coil or Steam Coil and R.A Controls

This diagram shows a classroom heating/cooling unit ventilator operating with ASHRAE cycle III. This system has a DX coil for the cooling and a steam, hot water or electric coil for the heating. The valves V-1 and V-2 as well as the heat controller S-1 are optional and not used together.

On a call for heat the OA damper is closed and the RA damper is open. The valve V-1 is open or the electric heat controller has the electric coil on. The unit is operating at full heating capacity. As the room temperature reaches set point the T-1 pressure increases and closes the valve V-1 to the coil. The low limit thermostat T-2 controls the OA and RA damper motor to maintain a minimum fan discharge temperature. The room thermostat controls the heat source without low limit interference. The EP switch E-1 closes the OA damper whenever the fan is not running.

When the OA can no longer provide cooling the return air thermostat T-1 increases its signal and the second circuit of the PE switch P-1 closes and that energizes E-2 which in turn operates AV-1. That action with a regulated supply from PR-1 to the damper returns the OA damper to a minimum position. As the signal from the T-1 increases, a second switch in the P-1 closes to start the refrigeration system. If the system uses the steam valve, the action of E-2 places full air pressure on V-2 to be sure it is off during the cooling cycle.

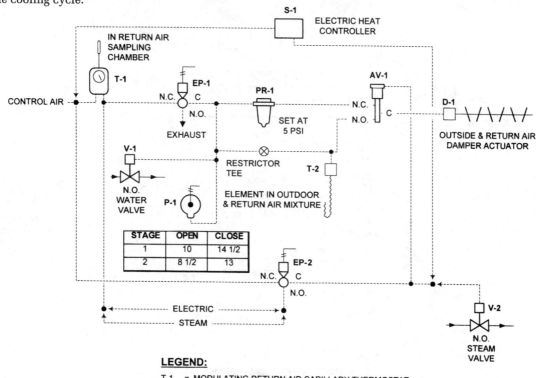

STAGE	OPEN	CLOSE
1	10	14 1/2
2	8 1/2	13

LEGEND:

T-1 = MODULATING RETURN AIR CAPILLARY THERMOSTAT
T-2 = LOW LIMIT THERMOSTAT
V-1 = N.O. WATER VALVE
V-2 = N.O. STEAM VALVE
EP-1 = SOLENOID AIR VALVE
EP-2 = SOLENOID AIR VALVE
D-1 = OUTSIDE & RETURN AIR DAMPER MOTOR
P-1 = PRESSURE/ELECTRIC SWITCH
PR-1 = PRESSURE REDUCING VALVE SET @ 5 PSI
AV-1 = 3-WAY SWITCHING AIR VALVE
S-1 = ELECTRIC STEP CONTROLLER FOR ELECTRIC HEAT COIL

NOTES:
S-1, V-1 & V-2 ARE NOT USED TOGETHER, ONLY ONE IS USED

Heating/Cooling UV (DX Cooling/HW) ASHRAE Cycle W Optional Electric Coil, or Steam Coil

This diagram shows a classroom heating/cooling unit ventilator operating with ASHRAE cycle W. This system has a DX coil for the cooling and a steam or hot water valve V-1 or electric coil for the heating. The valves V-1 as well as the heat controller S-1 are optional and not used together.

On a call for heat the OA damper is closed and the RA damper is open. The valve V-1 is open or the electric heat controller has the electric coil on. The unit is operating at full heating capacity. As the room temperature reaches set point the T-1 pressure increases and closes the valve V-1 to the coil. The low limit thermostat T-2 controls the OA and RA damper motor to maintain a minimum fan discharge temperature. The thermostat T-1 controls the heat source without low limit interference as it controls only the OA/RA damper motor. The EP switch E-1 closes the OA damper whenever the fan is not running.

When the OA can no longer provide cooling the thermostat T-1 increases its signal and the second circuit of the PE switch P-1 closes and that energizes E-2 which allows the pressure from the PRV-1 to bring the OA damper to a minimum position. As the signal from the T-1 increases, a second switch in the P-1 closes to start the refrigeration system.

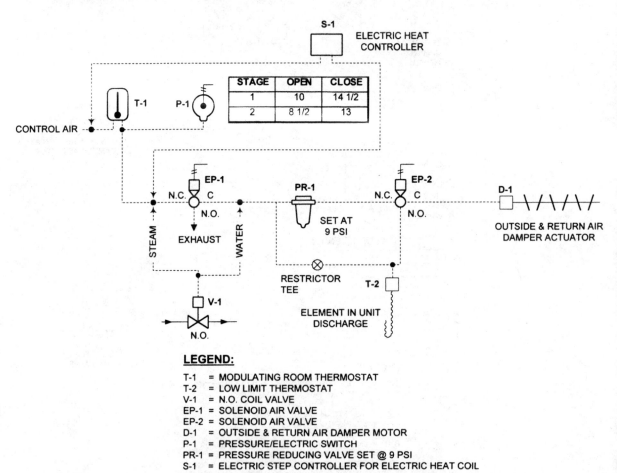

STAGE	OPEN	CLOSE
1	10	14 1/2
2	8 1/2	13

LEGEND:

T-1 = MODULATING ROOM THERMOSTAT
T-2 = LOW LIMIT THERMOSTAT
V-1 = N.O. COIL VALVE
EP-1 = SOLENOID AIR VALVE
EP-2 = SOLENOID AIR VALVE
D-1 = OUTSIDE & RETURN AIR DAMPER MOTOR
P-1 = PRESSURE/ELECTRIC SWITCH
PR-1 = PRESSURE REDUCING VALVE SET @ 9 PSI
S-1 = ELECTRIC STEP CONTROLLER FOR ELECTRIC HEAT COIL

NOTES:
    S-1 & V-1 ARE NOT USED TOGETHER, ONLY ONE IS USED

Heating/Cooling UV (DX Cooling/HW/Steam Htg) ASHRAE Cycle W Optional Electric or Steam Coil and R.A. Controls

This diagram shows a classroom heating/cooling unit ventilator operating with ASHRAE cycle W. This system has a DX coil for the cooling and a steam or hot water valve V-1 or electric coil for the heating. The valves V-1 as well as the heat controller S-1 are optional and not used together.

On a call for heat the OA damper is closed and the RA damper is open. The valve V-1 is open or the electric heat controller has the electric coil on. The unit is operating at full heating capacity. As the room temperature reaches set point the T-1 pressure increases and closes the valve V-1 to the coil. The low limit thermostat T-2 controls the OA and RA damper motor to maintain a minimum fan discharge temperature. The thermostat T-1 controls the heat source without low limit interference as it controls only the OA/RA damper motor. The EP switch E-1 closes the OA damper whenever the fan is not running.

When the OA can no longer provide cooling the thermostat T-1 increases its signal and the second circuit of the PE switch P-1 closes and that energizes E-2 which allows the pressure from the PRV-1 to bring the OA damper to a minimum position. As the signal from the T-1 increases, a second switch in the P-1 closes to start the refrigeration system.

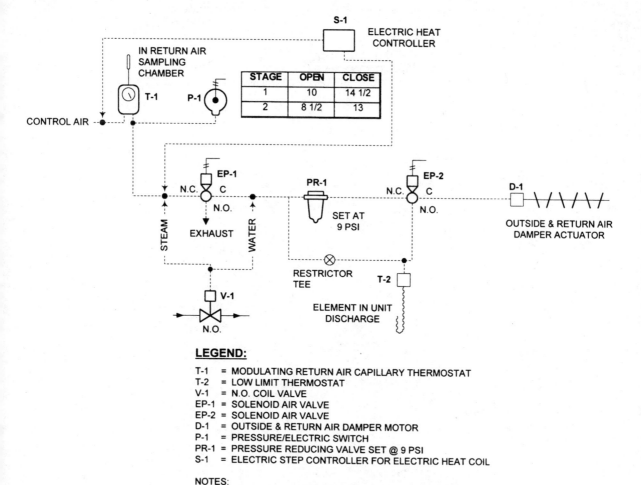

STAGE	OPEN	CLOSE
1	10	14 1/2
2	8 1/2	13

LEGEND:

T-1	=	MODULATING RETURN AIR CAPILLARY THERMOSTAT
T-2	=	LOW LIMIT THERMOSTAT
V-1	=	N.O. COIL VALVE
EP-1	=	SOLENOID AIR VALVE
EP-2	=	SOLENOID AIR VALVE
D-1	=	OUTSIDE & RETURN AIR DAMPER MOTOR
P-1	=	PRESSURE/ELECTRIC SWITCH
PR-1	=	PRESSURE REDUCING VALVE SET @ 9 PSI
S-1	=	ELECTRIC STEP CONTROLLER FOR ELECTRIC HEAT COIL

NOTES:
    S-1 & V-1 ARE NOT USED TOGETHER, ONLY ONE IS USED

Heating/Cooling Unit (HW Heating/CW Cooling) Single Coil as a Three-Pipe System

This diagram shows a system where a single coil is used with a 3-pipe system. Sequenced two-way valves supply hot or chilled water. The 3-way valve V-3 is used for changeover to return the hot water to the boiler (heat exchanger) or return the chilled water to the chiller.

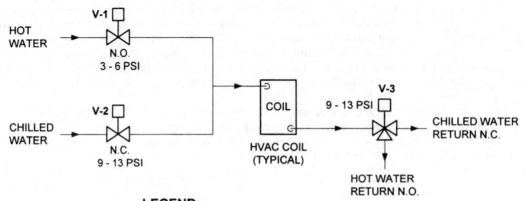

LEGEND:

V-1 = N.O. HW VALVE
V-2 = N.C. CW VALVE
V-3 = TWO POSITION RETURN LINE 3-WAY VALVE

Heating/Cooling UV (HW Heating/CW Cooling) Double Coil as a Four-Pipe System

This diagram shows a system where a single coil is used with a four-pipe system. Sequenced two-way valves supply hot or chilled water on two separate coils. The valves are sequenced so that they are never open at the same time.

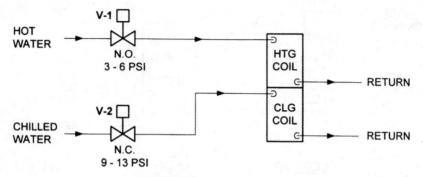

LEGEND:

V-1 = N.O. HW VALVE
V-2 = N.C. CW VALVE

Heating/Cooling UV (Electric Heating/CW Cooling) Double Coil Unit

This diagram shows a system where two coils (one electric and one chilled water) are controlled by a step controller for the heating coil and a valve V-1 on the chilled water coil.

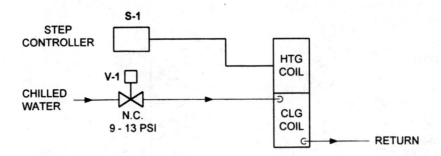

LEGEND:

V-1 = N.C. CW VALVE
S-1 = ELECTRIC STEP CONTROLLER FOR ELECTRIC HEAT COIL

Heating Only UV (HW Heating) Night-Setback, with Partial Day Restoration

This diagram shows a typical classroom unit ventilator controls where the unit is cycled at unoccupied periods (for night). The system requires a "dual" air supply pressure (15 psi for day and 20 psi for night). The cycle is a typical ASHRAE cycle II and the PE switch PE-1 with two circuits wired to the fan motor stops the fan when the supply air temperature reached 20 psi. The T-1 thermostat can close the circuit and restart the fan when the nighttime temperature falls below a certain point.

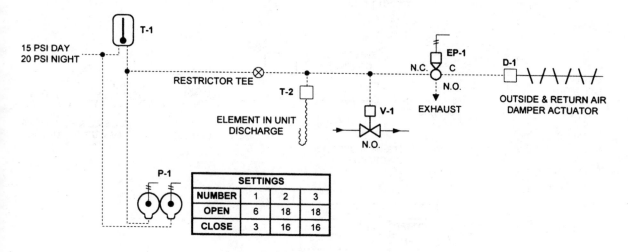

SETTINGS			
NUMBER	1	2	3
OPEN	6	18	18
CLOSE	3	16	16

LEGEND:

T-1 = MODULATING DAY/NIGHT ROOM THERMOSTAT
T-2 = LOW LIMIT THERMOSTAT
V-1 = N.O. HW HEATING VALVE
D-1 = OUTSIDE & RETURN AIR DAMPER MOTOR
EP-1 = SOLENOID AIR VALVE
PE-1 = PRESSURE/ELECTRIC SWITCH

Heating Only UV (Electric Heating) Night-Setback with Partial Day Restoration

This diagram shows a typical classroom unit ventilator controls where the unit is cycled at unoccupied periods (for night). The system requires a "dual" air supply pressure (15psi for day and 20psi for night). The cycle is a typical ASHRAE cycle II and the PE switch PE-1 with two circuits wired to the fan motor stops the fan when the supply air temperature reached 20 psi. The heating is done with an electric step controller on an electric coil. The T-1 thermostat can close the circuit and restart the fan when the nighttime temperature falls below a certain point.

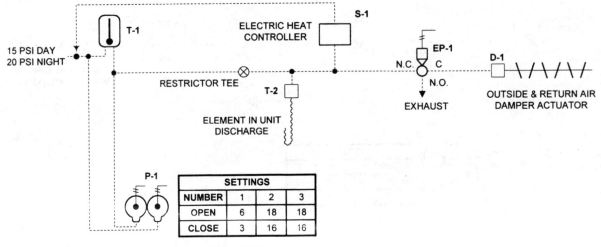

SETTINGS			
NUMBER	1	2	3
OPEN	6	18	18
CLOSE	3	16	16

LEGEND:

T-1	=	MODULATING DAY/NIGHT ROOM THERMOSTAT
T-2	=	LOW LIMIT THERMOSTAT
D-1	=	OUTSIDE & RETURN AIR DAMPER MOTOR
EP-1	=	SOLENOID AIR VALVE
PE-1	=	PRESSURE/ELECTRIC SWITCH
S-1	=	ELECTRIC STEP CONTROLLER FOR ELECTRIC HEAT COIL

Heating Only UV (HW Heating) Night-Setback with Partial Day Restoration and Face and Bypass Controls on Coil

This diagram shows a typical classroom unit ventilator controls where the unit is cycled at unoccupied periods (for night). The system requires a "dual" air supply pressure (15 psi for day and 20 psi for night). The cycle is a typical ASHRAE cycle II and the PE switch PE-1 with two circuits wired to the fan motor stops the fan when the supply air temperature reaches 20 psi. The heating is done with face & by-pass dampers across the heating coil. The T-1 thermostat can close the circuit and restart the fan when the nighttime temperature falls below a certain point.

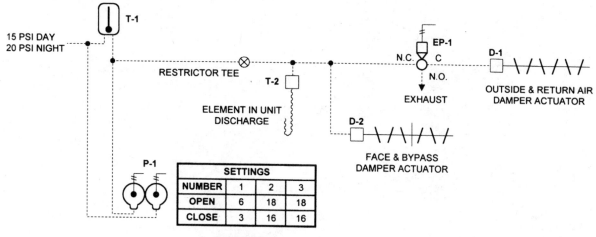

SETTINGS			
NUMBER	1	2	3
OPEN	6	18	18
CLOSE	3	16	16

LEGEND:

T-1	=	MODULATING DAY/NIGHT ROOM THERMOSTAT
T-2	=	LOW LIMIT THERMOSTAT
D-1	=	OUTSIDE & RETURN AIR DAMPER MOTOR
D-2	=	UNIT COIL FACE & BYPASS DAMPER MOTOR
EP-1	=	SOLENOID AIR VALVE
PE-1	=	PRESSURE/ELECTRIC SWITCH

CHAPTER 13

Unit Heaters

Unit Heaters (Steam Heating), Ceiling Hung, with No Valve on Coil, Electric Line Voltage Thermostat

This diagram shows a typical unit heater with a steam coil using a line voltage thermostat to cycle the fan. The strap-on thermostat T-2 prevents the unit fan from running when there is no steam.

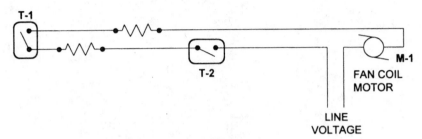

LEGEND:

T-1 = TWO POSITION ELECTRIC LINE VOLTAGE THERMOSTAT
T-2 = TWO POSITION ELECTRIC STRAP-ON THERMOSTAT
M-1 = UNIT HEATER FAN MOTOR

Unit Heaters (HW Heating) Ceiling Hung, with No Valve on Coil, Electric Line Voltage Thermostat

This diagram shows a typical unit heater with a water coil using a line voltage thermostat to cycle the fan. The strap-on thermostat T-2 prevents the unit fan from running when there is no steam.

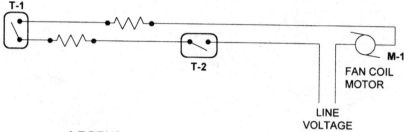

LEGEND:

T-1 = TWO POSITION ELECTRIC LINE VOLTAGE THERMOSTAT
T-2 = TWO POSITION ELECTRIC STRAP-ON THERMOSTAT
M-1 = UNIT HEATER FAN MOTOR

Unit Heaters (HW Heating) Cabinet Type, with No Valve on Coil, Electric Line Voltage Thermostat

This diagram shows a typical cabinet unit heater with a water coil using a line voltage thermostat to cycle the fan. The strap-on thermostat T-2 prevents the unit fan from running when there is no hot water.

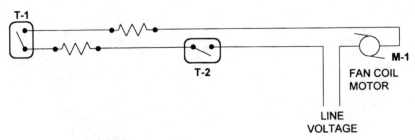

LEGEND:

T-1 = TWO POSITION ELECTRIC LINE VOLTAGE THERMOSTAT
T-2 = TWO POSITION ELECTRIC STRAP-ON THERMOSTAT
M-1 = UNIT HEATER FAN MOTOR

Unit Heaters (Steam Heating) Cabinet Type, with No Valve on Coil, Electric Line Voltage Thermostat

This diagram shows a typical cabinet unit heater with a steam coil using a line voltage thermostat to cycle the fan. The strap-on thermostat T-2 prevents the unit fan from running when there is no steam.

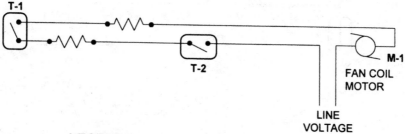

LEGEND:

T-1 = TWO POSITION ELECTRIC LINE VOLTAGE THERMOSTAT
T-2 = TWO POSITION ELECTRIC STRAP-ON THERMOSTAT
M-1 = UNIT HEATER FAN MOTOR

Unit Heaters (Steam or HW Heating) Ceiling Hung, with Valve on Coil

This diagram shows a typical unit heater with a steam coil and a valve on the coil controlled by a modulating room thermostat. The strap-on thermostat T-2 prevents the unit fan from running when there is no steam.

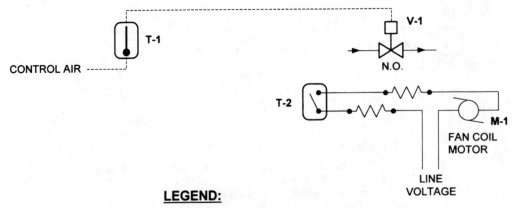

LEGEND:

T-1 = MODULATING ROOM THERMOSTAT
T-2 = TWO POSITION ELECTRIC STRAP-ON THERMOSTAT
V-1 = MODULATING STEAM COIL VALVE
M-1 = UNIT HEATER FAN MOTOR

Unit Heaters (Steam or HW Heating) Cabinet Type, with Valve on Coil

This diagram shows a typical cabinet type unit heater with a steam or hot water coil and a valve on the coil controlled by a modulating room thermostat. The strap-on thermostat T-2 prevents the unit fan from running when there is no steam or HW.

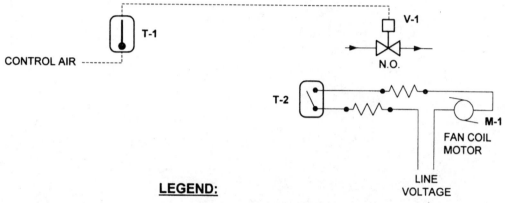

LEGEND:

T-1 = MODULATING ROOM THERMOSTAT
T-2 = TWO POSITION ELECTRIC STRAP-ON THERMOSTAT
V-1 = MODULATING STEAM COIL VALVE
M-1 = UNIT HEATER FAN MOTOR

CHAPTER 14

Radiant Systems

HW Radiant Floor Heating Systems, with Room Controls

This diagram shows a typical hot water floor radiant panel system. The room thermostat T-1 modulates the 3-way valve V-1 feeding the floor coils.

Typically the systems are zoned in schools where the children are on the floor a lot.

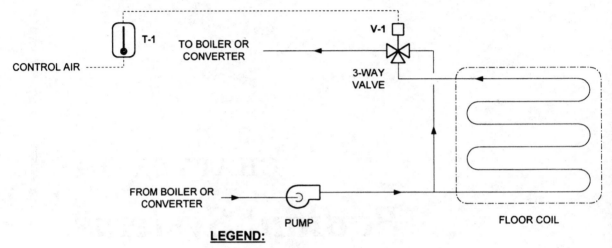

LEGEND:

T-1   = MODULATING ROOM THERMOSTAT
V-1   = MODULATING 3-WAY HW VALVE FOR ZONE

NOTE:
SEE SECTION I FOR PRIMARY CONTROL OF HW SUPPLIED
TO ZONE CONTROL VALVES

Chilled Water Radiant Cooling Systems with Coils in Ceiling

This diagram shows a typical chilled water ceiling radiant panel system. The room thermostat T-1 modulates the 3-way valve V-1 feeding the ceiling coils. The thermostat T-2 is a special "dew point" thermostat that changes the signal to the 3-way valve should "sweating" on the pipes develop.

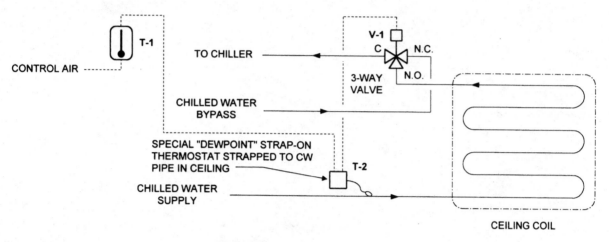

LEGEND:

T-1 = MODULATING ROOM THERMOSTAT
T-2 = SPECIAL DEWPOINT THERMOSTAT
V-1 = MODULATING 3-WAY VALVE ON CW

CHAPTER 15

VAV Box Controls

VAV Box Controls, Pressure Dependent (Cooling Only)

This diagram shows a typical cutoff VAV box that is pressure dependent (no pressure controls at the box). The room thermostat T-1 controls the box damper to reduce the volume of air going into the space. This box is cooling only.

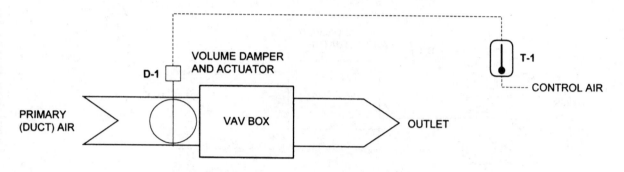

LEGEND:

T-1 = MODULATING ROOM THERMOSTAT
D-1 = VAV BOX DAMPER MOTOR

VAV Box Controls, Pressure Independent (Cooling Only)

This diagram shows a typical cutoff VAV box that is pressure independent. The room thermostat T-1 controls the box damper to reduce the volume of air going into the space. This box is cooling only. The flow controller F-1 at the box maintains a minimum flow to the space.

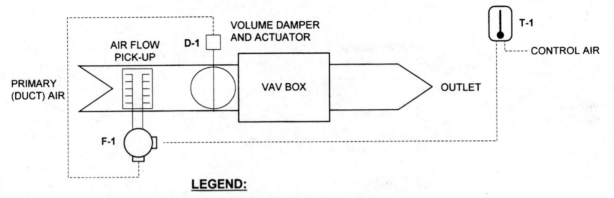

LEGEND:

T-1   = MODULATING ROOM THERMOSTAT
D-1   = VAV BOX DAMPER MOTOR
F-1   = FLOW CONTROLLER

VAV Box Controls, Pressure Dependent, with Reheat HW Coil

This diagram shows a typical cutoff VAV box that is pressure dependent. The room thermostat T-1 controls the box damper and valve V-1 (reheat valve) in sequence. This box is cooling with reheat.

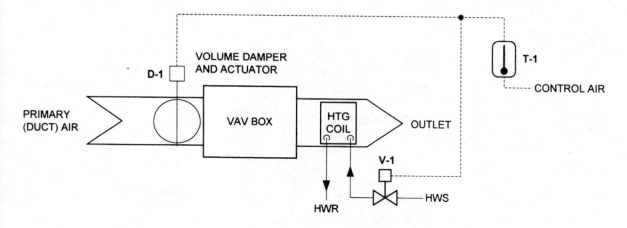

LEGEND:

T-1 = MODULATING ROOM THERMOSTAT
V-1 = MODULATING HW COIL VALVE
D-1 = VAV BOX DAMPER MOTOR

VAV Box Controls, Pressure Dependent, with Reheat Steam Coil

This diagram shows a typical cutoff VAV box that is pressure dependent. The room thermostat T-1 controls the box damper and valve V-1 (reheat valve) in sequence. This box is cooling with reheat.

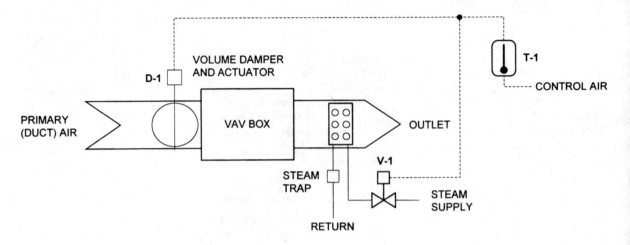

LEGEND:

T-1   = MODULATING ROOM THERMOSTAT
V-1   = MODULATING STEAM COIL VALVE
D-1   = VAV BOX DAMPER MOTOR

VAV Box Controls, Pressure Dependent, with Electric Reheat Coil

This diagram shows a typical cutoff VAV box that is pressure dependent. The room thermostat T-1 controls the box damper and step controller on the electric coil. This box is cooling with reheat.

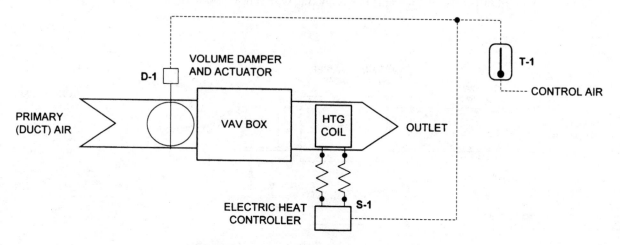

LEGEND:

T-1 = MODULATING ROOM THERMOSTAT
D-1 = VAV BOX DAMPER MOTOR
S-1 = ELECTRIC STEP CONTROLLER FOR ELECTRIC HEAT COIL

VAV Box Controls, Pressure Independent, with HW Reheat Coil

This diagram shows a typical cutoff VAV box that is pressure independent. The room thermostat T-1 controls the box damper and valve V-1 on the reheat coil. The flow controller F-1 operates the damper to prevent airflow that is too low. This box is cooling with reheat.

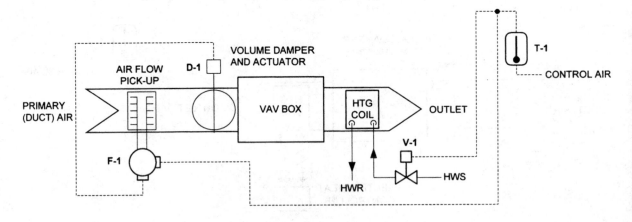

LEGEND:

T-1 = MODULATING ROOM THERMOSTAT
V-1 = MODULATING HW COIL VALVE
D-1 = VAV BOX DAMPER MOTOR

VAV Box Controls, Pressure Independent, with Steam Reheat Coil

This diagram shows a typical cutoff VAV box that is pressure independent. The room thermostat T-1 controls the box damper and valve V-1 on the reheat coil. The flow controller F-1 operates the damper to prevent airflow that is too low. This box is cooling with reheat.

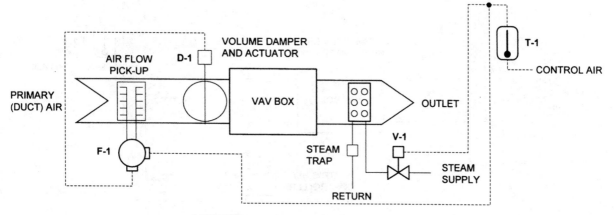

LEGEND:

T-1 = MODULATING ROOM THERMOSTAT
V-1 = MODULATING STEAM COIL VALVE
D-1 = VAV BOX DAMPER MOTOR
F-1 = FLOW CONTROLLER

VAV Box Controls, Pressure Independent, with Electric Reheat Coil

This diagram shows a typical cutoff VAV box that is pressure independent. The room thermostat T-1 controls the box damper and step controller on the electric reheat coil. The flow controller F-1 operates the damper to prevent airflow that is too low. This box is cooling with reheat.

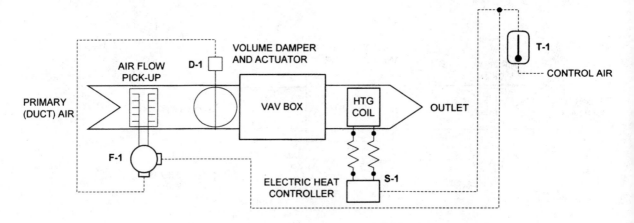

LEGEND:

T-1 = MODULATING ROOM THERMOSTAT
D-1 = VAV BOX DAMPER MOTOR
F-1 = FLOW CONTROLLER
S-1 = ELECTRIC STEP CONTROLLER FOR ELECTRIC HEAT COIL

VAV Box Controls, Pressure Dependent, Bypass Style (Cooling Only)

This diagram shows a typical bypass VAV box where the room thermostat T-1 operates the box damper as a cutoff box that is pressure dependent. When the system is satisfied the damper D-1 returns some of the air to the return plenum.

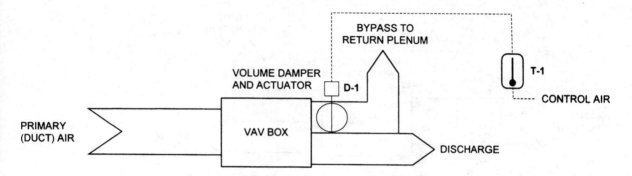

LEGEND:

T-1   = MODULATING ROOM THERMOSTAT
D-1   = VAV BOX DAMPER MOTOR

VAV Box Controls, Pressure Independent, Bypass Style (Cooling Only)

This diagram shows a typical bypass VAV box where the room thermostat T-1 operates the box damper as a cutoff box that is pressure independent. When the system is satisfied the damper D-1 returns some of the air to the return plenum. The flow controller F-1 limits the minimum airflow in to the space.

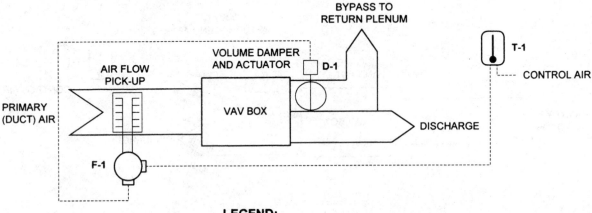

LEGEND:

T-1 = MODULATING ROOM THERMOSTAT
D-1 = VAV BOX DAMPER MOTOR
F-1 = FLOW CONTROLLER

VAV Box Controls, Pressure Dependent, Induction Type (Cooling Only)

This diagram shows a typical induction type VAV box where the room thermostat T-1 operates the box damper as well as the bypass dampers that inject return air from a typical ceiling into the space.

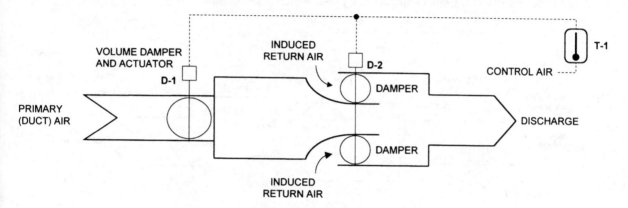

LEGEND:

T-1	= MODULATING ROOM THERMOSTAT
D-1	= VAV BOX DAMPER MOTOR
D-2	= VAV R.A. INDUCED DAMPER MOTOR

VAV Box Controls, Pressure Independent, Induction Type (Cooling Only)

This diagram shows a typical induction type VAV box where the room thermostat T-1 operates the box damper as well as the bypass dampers that inject return air from a typical ceiling into the space. Flow controller F-I allows the box to operate as a pressure independent unit.

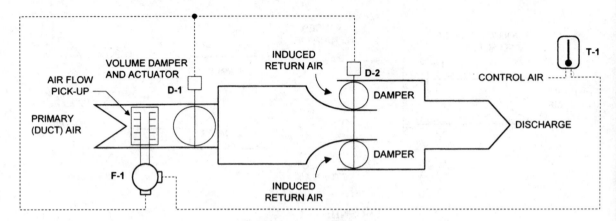

LEGEND:

T-1 = MODULATING ROOM THERMOSTAT
D-1 = VAV BOX DAMPER MOTOR
D-2 = VAV R.A. INDUCED DAMPER MOTOR
F-1 = FLOW CONTROLLER

VAV Box Controls, Pressure Dependent, with Parallel Fan (Cooling Only)

This diagram shows a typical parallel fan type VAV box where the room thermostat T-1 operates the box damper as well as a parallel fan to move air when the damper is closed. T-1 modulates D-1 and starts and stops the fan through PE-1. The unit is pressure dependent.

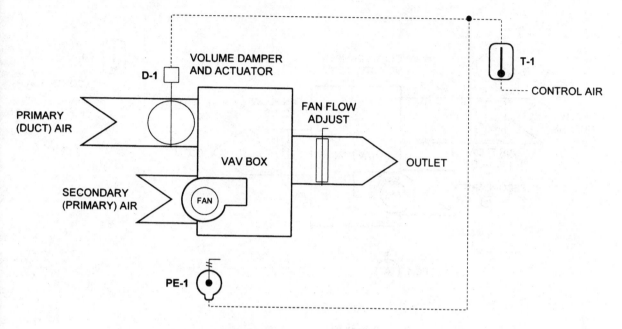

LEGEND:

T-1   = MODULATING ROOM THERMOSTAT
D-1   = VAV PRIMARY AIR DAMPER MOTOR
PE-1 = PRESSURE/ELECTRIC SWITCH

VAV Box Controls, Pressure Independent, with Parallel Fan (Cooling Only)

This diagram shows a typical parallel fan type VAV box where the room thermostat T-1 operates the box damper as well as a parallel fan to move air when the damper is closed. T-1 modulates D-1 and starts and stops the fan through PE-1. The unit is pressure independent.

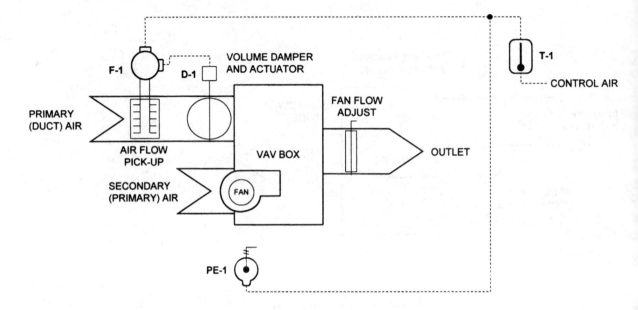

LEGEND:

T-1   = MODULATING ROOM THERMOSTAT
D-1   = VAV PRIMARY AIR DAMPER MOTOR
PE-1 = PRESSURE/ELECTRIC SWITCH
F-1   = FLOW CONTROLLER

VAV Box Controls, Pressure Independent, with Parallel Fan and Steam Reheat Coil

This diagram shows a typical parallel fan type VAV box where the room thermostat T-1 operates the box damper as well as a parallel fan to move air when the damper is closed. T-1 modulates D-1 and starts and stops the fan through PE-1. It also sequences the reheat coil valve V-1. The unit is pressure independent. The flow controller F-1 modulates the damper to limit the amount of air entering the space below a certain level.

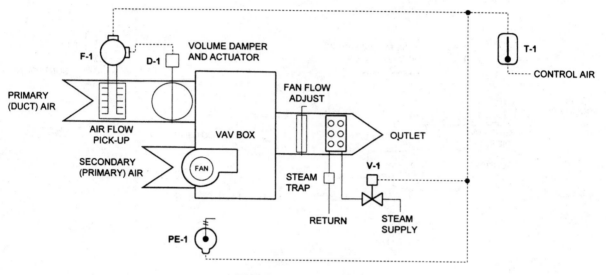

LEGEND:

T-1	=	MODULATING ROOM THERMOSTAT
V-1	=	MODULATING STEAM COIL VALVE
D-1	=	VAV PRIMARY AIR DAMPER MOTOR
PE-1	=	PRESSURE/ELECTRIC SWITCH
F-1	=	FLOW CONTROLLER

VAV Box Controls, Pressure Independent, with Parallel Fan and Hot Water Reheat Coil

This diagram shows a typical parallel fan type VAV box where the room thermostat T-1 operates the box damper as well as a parallel fan to move air when the damper is closed. T-1 modulates D-1 and starts and stops the fan through PE-1. It also sequences the reheat coil valve V-1. The unit is pressure independent. The flow controller F-1 modulates the damper to limit the amount of air entering the space below a certain level.

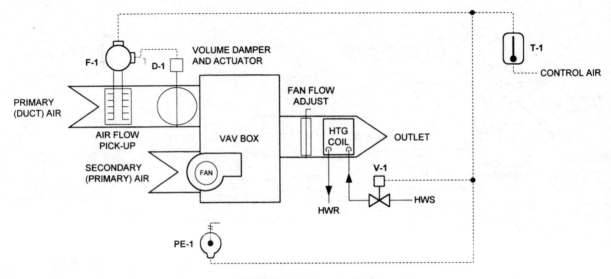

LEGEND:

T-1 = MODULATING ROOM THERMOSTAT
V-1 = MODULATING HW COIL VALVE
D-1 = VAV PRIMARY AIR DAMPER MOTOR
PE-1 = PRESSURE/ELECTRIC SWITCH
F-1 = FLOW CONTROLLER

VAV Box Controls, Pressure Independent, with Parallel Fan and Electric Reheat Coil

This diagram shows a typical parallel fan type VAV box where the room thermostat T-1 operates the box damper as well as a parallel fan to move air when the damper is closed. T-1 modulates D-1 and starts and stops the fan through PE-1. It also sequences the electric reheat coil through the step controller S-1. The unit is pressure independent. The flow controller F-1 modulates the damper to limit the amount of air entering the space below a certain level.

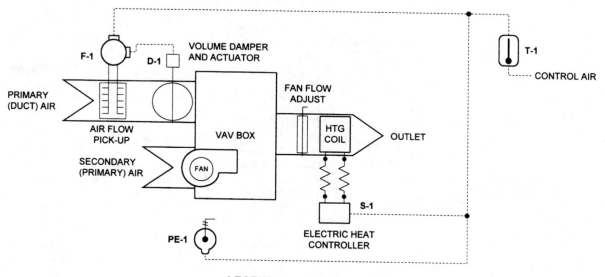

LEGEND:

T-1 = MODULATING ROOM THERMOSTAT
D-1 = VAV PRIMARY AIR DAMPER MOTOR
PE-1 = PRESSURE/ELECTRIC SWITCH
F-1 = FLOW CONTROLLER
S-1 = ELECTRIC STEP CONTROLLER FOR ELECTRIC HEAT COIL

VAV Box Controls, Pressure Dependent, with Parallel Fan and Steam Reheat Coil

This diagram shows a typical parallel fan type VAV box where the room thermostat T-1 operates the box damper as well as a parallel fan to move air when the damper is closed. T-1 modulates D-1 and starts and stops the fan through PE-1. It also sequences the steam reheat coil. The unit is pressure dependent.

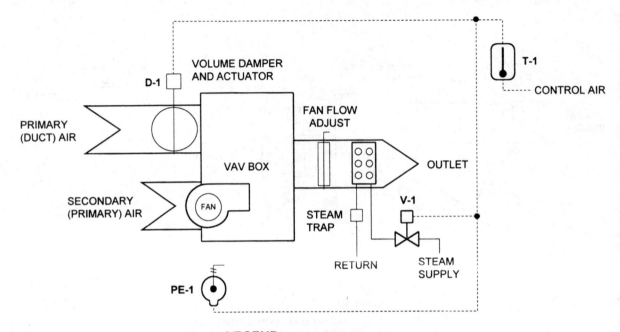

LEGEND:

T-1 = MODULATING ROOM THERMOSTAT
V-1 = MODULATING STEAM COIL VALVE
D-1 = VAV PRIMARY AIR DAMPER MOTOR
PE-1 = PRESSURE/ELECTRIC SWITCH

VAV Box Controls, Pressure Dependent, with Parallel Fan and Hot Water Reheat Coil

This diagram shows a typical parallel fan type VAV box where the room thermostat T-1 operates the box damper as well as a parallel fan to move air when the damper is closed. T-1 modulates D-1 and starts and stops the fan through PE-1. It also sequences the HW reheat coil. The unit is pressure dependent.

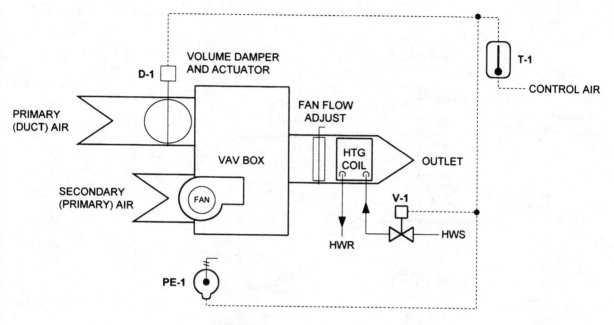

LEGEND:

T-1 = MODULATING ROOM THERMOSTAT
V-1 = MODULATING HW COIL VALVE
D-1 = VAV PRIMARY AIR DAMPER MOTOR
PE-1 = PRESSURE/ELECTRIC SWITCH

VAV Box Controls, Pressure Dependent, with Parallel Fan and Electric Reheat Coil

This diagram shows a typical parallel fan type VAV box where the room thermostat T-1 operates the box damper as well as a parallel fan to move air when the damper is closed. T-1 modulates D-1 and starts and stops the fan through PE-1. It also sequences the electric reheat coil through the step controller S-1. The unit is pressure dependent.

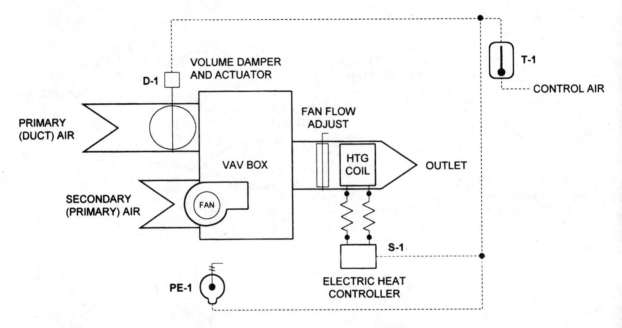

LEGEND:

T-1 = MODULATING ROOM THERMOSTAT
D-1 = VAV PRIMARY AIR DAMPER MOTOR
PE-1 = PRESSURE/ELECTRIC SWITCH
S-1 = ELECTRIC STEP CONTROLLER FOR ELECTRIC HEAT COIL

VAV Box Controls, Pressure Independent, Dual Duct Type with Steam Reheat Coil

This diagram shows a typical double duct type VAV box where the room thermostat T-1 operates the box dampers as well as a reheat coil valve V-1 in sequence. The hot and cold duct dampers operate equal and opposite to each other. The unit is pressure independent, and uses a flow controller F-1.

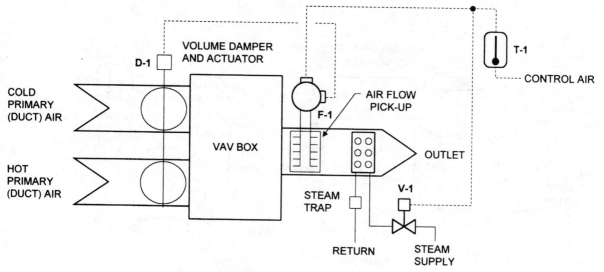

LEGEND:

T-1 = MODULATING ROOM THERMOSTAT
V-1 = MODULATING STEAM COIL VALVE
D-1 = VAV HOT/COLD DAMPER MOTOR
F-1 = FLOW CONTROLLER

VAV Box Controls, Pressure Independent, Dual Duct Type with HW Reheat Coil

This diagram shows a typical double duct type VAV box where the room thermostat T-1 operates the box dampers as well as a reheat coil valve V-1 in sequence. The hot and cold duct dampers operate equal and opposite to each other. The unit is pressure independent, and uses a flow controller

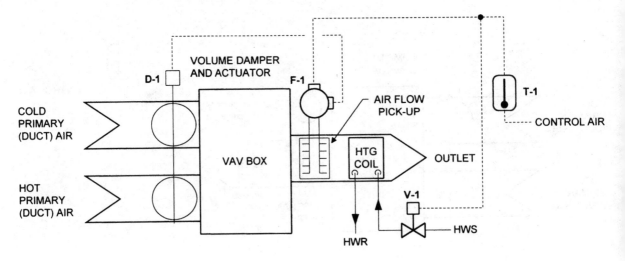

LEGEND:

T-1 = MODULATING ROOM THERMOSTAT
V-1 = MODULATING HW COIL VALVE
D-1 = VAV HOT/COLD DAMPER MOTOR
F-1 = FLOW CONTROLLER

VAV Box Controls, Pressure Dependent, with Electric Reheat Coil

This diagram shows a typical double duct type VAV box where the room thermostat T-1 operates the box dampers as well as an electric reheat coil through step controller S-1. The hot and cold duct dampers operate equal and opposite to each other. The unit is pressure dependent.

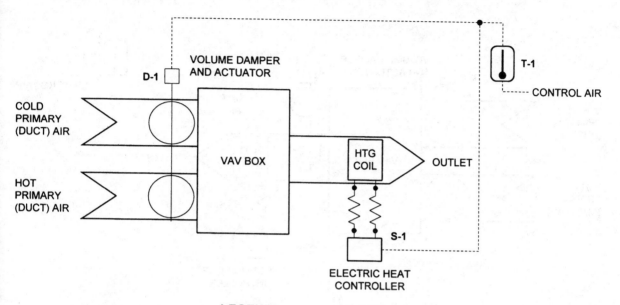

LEGEND:

T-1 = MODULATING ROOM THERMOSTAT
D-1 = VAV HOT/COLD DAMPER MOTOR
S-1 = ELECTRIC STEP CONTROLLER FOR ELECTRIC HEAT COIL

VAV Box Controls, Pressure Independent, with Electric Reheat Coil

This diagram shows a typical double duct type VAV box where the room thermostat T-1 operates the box dampers as well as an electric reheat coil through step controller S-1. The hot and cold duct dampers operate equal and opposite to each other. The unit is pressure independent as it has a flow controller F-1.

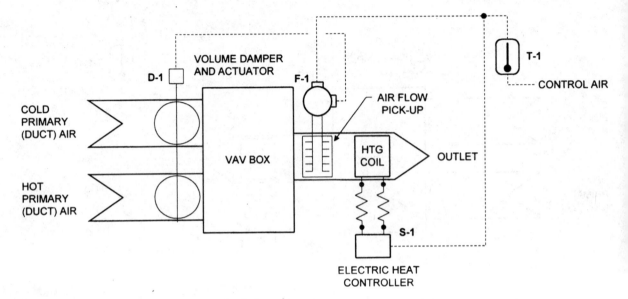

LEGEND:

T-1	=	MODULATING ROOM THERMOSTAT
D-1	=	VAV HOT/COLD DAMPER MOTOR
F-1	=	FLOW CONTROLLER
S-1	=	ELECTRIC STEP CONTROLLER FOR ELECTRIC HEAT COIL

VAV Box Controls, Variable-Constant Volume, Pressure Independent

This diagram shows a typical double duct type VAV box that is variable volume and constant volume with two flow controllers. The room thermostat T-1 operates the box dampers through the flow controllers F-1 and F-2. The hot and cold duct dampers operate equal and opposite to each other. The unit is pressure independent as it has flow controllers F-1 and F-2 in the hot and cold ducts.

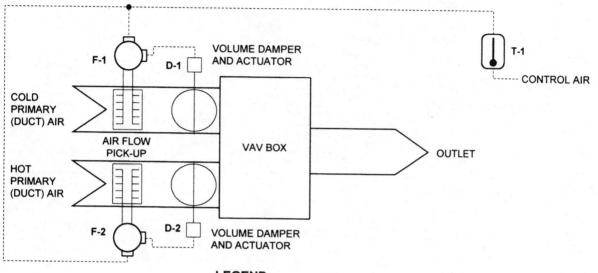

LEGEND:

T-1 = MODULATING ROOM THERMOSTAT
D-1 = VAV COLD DAMPER MOTOR
D-2 = VAV HOT DAMPER MOTOR
F-1 = FLOW CONTROLLER
F-2 = FLOW CONTROLLER

Air-Handling Units

CHAPTER 16

Single-Path Units

100% O.A. Unit, Ventilation Only, with Steam Preheat Coil

Whenever the fan runs, the EP-1 is energized and the O.A. damper opens. If the unit is large, arrangements need to be made to be sure the damper is open before the fan runs.

Thermostat T-1 opens the steam coil valve on the preheat coil whenever the O.A. temperature is below the setting of the thermostat.

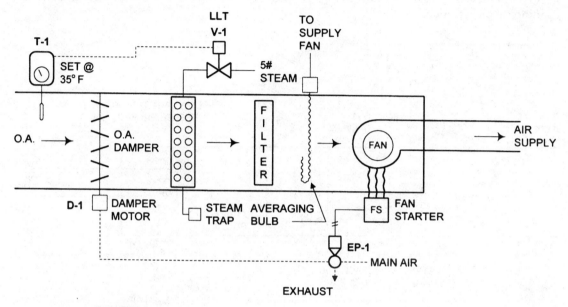

LEGEND:

T-1   = TWO POSITION CAPILLARY THERMOSTAT
V-1   = PREHEAT COIL STEAM VALVE
D-1   = TWO POSITION OA DAMPER MOTOR
EP-1 = SOLENOID AIR VALVE
LLT   = LOW LIMIT ELECTRIC FREEZE PROTECTION THERMOSTAT

100% O.A. Unit Ventilation Only, with Steam Preheat Coil and Face and Bypass Dampers

Whenever the fan runs, the EP-1 is energized and the O.A. damper opens. If the unit is large, arrangements need to be made to be sure the damper is open before the fan runs.

Thermostat T-1 opens the steam coil valve on the preheat coil whenever the O.A. temperature is below the setting of the thermostat. Thermostat T-2 modulates the face and bypass dampers on the preheat coil to prevent overheating downstream. Low-limit thermostat LLT stops the fan on a freezing condition at that location.

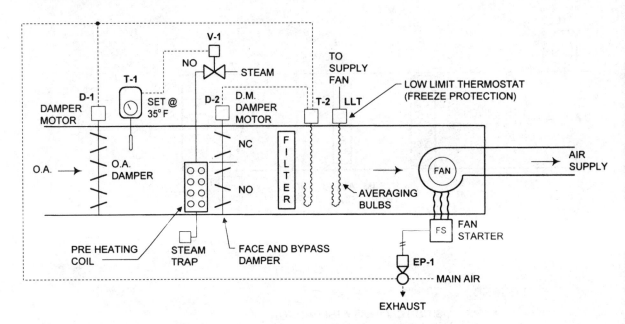

LEGEND:

T-1 = TWO POSITION CAPILLARY THERMOSTAT
T-2 = MODULATING CAPILLARY THERMOSTAT
D-1 = TWO POSITION OA DAMPER MOTOR
D-2 = MODULATING FACE & BYPASS DAMPER MOTOR
V-1 = TWO POSITION STEAM VALVE
EP-1 = SOLENOID AIR VALVE
LLT = LOW LIMIT ELECTRIC FREEZE PROTECTION THERMOSTAT

100% O.A. Unit, Ventilation Only, with Hot Water Preheat Coil

Whenever the fan runs, the EP-1 is energized and the O.A. damper opens. If the unit is large, arrangements need to be made to be sure the damper is open before the fan runs.

Thermostat T-1 modulates the three-way valve on the hot water preheat coil. Low-limit thermostat LLT stops the fan on a freezing condition at that location.

The pump must run whenever the fan runs, and an alarm system is recommended should the pump stop when O.A. freezing conditions exist.

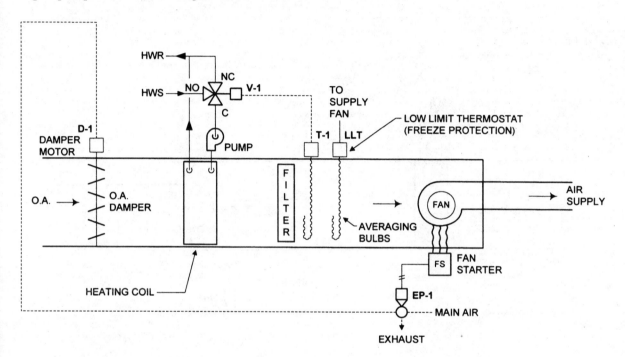

LEGEND:

T-1 = MODULATING CAPILLARY THERMOSTAT
V-1 = MODULATING 3-WAY HOT WATER VALVE
D-1 = TWO POSITION OA DAMPER MOTOR
EP-1 = SOLENOID AIR VALVE
LLT = LOW LIMIT ELECTRIC FREEZE PROTECTION THERMOSTAT

100% O.A. Unit, Ventilation Only, with Hot Water Preheat Coil

Whenever the fan runs, the EP-1 is energized and the O.A. damper opens. If the unit is large, arrangements need to be made to be sure the damper is open before the fan runs.

Thermostat T-1 modulates the two-way water valve on the hot water preheat coil. Low-limit thermostat LLT stops the fan on a freezing condition at that location.

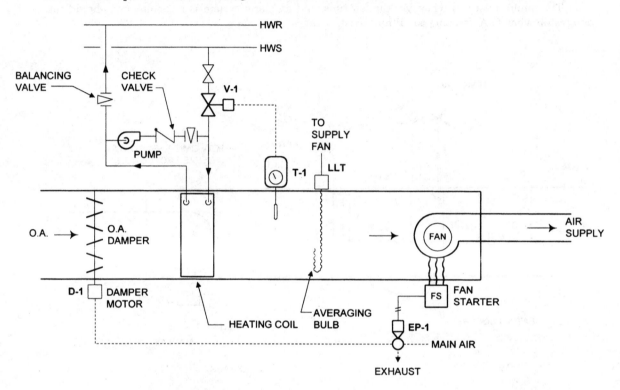

LEGEND:

T-1 = MODULATING CAPILLARY THERMOSTAT
V-1 = MODULATING HOT WATER VALVE
D-1 = TWO POSITION OA DAMPER MOTOR
EP-1 = SOLENOID AIR VALVE
LLT = LOW LIMIT ELECTRIC FREEZE PROTECTION THERMOSTAT

100% O.A. Unit, Ventilation Only, with Steam Preheat Coil and Hot Water or Steam Reheat Coil

Whenever the fan runs, the EP-1 is energized and the O.A. damper opens. If the unit is large, arrangements need to be made to be sure the damper is open before the fan runs.

Thermostat T-1 opens the steam coil valve on the preheat coil whenever the O.A. temperature is below the setting of the thermostat. Room thermostat T-2 modulates the steam or hot water valve on the reheat coil(s).

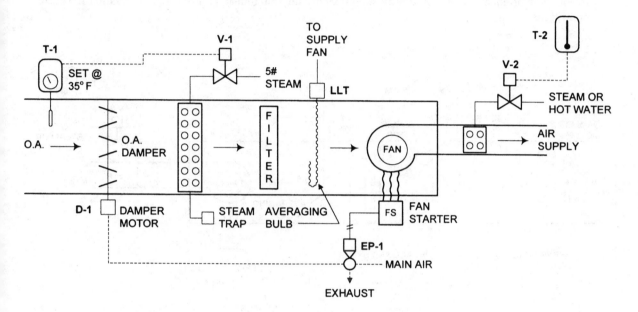

LEGEND:

T-1 = TWO POSITION CAPILLARY THERMOSTAT
T-2 = MODULATING ROOM THERMOSTAT
V-1 = PREHEAT COIL STEAM VALVE
V-2 = MODULATING REHEAT STEAM/HOT WATER COIL VALVE
D-1 = TWO POSITION OA DAMPER MOTOR
EP-1 = SOLENOID AIR VALVE
LLT = LOW LIMIT ELECTRIC FREEZE PROTECTION THERMOSTAT

100% O.A. Unit, Ventilation Only, with Steam Preheat Coil and Face and Bypass Dampers, and Steam or Hot Water Reheat Coil(s)

Whenever the fan runs, the EP-1 is energized and the O.A. damper opens. If the unit is large, arrangements need to be made to be sure the damper is open before the fan runs.

Thermostat T-1 opens the steam coil valve on the preheat coil whenever the O.A. temperature is below the setting of the thermostat. Thermostat T-2 modulates the face and bypass dampers on the preheat coil to prevent overheating downstream. Low-limit thermostat LLT stops the fan on a freezing condition at that location. Room thermostat(s) modulate the reheat coil valve(s) to maintain space conditions.

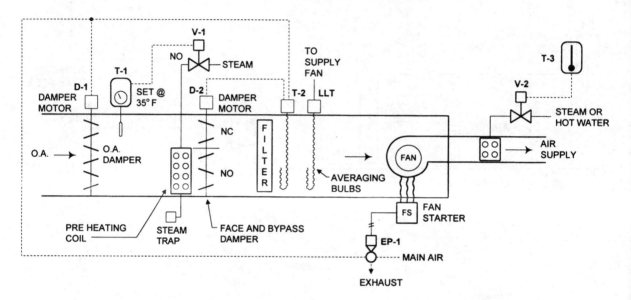

LEGEND:

T-1	=	TWO POSITION CAPILLARY THERMOSTAT
T-2	=	MODULATING CAPILLARY THERMOSTAT
T-3	=	MODULATING ROOM THERMOSTAT
V-1	=	TWO POSITION STEAM VALVE
V-2	=	MODULATING STEAM/HOT WATER REHEAT COIL VALVE
D-1	=	TWO POSITION OA DAMPER MOTOR
D-2	=	MODULATING FACE & BYPASS DAMPER MOTOR
EP-1	=	SOLENOID AIR VALVE
LLT	=	LOW LIMIT ELECTRIC FREEZE PROTECTION THERMOSTAT

100% O.A. Unit, Ventilation Only, with Hot Water Preheat Coil, and Hot Water or Steam Reheat Coil(s)

Whenever the fan runs, the EP-1 is energized and the O.A. damper opens. If the unit is large, arrangements need to be made to be sure the damper is open before the fan runs.

Thermostat T-1 modulates three-way valve V-1 on the hot water preheat coil. Low-limit thermostat LLT stops the fan on a freezing condition at that location. Room thermostat(s) T-2 modulate valve(s) V-2 on steam or hot water reheat coils.

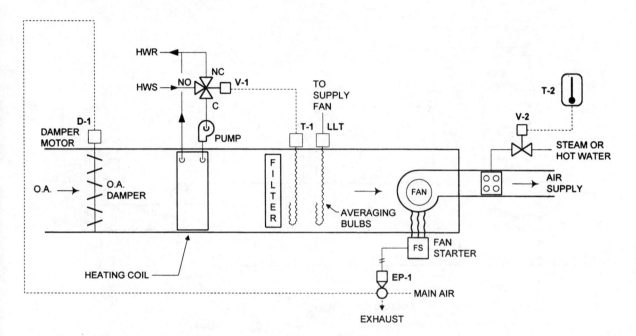

LEGEND:

T-1 = MODULATING CAPILLARY THERMOSTAT
T-2 = MODULATING ROOM THERMOSTAT
V-1 = MODULATING 3-WAY HOT WATER VALVE
V-2 = MODULATING STEAM/HOT WATER REHEAT COIL VALVE
D-1 = TWO POSITION OA DAMPER MOTOR
EP-1 = SOLENOID AIR VALVE
LLT = LOW LIMIT ELECTRIC FREEZE PROTECTION THERMOSTAT

100% O.A. Unit, Ventilation Only, with Hot Water Preheat Coil and Hot Water or Steam Reheat Coil(s)

Whenever the fan runs, the EP-1 is energized and the O.A. damper opens. If the unit is large, arrangements need to be made to be sure the damper is open before the fan runs.

Thermostat T-1 modulates the two-way valve V-1 on the hot water preheat coil. Low-limit thermostat LLT stops the fan on a freezing condition at that location. Room thermostat(s) T-2 modulate valve(s) V-2 on steam or hot water reheat coils.

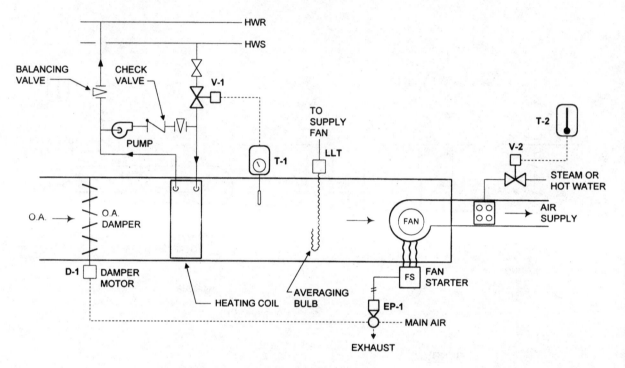

LEGEND:

T-1 = MODULATING CAPILLARY THERMOSTAT
T-2 = MODULATING ROOM THERMOSTAT
V-1 = MODULATING HOT WATER VALVE
V-2 = MODULATING STEAM/HOT WATER REHEAT COIL VALVE
D-1 = TWO POSITION OA DAMPER MOTOR
EP-1 = SOLENOID AIR VALVE
LLT = LOW LIMIT ELECTRIC FREEZE PROTECTION THERMOSTAT

100% O.A. Unit, Ventilation Only, with Steam Preheat Coil and Steam Humidifier

Whenever the fan runs, the EP-1 is energized and the O.A. damper opens. If the unit is large, arrangements need to be made to be sure the damper is open before the fan runs.

Thermostat T-1 opens valve V-1 whenever freeze-up temperatures exist. Room humidistat H-1 modulates valve V-2 on the steam humidifier through the high-limit duct humidistat.

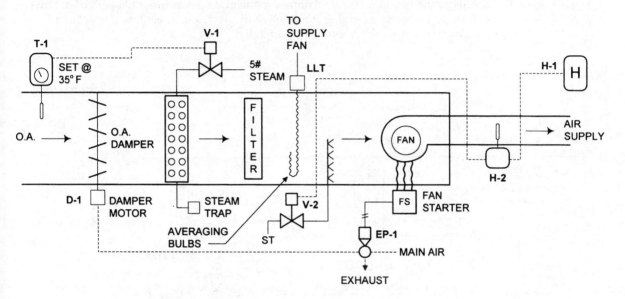

LEGEND:

T-1 = TWO POSITION CAPILLARY THERMOSTAT
H-1 = MODULATING ROOM HUMIDISTAT
H-2 = HIGH LIMIT DUCT HUMIDISTAT
V-1 = PREHEAT COIL STEAM VALVE
V-2 = N.C. MODULATING HUMIDIFIER STEAM VALVE
D-1 = TWO POSITION OA DAMPER MOTOR
EP-1 = SOLENOID AIR VALVE
LLT = LOW LIMIT ELECTRIC FREEZE PROTECTION THERMOSTAT

100% O.A. Unit, Ventilation Only, with Steam Preheat Coil, Coil Face-Bypass Dampers and Steam Humidifier

Whenever the fan runs, the EP-1 is energized and the O.A. damper opens. If the unit is large, arrangements need to be made to be sure the damper is open before the fan runs.

Thermostat T-1 opens valve V-1 whenever the condition exists indicating freeze-up temperatures. Thermostat T-2 modulates the face and by-pass dampers to maintain downstream temperatures. Low-limit thermostat LLT stops the fan on a freezing condition at that location. Room humidistat H-1 modulates valve(s) V-2 on the steam humidifier through the high-limit duct humidistat.

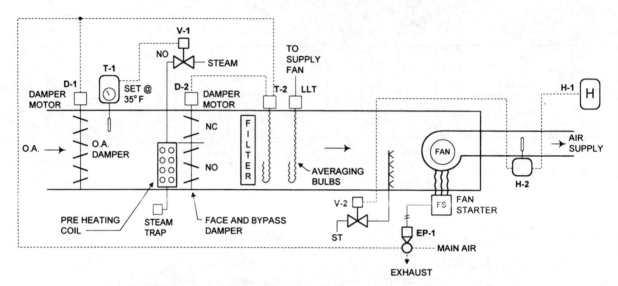

LEGEND:

T-1   = TWO POSITION CAPILLARY THERMOSTAT
T-2   = MODULATING CAPILLARY THERMOSTAT
H-1   = MODULATING ROOM HUMIDISTAT
H-2   = HIGH LIMIT DUCT HUMIDISTAT
V-1   = TWO POSITION STEAM VALVE
V-2   = N.C. MODULATING HUMIDIFIER STEAM VALVE
D-1   = TWO POSITION OA DAMPER MOTOR
D-2   = MODULATING FACE & BYPASS DAMPER MOTOR
EP-1 = SOLENOID AIR VALVE
LLT   = LOW LIMIT ELECTRIC FREEZE PROTECTION THERMOSTAT

100% O.A. Unit, Ventilation Only, with Hot Water Preheat Coil and Steam Humidifier

Whenever the fan runs, the EP-1 is energized and the O.A. damper opens. If the unit is large, arrangements need to be made to be sure the damper is open before the fan runs.

Thermostat T-1 modulates valve V-1 on the hot water preheat coil. Low-limit thermostat LLT stops the fan whenever freezing conditions exist at its location. Room humidistat H-1 modulates valve V-2 on the steam humidifier through the high-limit duct humidistat H-2.

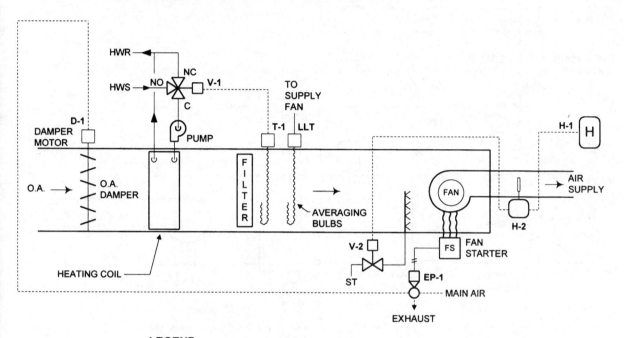

LEGEND:

T-1	=	MODULATING CAPILLARY THERMOSTAT
H-1	=	MODULATING ROOM HUMIDISTAT
H-2	=	HIGH LIMIT DUCT HUMIDISTAT
V-1	=	MODULATING 3-WAY HOT WATER VALVE
V-2	=	N.C. MODULATING HUMIDIFIER STEAM VALVE
D-1	=	TWO POSITION OA DAMPER MOTOR
EP-1	=	SOLENOID AIR VALVE
LLT	=	LOW LIMIT ELECTRIC FREEZE PROTECTION THERMOSTAT

100% O.A. Unit, Ventilation Only, with Hot Water Preheat Coil and Steam Humidifier

Whenever the fan runs, the EP-1 is energized and the O.A. damper opens. If the unit is large, arrangements need to be made to be sure the damper is open before the fan runs.

Thermostat T-1 modulates valve V-1 on the hot water preheat coil. Room humidistat H-1 modulates valve V-2 on the steam humidifier through the high-limit duct humidistat H-2.

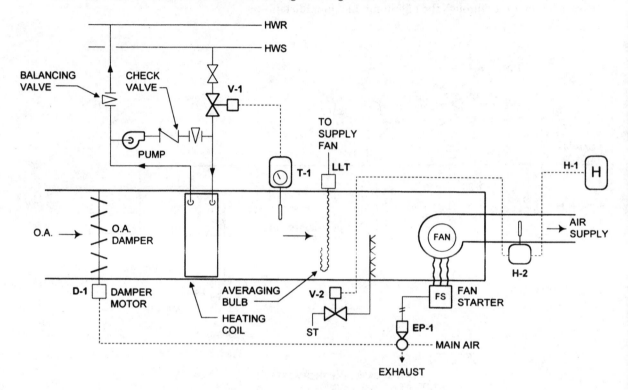

LEGEND:

T-1	=	MODULATING CAPILLARY THERMOSTAT
H-1	=	MODULATING ROOM HUMIDISTAT
H-2	=	HIGH LIMIT DUCT HUMIDISTAT
V-1	=	MODULATING HOT WATER VALVE
V-2	=	N.C. MODULATING HUMIDIFIER STEAM VALVE
D-1	=	TWO POSITION OA DAMPER MOTOR
EP-1	=	SOLENOID AIR VALVE

100% O.A. Unit, Ventilation Only, with Steam Preheat Coil, Steam Humidifier, and Hot Water or Steam Reheat Coil(s)

Whenever the fan runs, the EP-1 is energized and the O.A. damper opens. If the unit is large, arrangements need to be made to be sure the damper is open before the fan runs.

Thermostat T-1 opens valve V-1 on the steam preheat coil. Room humidistat H-1 modulates valve V-3 on the steam humidifier through the high-limit duct humidistat H-2. Room thermostat T-2 modulates valve V-2 on the reheat coil(s).

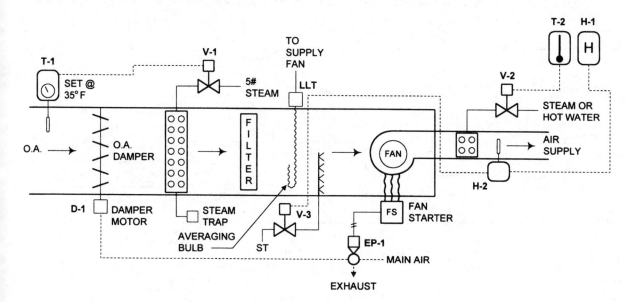

LEGEND:

T-1 = TWO POSITION CAPILLARY THERMOSTAT
T-2 = MODULATING ROOM THERMOSTAT
H-1 = MODULATING ROOM HUMIDISTAT
H-2 = HIGH LIMIT DUCT HUMIDISTAT
V-1 = PREHEAT COIL STEAM VALVE
V-2 = MODULATING REHEAT STEAM/HOT WATER COIL VALVE
V-3 = N.C. MODULATING HUMIDIFIER STEAM VALVE
D-1 = TWO POSITION OA DAMPER MOTOR
EP-1 = SOLENOID AIR VALVE
LLT = LOW LIMIT ELECTRIC FREEZE PROTECTION THERMOSTAT

100% O.A. Unit, Ventilation Only, with Steam Preheat Coil, Face and Bypass Dampers, Steam Humidifier, and Hot Water or Steam Reheat Coil(s)

Whenever the fan runs, the EP-1 is energized and the O.A. damper opens. If the unit is large, arrangements need to be made to be sure the damper is open before the fan runs.

Thermostat T-1 opens valve V-1 on the steam preheat coil. Thermostat T-2 modulates the coil face and bypass dampers on the preheat coil. Low-limit thermostat LLT stops the fan on evidence of freezing conditions at its location. Room humidistat H-1 modulates valve V-3 on the steam humidifier through the high-limit duct humidistat H-2. Room thermostat T-3 modulates valve V-2 on the reheat coil(s).

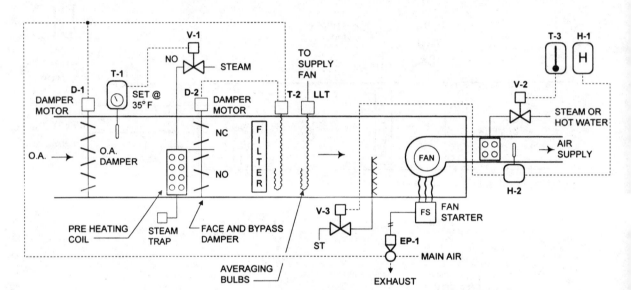

LEGEND:

T-1	=	TWO POSITION CAPILLARY THERMOSTAT
T-2	=	MODULATING CAPILLARY THERMOSTAT
T-3	=	MODULATING ROOM THERMOSTAT
H-1	=	MODULATING ROOM HUMIDISTAT
H-2	=	HIGH LIMIT DUCT HUMIDISTAT
V-1	=	TWO POSITION STEAM VALVE
V-2	=	MODULATING REHEAT STEAM/HOT WATER COIL VALVE
V-3	=	N.C. MODULATING HUMIDIFIER STEAM VALVE
D-1	=	TWO POSITION OA DAMPER MOTOR
D-2	=	MODULATING FACE & BYPASS DAMPER MOTOR
EP-1	=	SOLENOID AIR VALVE
LLT	=	LOW LIMIT ELECTRIC FREEZE PROTECTION THERMOSTAT

100% O.A. Unit, Ventilation Only, with Hot Water Preheat Coil, Steam Humidifier, and Hot Water or Steam Reheat Coil(s)

Whenever the fan runs, the EP-1 is energized and the O.A. damper opens. If the unit is large, arrangements need to be made to be sure the damper is open before the fan runs.

Thermostat T-1 modulates valve V-1 on the hot water preheat coil. Low-limit thermostat LLT stops the fan on evidence of freezing conditions at its location. Room humidistat H-1 modulates valve V-3 on the steam humidifier through the high-limit duct humidistat H-2. Room thermostat T-2 modulates valve V-2 on the reheat coil(s).

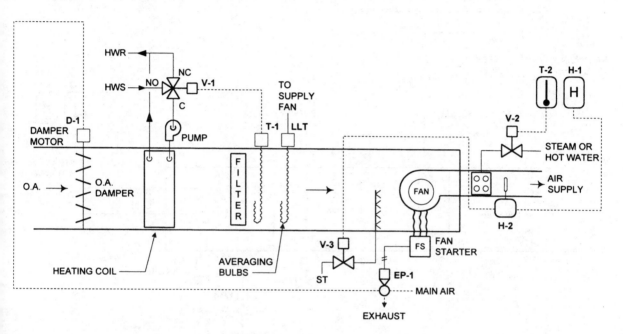

LEGEND:

T-1	= MODULATING CAPILLARY THERMOSTAT
T-2	= MODULATING ROOM THERMOSTAT
H-1	= MODULATING ROOM HUMIDISTAT
H-2	= HIGH LIMIT DUCT HUMIDISTAT
V-1	= MODULATING 3-WAY HOT WATER VALVE
V-2	= MODULATING REHEAT STEAM/HOT WATER COIL VALVE
V-3	= N.C. MODULATING HUMIDIFIER STEAM VALVE
D-1	= TWO POSITION OA DAMPER MOTOR
EP-1	= SOLENOID AIR VALVE
LLT	= LOW LIMIT ELECTRIC FREEZE PROTECTION THERMOSTAT

100% O.A. Unit, Ventilation Only, with Hot Water Preheat Coil, Steam Humidifier, and Hot Water or Steam Reheat Coil(s)

Whenever the fan runs, the EP-1 is energized and the O.A. damper opens. If the unit is large, arrangements need to be made to be sure the damper is open before the fan runs.

Thermostat T-1 modulates valve V-1 on the hot water preheat coil. Room humidistat H-1 modulates valve V-3 on the steam humidifier through the high-limit duct humidistat H-2. Room thermostat T-2 modulates valve V-2 on the reheat coil(s).

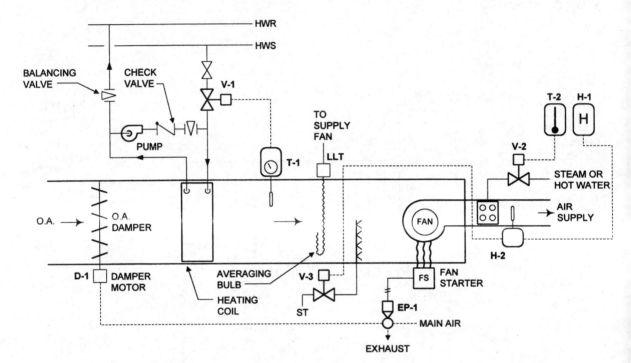

LEGEND:

T-1 = MODULATING CAPILLARY THERMOSTAT
T-2 = MODULATING ROOM THERMOSTAT
H-1 = MODULATING ROOM HUMIDISTAT
H-2 = HIGH LIMIT DUCT HUMIDISTAT
V-1 = MODULATING HOT WATER VALVE
V-2 = MODULATING REHEAT STEAM/HOT WATER COIL VALVE
V-3 = N.C. MODULATING HUMIDIFIER STEAM VALVE
D-1 = TWO POSITION OA DAMPER MOTOR
EP-1 = SOLENOID AIR VALVE
LLT = LOW LIMIT ELECTRIC FREEZE PROTECTION THERMOSTAT

100% O.A. Unit, Ventilation Heating and Cooling, with Steam Preheat Coil, Hot Water Heating, and Chilled Water Cooling Coils

Whenever the fan runs, the EP-1 is energized and the O.A. damper opens. If the unit is large, arrangements need to be made to be sure the damper is open before the fan runs.

Thermostat T-1 opens valve V-1 on the steam preheat coil, when freezing conditions exist. Room thermostat T-2 modulates valves V-2 and V-3 on heating and cooling coils in sequence to maintain space temperatures.

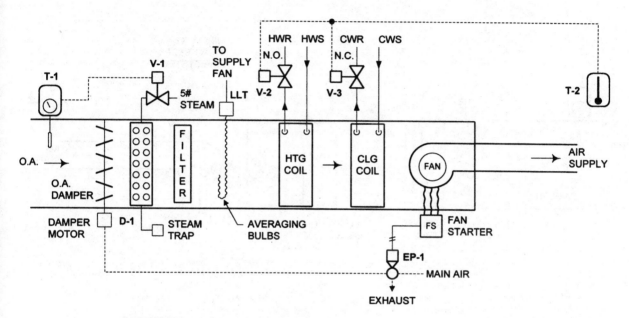

LEGEND:

T-1 = TWO POSITION CAPILLARY THERMOSTAT
T-2 = MODULATING ROOM THERMOSTAT
V-1 = TWO POSITION PREHEAT STEAM COIL VALVE
V-2 = N.O. MODULATING 2-WAY HEATING COIL VALVE
V-3 = N.C. MODULATING 2-WAY COOLING COIL VALVE
D-1 = TWO POSITION OA DAMPER MOTOR
EP-1 = SOLENOID AIR VALVE
LLT = LOW LIMIT ELECTRIC FREEZE PROTECTION THERMOSTAT

100% O.A. Unit, Ventilation Heating and Cooling, with Steam Preheat coil, Hot Water Heating and Cooling Coils

Whenever the fan runs, the EP-1 is energized and the O.A. damper opens. If the unit is large, arrangements need to be made to be sure the damper is open before the fan runs.

Thermostat T-1 opens valve V-1 on the steam preheat coil, when freezing conditions exist. Room thermostat T-2 modulates valves V-2 and also controls PE-1 on the DX cooling coil sequence to maintain space temperatures.

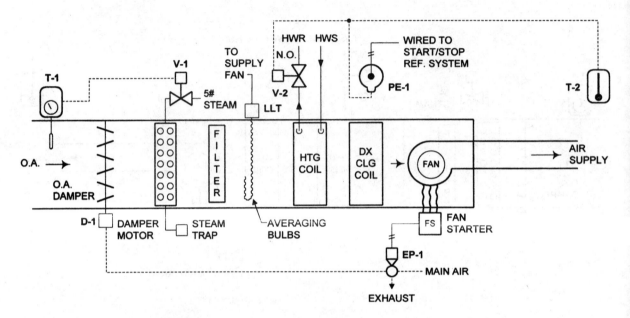

LEGEND:

T-1 = TWO POSITION CAPILLARY THERMOSTAT
T-2 = MODULATING ROOM THERMOSTAT
V-1 = TWO POSITION PREHEAT STEAM COIL VALVE
V-2 = N.O. MODULATING 2-WAY HEATING COIL VALVE
D-1 = TWO POSITION OA DAMPER MOTOR
EP-1 = SOLENOID AIR VALVE
PE-1 = PRESSURE/ELECTRIC SWITCH
LLT = LOW LIMIT ELECTRIC FREEZE PROTECTION THERMOSTAT

100% O.A. Unit, Ventilation Heating and Cooling, with Steam Preheat Coil, Face and Bypass Dampers on the Coil, and Heating and Chilled Water Cooling Coils

Whenever the fan runs, the EP-1 is energized and the O.A. damper opens. If the unit is large, arrangements need to be made to be sure the damper is open before the fan runs.

Thermostat T-1 opens valve V-1 on the steam preheat coil, when freezing conditions exist. Thermostat T-2 modulates the face and bypass dampers on the coil to control downstream temperature. Low-limit thermostat LLT stops the fan on freezing conditions at that location. Room thermostat T-3 modulates valves V-2 and V-3 on the cooling and heating coils in sequence to maintain space temperatures.

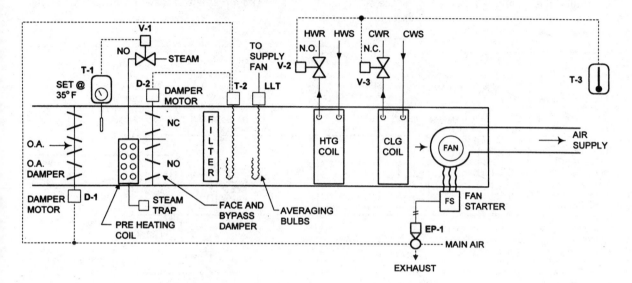

LEGEND:

T-1	=	TWO POSITION CAPILLARY THERMOSTAT
T-2	=	MODULATING CAPILLARY THERMOSTAT
T-3	=	MODULATING ROOM THERMOSTAT
V-1	=	TWO POSITION PREHEAT STEAM COIL VALVE
V-2	=	N.O. MODULATING 2-WAY HEATING COIL VALVE
V-3	=	N.C. MODULATING 2-WAY COOLING COIL VALVE
D-1	=	TWO POSITION OA DAMPER MOTOR
D-2	=	MODULATING FACE & BYPASS DAMPER MOTOR
EP-1	=	SOLENOID AIR VALVE
LLT	=	LOW LIMIT ELECTRIC FREEZE PROTECTION THERMOSTAT

100% O.A. Unit, Ventilation Heating and Cooling, with Steam Preheat Coil, Face and Bypass Dampers on the Coil and Hot Water Heating and DX Cooling Coil with Face and Bypass Dampers

Whenever the fan runs, the EP-1 is energized and the O.A. damper opens. If the unit is large, arrangements need to be made to be sure the damper is open before the fan runs.

Thermostat T-1 opens valve V-1 on the steam preheat coil, when freezing conditions exist. Thermostat T-2 modulates the face and bypass dampers on the coil to control downstream temperature. Low-limit thermostat LLT stops the fan on freezing conditions at that location. Room thermostat T-3 modulates valve V-2 and DX coil face and bypass dampers D-3 along with PE switch PE-1. PE-1 stops the refrigeration when the face dampers are closed.

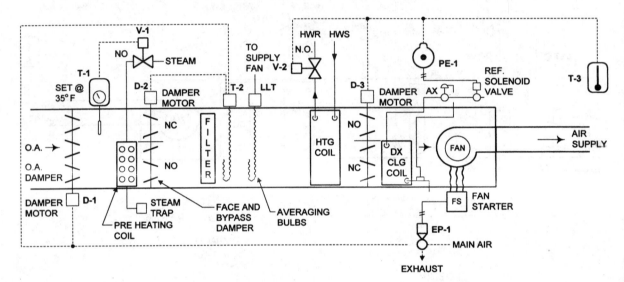

LEGEND:

T-1	= TWO POSITION CAPILLARY THERMOSTAT
T-2	= MODULATING CAPILLARY THERMOSTAT
T-3	= MODULATING ROOM THERMOSTAT
V-1	= TWO POSITION PREHEAT STEAM COIL VALVE
V-2	= N.O. MODULATING 2-WAY HEATING COIL VALVE
D-1	= TWO POSITION OA DAMPER MOTOR
D-2	= MODULATING FACE & BYPASS DAMPER MOTOR
D-3	= FACE & BYPASS DAMPER MOTOR
EP-1	= SOLENOID AIR VALVE
PE-1	= PRESSURE/ELECTRIC SWITCH
LLT	= LOW LIMIT ELECTRIC FREEZE PROTECTION THERMOSTAT

100% O.A. Unit, Ventilation Heating and Cooling, with Hot Water Preheat Coil and Hot Water Heating and DX Cooling Coil

Whenever the fan runs, the EP-1 is energized and the O.A. damper opens. If the unit is large, arrangements need to be made to be sure the damper is open before the fan runs.

Thermostat T-1 controls valve V-1 on the hot water preheat coil. Low-limit thermostat LLT stops the fan on freezing conditions at that location. Room thermostat T-2 modulates valve V-2 along with PE switch PE-1. PE-1 stops the refrigeration when room conditions are satisfied.

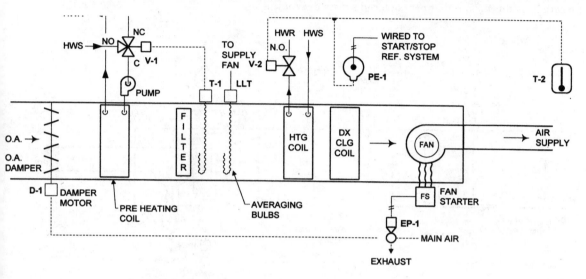

LEGEND:

T-1	=	MODULATING CAPILLARY THERMOSTAT
T-2	=	MODULATING ROOM THERMOSTAT
V-1	=	MODULATING 3-WAY HOT WATER VALVE
V-2	=	N.O. MODULATING 2-WAY HEATING COIL VALVE
D-1	=	TWO POSITION OA DAMPER MOTOR
EP-1	=	SOLENOID AIR VALVE
PE-1	=	PRESSURE/ELECTRIC SWITCH
LLT	=	LOW LIMIT ELECTRIC FREEZE PROTECTION THERMOSTAT

100% O.A. Unit, Ventilation Heating and Cooling, with Hot Water Preheat Coil, Hot Water Heating Coil, and DX Cooling Using Face and Bypass Dampers

Whenever the fan runs, the EP-1 is energized and the O.A. damper opens. If the unit is large, arrangements need to be made to be sure the damper is open before the fan runs.

Thermostat T-1 controls valve V-1 on the hot water preheat coil. Low-limit thermostat LLT stops the fan on freezing conditions at that location. Room thermostat T-2 modulates valve V-2 along with face and bypass damper motor D-2 and PE switch PE-1. PE-1 stops the refrigeration when room conditions are satisfied and the face damper is closed across the cooling coil.

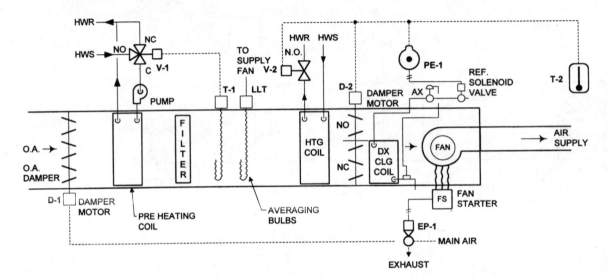

LEGEND:

T-1 = MODULATING CAPILLARY THERMOSTAT
T-2 = MODULATING ROOM THERMOSTAT
V-1 = MODULATING 3-WAY HOT WATER VALVE
V-2 = N.O. MODULATING 2-WAY HEATING COIL VALVE
D-1 = TWO POSITION OA DAMPER MOTOR
D-2 = FACE & BYPASS DAMPER MOTOR
EP-1 = SOLENOID AIR VALVE
PE-1 = PRESSURE/ELECTRIC SWITCH
LLT = LOW LIMIT ELECTRIC FREEZE PROTECTION THERMOSTAT

100% O.A. Unit, Ventilation Heating and Cooling, Hot Water Preheat Coil, and Hot Water Heating and DX Cooling

Whenever the fan runs, the EP-1 is energized and the O.A. damper opens. If the unit is large, arrangements need to be made to be sure the damper is open before the fan runs.

Thermostat T-1 controls valve V-1 on the hot water preheat coil. Low-limit thermostat LLT stops the fan on freezing conditions at that location. Room thermostat T-2 modulates valve V-2 along with PE switch PE-1. PE-1 stops the refrigeration when room conditions are satisfied.

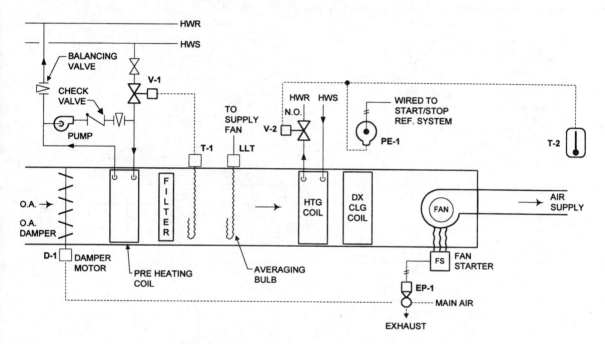

LEGEND:

T-1 = MODULATING CAPILLARY THERMOSTAT
T-2 = MODULATING ROOM THERMOSTAT
V-1 = MODULATING HOT WATER VALVE
V-2 = N.O. MODULATING 2-WAY HEATING COIL VALVE
D-1 = TWO POSITION OA DAMPER MOTOR
EP-1 = SOLENOID AIR VALVE
PE-1 = PRESSURE/ELECTRIC SWITCH
LLT = LOW LIMIT ELECTRIC FREEZE PROTECTION THERMOSTAT

100% O.A. Unit, Ventilation Heating and Cooling, with Hot Water Preheat Coil, Hot Water Heating and DX Cooling Using Face and Bypass Dampers

Whenever the fan runs, the EP-1 is energized and the O.A. damper opens. If the unit is large, arrangements need to be made to be sure the damper is open before the fan runs.

Thermostat T-1 controls valve V-1 on the hot water preheat coil. Low-limit thermostat LLT stops the fan on freezing conditions at that location. Room thermostat T-2 modulates valve V-2 along with face and bypass damper motor D-2 and PE switch PE-1. PE-1 stops the refrigeration when room conditions are satisfied and when face damper is closed across the cooling coil.

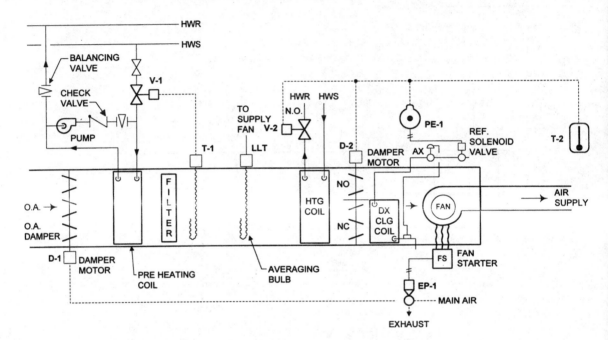

LEGEND:

T-1 = MODULATING CAPILLARY THERMOSTAT
T-2 = MODULATING ROOM THERMOSTAT
V-1 = MODULATING HOT WATER VALVE
V-2 = N.O. MODULATING 2-WAY HEATING COIL VALVE
D-1 = TWO POSITION OA DAMPER MOTOR
D-2 = FACE & BYPASS DAMPER MOTOR
EP-1 = SOLENOID AIR VALVE
PE-1 = PRESSURE/ELECTRIC SWITCH
LLT = LOW LIMIT ELECTRIC FREEZE PROTECTION THERMOSTAT

100% O.A. Unit, Heating and Cooling, with Hot Water Preheat Coil, Hot Water Heating, and Chilled Water Cooling

Whenever the fan runs, the EP-1 is energized and the O.A. damper opens. If the unit is large, arrangements need to be made to be sure the damper is open before the fan runs.

Thermostat T-1 controls valve V-1 on the hot water preheat coil. Low-limit thermostat LLT stops the fan on freezing conditions at that location. Room thermostat T-2 modulates valve V-2 along with valve V-3 in sequence.

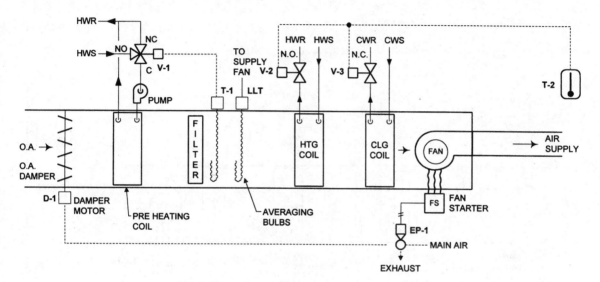

LEGEND:

T-1 = MODULATING CAPILLARY THERMOSTAT
T-2 = MODULATING ROOM THERMOSTAT
V-1 = MODULATING 3-WAY HOT WATER VALVE
V-2 = N.O. MODULATING 2-WAY HEATING COIL VALVE
V-3 = N.C. MODULATING 2-WAY COOLING COIL VALVE
D-1 = TWO POSITION OA DAMPER MOTOR
EP-1 = SOLENOID AIR VALVE
LLT = LOW LIMIT ELECTRIC FREEZE PROTECTION THERMOSTAT

100% O.A. Unit, Ventilation Heating and Cooling, with Hot Water Preheat Coil, Hot Water Heating, and Chilled Water Cooling

Whenever the fan runs, the EP-1 is energized and the O.A. damper opens. If the unit is large, arrangements need to be made to be sure the damper is open before the fan runs.

Thermostat T-1 controls valve V-1 on the hot water preheat coil. Low-limit thermostat LLT stops the fan on freezing conditions at that location. Room thermostat T-2 modulates valve V-2 along with valve V-3 in sequence.

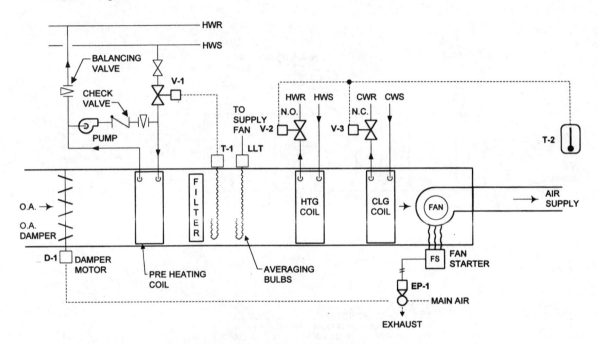

LEGEND:

T-1 = MODULATING CAPILLARY THERMOSTAT
T-2 = MODULATING ROOM THERMOSTAT
V-1 = MODULATING HOT WATER VALVE
V-2 = N.O. MODULATING 2-WAY HEATING COIL VALVE
V-3 = N.C. MODULATING 2-WAY COOLING COIL VALVE
D-1 = TWO POSITION OA DAMPER MOTOR
EP-1 = SOLENOID AIR VALVE
LLT = LOW LIMIT ELECTRIC FREEZE PROTECTION THERMOSTAT

100% O.A. Unit, Ventilation Heating and Cooling, wIth Steam Preheat Coil, Chilled Water Cooling, and Reheat Coil from Room Thermostat

Whenever the fan runs, the EP-1 is energized and the O.A. damper opens. If the unit is large, arrangements need to be made to be sure the damper is open before the fan runs.

Thermostat T-1 controls valve V-1 on the steam preheat coil. Room thermostat T-2 modulates valve(s) V-2 on the reheat coil(s). Duct thermostat T-3 controls valve V-3 on the cooling coil. There may be more than one reheat coil in the system. The low-limit thermostat LLT stops the fan on an indication of freezing conditions at its location.

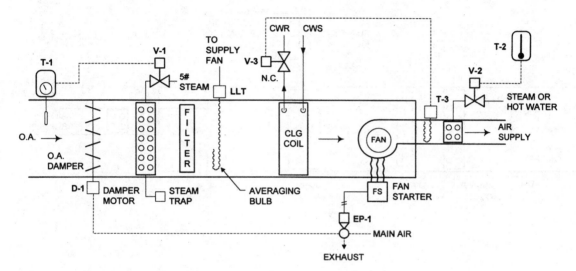

LEGEND:

T-1 = TWO POSITION CAPILLARY THERMOSTAT
T-2 = MODULATING ROOM THERMOSTAT
T-3 = MODULATING CAPILLARY DUCT THERMOSTAT
V-1 = TWO POSITION STEAM VALVE
V-2 = MODULATING STEAM/HOT WATER REHEAT COIL VALVE
V-3 = N.C. MODULATING 2-WAY COOLING COIL VALVE
D-1 = TWO POSITION OA DAMPER MOTOR
EP-1 = SOLENOID AIR VALVE

100% O.A. Unit, Heating and Cooling, with Steam Preheat Coil, DX Cooling Coil, and Reheat Coil From Room Thermostat

Whenever the fan runs, the EP-1 is energized and the O.A. damper opens. If the unit is large, arrangements need to be made to be sure the damper is open before the fan runs.

Thermostat T-1 controls valve V-1 on the steam preheat coil. Room thermostat T-2 modulates valve(s) V-2 on the reheat coil(s). Duct thermostat T-3 controls PE switch PE-1 to start and stop the refrigeration systems and satisfy room conditions. There may be more than one reheat coil in the system.

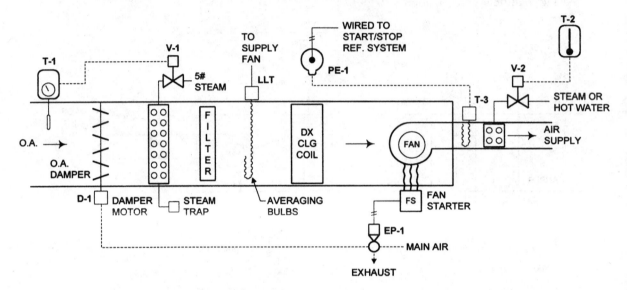

LEGEND:

T-1 = TWO POSITION CAPILLARY THERMOSTAT
T-2 = MODULATING ROOM THERMOSTAT
T-3 = MODULATING CAPILLARY DUCT THERMOSTAT
V-1 = TWO POSITION PREHEAT STEAM COIL VALVE
V-2 = MODULATING STEAM/HOT WATER REHEAT COIL VALVE
D-1 = TWO POSITION OA DAMPER MOTOR
EP-1 = SOLENOID AIR VALVE
PE-1 = PRESSURE/ELECTRIC SWITCH
LLT = LOW LIMIT ELECTRIC FREEZE PROTECTION THERMOSTAT

100% O.A. Unit, Ventilation Heating and Cooling, with Steam Preheat Coil, Hot Water Heating, Chilled Water Cooling, and Reheat Coil from Room Thermostat

Whenever the fan runs, the EP-1 is energized and the O.A. damper opens. If the unit is large, arrangements need to be made to be sure the damper is open before the fan runs.

Thermostat T-1 controls valve V-1 on the steam preheat coil. Thermostat LLT stops the fan on freezing conditions at that location. Room thermostat T-2 modulates valve(s) V-4 on the reheat coil(s). Duct thermostat T-3 controls in sequence valves V-2 and V-3 of the heating and cooling coils. There may be more than one reheat coil in the system.

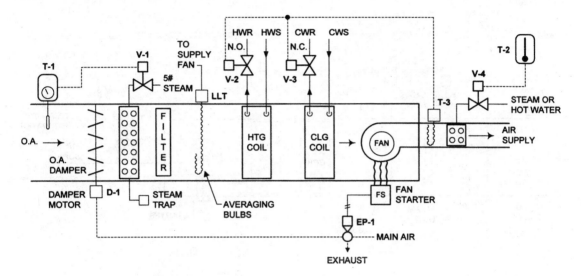

LEGEND:

T-1 = TWO POSITION CAPILLARY THERMOSTAT
T-2 = MODULATING ROOM THERMOSTAT
T-3 = MODULATING CAPILLARY DUCT THERMOSTAT
V-1 = TWO POSITION PREHEAT STEAM COIL VALVE
V-2 = N.O. MODULATING 2-WAY HEATING COIL VALVE
V-3 = N.C. MODULATING 2-WAY COOLING COIL VALVE
V-4 = MODULATING STEAM/HOT WATER REHEAT COIL VALVE
D-1 = TWO POSITION OA DAMPER MOTOR
EP-1 = SOLENOID AIR VALVE
LLT = LOW LIMIT ELECTRIC FREEZE PROTECTION THERMOSTAT

100% O.A. Unit, Heating and Cooling, with Steam Preheat Coil, Hot Water Heating, DX Cooling, and Reheat Coil from Room Thermostat

Whenever the fan runs, the EP-1 is energized and the O.A. damper opens. If the unit is large, arrangements need to be made to be sure the damper is open before the fan runs.

Thermostat T-1 controls valve V-1 on the steam preheat coil. Thermostat LLT stops the fan on freezing conditions at that location. Room thermostat T-2 modulates valve(s) V-3 on the reheat coil(s). Duct thermostat T-3 controls in sequence valve V-2 and PE switch PE-1 on the cooling coil. When the room conditions are satisfied, the cooling coil is shut down. There may be more than one reheat coil in the system.

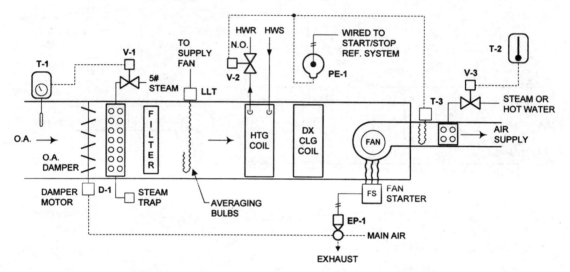

LEGEND:

T-1 = TWO POSITION CAPILLARY THERMOSTAT
T-2 = MODULATING ROOM THERMOSTAT
T-3 = MODULATING CAPILLARY DUCT THERMOSTAT
V-1 = TWO POSITION PREHEAT STEAM COIL VALVE
V-2 = N.O. MODULATING 2-WAY HEATING COIL VALVE
V-3 = MODULATING STEAM/HOT WATER REHEAT COIL VALVE
D-1 = TWO POSITION OA DAMPER MOTOR
EP-1 = SOLENOID AIR VALVE
PE-1 = PRESSURE/ELECTRIC SWITCH
LLT = LOW LIMIT ELECTRIC FREEZE PROTECTION THERMOSTAT

100% O.A. Unit, Heating and Cooling, with Steam Preheat Coil, Hot Water Heating, Chilled Water Cooling, and Reheat Coil from Room Thermostat

Whenever the fan runs, the EP-1 is energized and the O.A. damper opens. If the unit is large, arrangements need to be made to be sure the damper is open before the fan runs.

Thermostat T-1 controls valve V-1 on the steam preheat coil. Thermostat LLT stops the fan on freezing conditions at that location. Duct thermostat T-2 controls face and bypass dampers on the steam preheat coil. Room thermostat T-3 modulates valve(s) V-4 on the reheat coil(s). Duct thermostat T-4 controls in sequence valve V-2 and V-3 on the heating and cooling coil. There may be more than one reheat coil in the system.

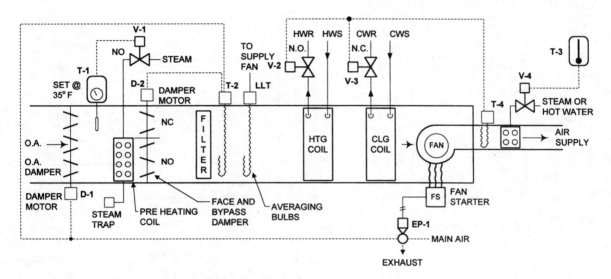

LEGEND:

T-1 = TWO POSITION CAPILLARY THERMOSTAT
T-2 = MODULATING CAPILLARY THERMOSTAT
T-3 = MODULATING ROOM THERMOSTAT
T-4 = MODULATING CAPILLARY DUCT THERMOSTAT
V-1 = TWO POSITION PREHEAT STEAM COIL VALVE
V-2 = N.O. MODULATING 2-WAY HEATING COIL VALVE
V-3 = N.C. MODULATING 2-WAY COOLING COIL VALVE
V-4 = MODULATING STEAM/HOT WATER REHEAT COIL VALVE
D-1 = TWO POSITION OA DAMPER MOTOR
D-2 = MODULATING FACE & BYPASS DAMPER MOTOR
EP-1 = SOLENOID AIR VALVE
LLT = LOW LIMIT ELECTRIC FREEZE PROTECTION THERMOSTAT

100% O.A. Unit, Heating and Cooling, with Steam Preheat Coil, Hot Water Heating, DX Cooling Using Face and Bypass Dampers, and Reheat Coil from Room Thermostat

Whenever the fan runs, the EP-1 is energized and the O.A. damper opens. If the unit is large, arrangements need to be made to be sure the damper is open before the fan runs.

Thermostat T-1 controls two-position valve V-1 on the steam preheat coil. Thermostat LLT stops the fan on freezing conditions at that location. Duct thermostat T-2 controls face and bypass dampers D-2 on the steam preheat coil. Room thermostat T-3 modulates valve(s) V-3 on the reheat coil(s). Duct thermostat T-4 controls in-sequence valve V-2 and damper motor D-3, which operates the face and bypass dampers and PE switch PE-1. PE-1 stops the cooling when space conditions are satisfied. There may be more than one reheat coil in the system.

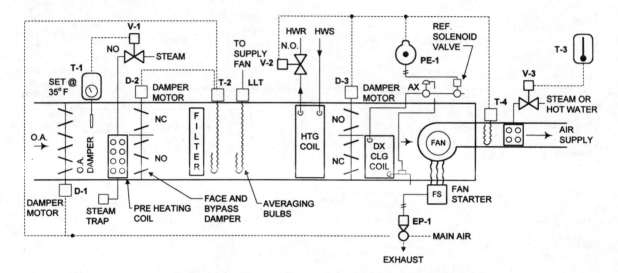

LEGEND:

T-1	=	TWO POSITION CAPILLARY THERMOSTAT
T-2	=	MODULATING CAPILLARY THERMOSTAT
T-3	=	MODULATING ROOM THERMOSTAT
T-4	=	MODULATING CAPILLARY DUCT THERMOSTAT
V-1	=	TWO POSITION PREHEAT STEAM COIL VALVE
V-2	=	N.O. MODULATING 2-WAY HEATING COIL VALVE
V-3	=	MODULATING STEAM/HOT WATER REHEAT COIL VALVE
D-1	=	TWO POSITION OA DAMPER MOTOR
D-2	=	MODULATING FACE & BYPASS DAMPER MOTOR
D-3	=	FACE & BYPASS DAMPER MOTOR
EP-1	=	SOLENOID AIR VALVE
PE-1	=	PRESSURE/ELECTRIC SWITCH
LLT	=	LOW LIMIT ELECTRIC FREEZE PROTECTION THERMOSTAT

100% O.A. Unit, Heating and Cooling, with Hot Water Preheat Coil, Hot Water Heating, DX Cooling, and Reheat Coil from Room Thermostat

Whenever the fan runs, the EP-1 is energized and the O.A. damper opens. If the unit is large, arrangements need to be made to be sure the damper is open before the fan runs.

Thermostat T-1 controls valve V-1 on the hot water preheat coil. Thermostat LLT stops the fan on freezing conditions at that location. Duct thermostat T-3 controls valve(s) V-2 and PE switch PE-1 in sequence. Room thermostat T-2 modulates valve(s) V-3 on the reheat coil(s). There may be more than one reheat coil in the system.

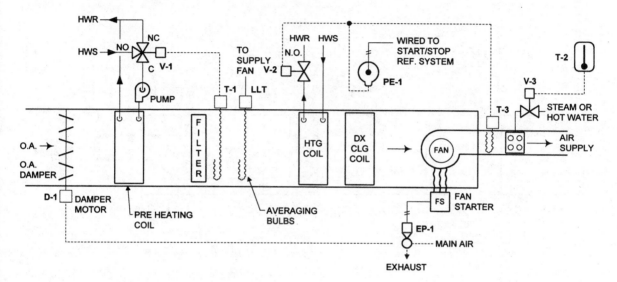

LEGEND:

T-1	= MODULATING CAPILLARY THERMOSTAT
T-2	= MODULATING ROOM THERMOSTAT
T-3	= MODULATING CAPILLARY DUCT THERMOSTAT
V-1	= MODULATING 3-WAY HOT WATER VALVE
V-2	= N.O. MODULATING 2-WAY HEATING COIL VALVE
V-3	= MODULATING STEAM/HOT WATER REHEAT COIL VALVE
D-1	= TWO POSITION OA DAMPER MOTOR
EP-1	= SOLENOID AIR VALVE
PE-1	= PRESSURE/ELECTRIC SWITCH
LLT	= LOW LIMIT ELECTRIC FREEZE PROTECTION THERMOSTAT

100% O.A. Unit, Heating and Cooling, with Hot Water Preheat Coil, Hot Water Heating, DX Cooling Using Face and Bypass Dampers, and Reheat Coil from Room Thermostat

Whenever the fan runs, the EP-1 is energized and the O.A. damper opens. If the unit is large, arrangements need to be made to be sure the damper is open before the fan runs.

Thermostat T-1 controls valve V-1 on the hot water preheat coil. Thermostat LLT stops the fan on freezing conditions at that location. Duct thermostat T-3 controls valve V-2 and PE switch PE-1 in sequence with damper motor D-2 on cooling coil face and bypass dampers. Room thermostat T-2 modulates valve(s) V-3 on the reheat coil(s). There may be more than one reheat coil in the system.

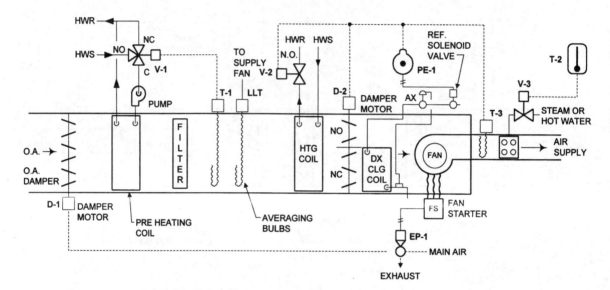

LEGEND:

T-1	=	MODULATING CAPILLARY THERMOSTAT
T-2	=	MODULATING ROOM THERMOSTAT
T-3	=	MODULATING CAPILLARY DUCT THERMOSTAT
V-1	=	MODULATING 3-WAY HOT WATER VALVE
V-2	=	N.O. MODULATING 2-WAY HEATING COIL VALVE
V-3	=	MODULATING STEAM/HOT WATER REHEAT COIL VALVE
D-1	=	TWO POSITION OA DAMPER MOTOR
D-2	=	FACE & BYPASS DAMPER MOTOR
EP-1	=	SOLENOID AIR VALVE
PE-1	=	PRESSURE/ELECTRIC SWITCH
LLT	=	LOW LIMIT ELECTRIC FREEZE PROTECTION THERMOSTAT

100% O.A. Unit, Heating and Cooling, with Hot Water Preheat Coil, Hot Water Heating, DX Cooling, and Reheat Coil from Room Thermostat

Whenever the fan runs, the EP-1 is energized and the O.A. damper opens. If the unit is large, arrangements need to be made to be sure the damper is open before the fan runs.

Thermostat T-1 controls valve V-1 on the hot water preheat coil. Thermostat LLT stops the fan on freezing conditions at that location. Duct thermostat T-3 controls valve V-2 and PE switch PE-1 in sequence. Room thermostat T-2 modulates valve(s) V-3 on the reheat coil(s). There may be more than one reheat coil in the system.

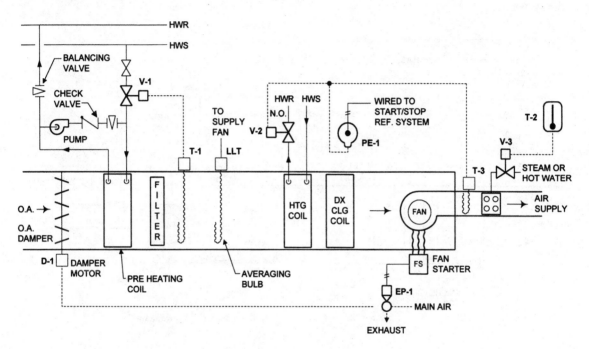

LEGEND:

T-1 = MODULATING CAPILLARY THERMOSTAT
T-2 = MODULATING ROOM THERMOSTAT
T-3 = MODULATING CAPILLARY DUCT THERMOSTAT
V-1 = MODULATING HOT WATER VALVE
V-2 = N.O. MODULATING 2-WAY HEATING COIL VALVE
V-3 = MODULATING STEAM/HOT WATER REHEAT COIL VALVE
D-1 = TWO POSITION OA DAMPER MOTOR
EP-1 = SOLENOID AIR VALVE
PE-1 = PRESSURE/ELECTRIC SWITCH
LLT = LOW LIMIT ELECTRIC FREEZE PROTECTION

100% O.A. Unit, with Hot Water Preheat Coil, Hot Water Heating, DX Cooling Using Face and Bypass Dampers, and Reheat Coil from Room Thermostat

Whenever the fan runs, the EP-1 is energized and the O.A. damper opens. If the unit is large, arrangements need to be made to be sure the damper is open before the fan runs.

Thermostat T-1 controls valve V-1 on the hot water preheat coil. Thermostat LLT stops the fan on freezing conditions at that location. Duct thermostat T-3 controls valve V-2, damper motor D-2, and PE switch PE-1 in sequence. PE switch PE-1 shuts down the cooling when the face damper is closed. Room thermostat(s) T-2 modulate valve(s) V-3 on the reheat coil(s). There may be more than one reheat coil in the system.

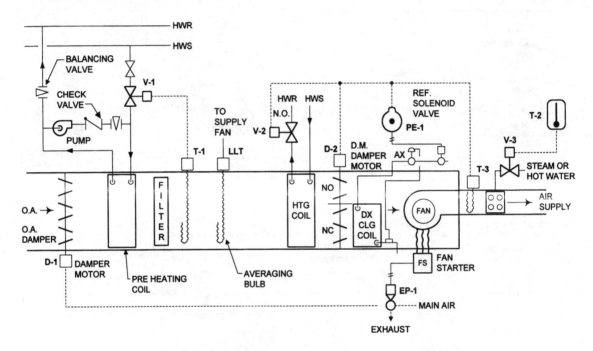

LEGEND:

T-1 = MODULATING CAPILLARY THERMOSTAT
T-2 = MODULATING ROOM THERMOSTAT
T-3 = MODULATING CAPILLARY DUCT THERMOSTAT
V-1 = MODULATING HOT WATER VALVE
V-2 = N.O. MODULATING 2-WAY HEATING COIL VALVE
V-3 = MODULATING STEAM/HOT WATER REHEAT COIL VALVE
D-1 = TWO POSITION OA DAMPER MOTOR
D-2 = FACE & BYPASS DAMPER MOTOR
EP-1 = SOLENOID AIR VALVE
PE-1 = PRESSURE/ELECTRIC SWITCH
LLT = LOW LIMIT ELECTRIC FREEZE PROTECTION

100% O.A. Unit, Heating and Cooling, with Hot Water Preheat Coil, Hot Water Heating, Chilled Water Cooling, and Reheat Coil from Room Thermostat

Whenever the fan runs, the EP-1 is energized and the O.A. damper opens. If the unit is large, arrangements need to be made to be sure the damper is open before the fan runs.

Thermostat T-1 controls valve V-1 on the hot water preheat coil. Thermostat LLT stops the fan on freezing conditions at that location. Duct thermostat T-3 controls valve V-2 and valve V-3 in sequence. Room thermostat(s) T-2 modulate valve(s) V-4 on the reheat coil(s). There may be more than one reheat coil in the system.

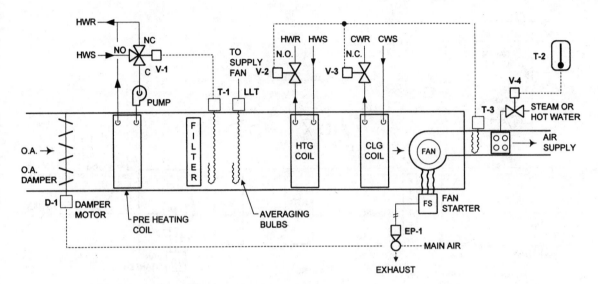

LEGEND:

T-1	=	MODULATING CAPILLARY THERMOSTAT
T-2	=	MODULATING ROOM THERMOSTAT
T-3	=	MODULATING CAPILLARY DUCT THERMOSTAT
V-1	=	MODULATING HOT WATER VALVE
V-2	=	N.O. MODULATING 2-WAY HEATING COIL VALVE
V-3	=	N.C. MODULATING 2-WAY COOLING COIL VALVE
V-4	=	MODULATING STEAM/HOT WATER REHEAT COIL VALVE
D-1	=	TWO POSITION OA DAMPER MOTOR
EP-1	=	SOLENOID AIR VALVE
LLT	=	LOW LIMIT ELECTRIC FREEZE PROTECTION

100% O.A. Unit, Heating and Cooling, with Hot Water Preheat Coil, Hot Water Heating, Chilled Water Cooling, and Reheat Coil from Room Thermostat

Whenever the fan runs, the EP-1 is energized and the O.A. damper opens. If the unit is large, arrangements need to be made to be sure the damper is open before the fan runs.

Thermostat T-1 controls valve V-1 on the hot water preheat coil. Thermostat LLT stops the fan on freezing conditions at that location. Duct thermostat T-3 controls valve V-2 and valve V-3 in sequence. Room thermostat(s) T-2 modulate valve(s) V-4 on the reheat coil(s). There may be more than one reheat coil in the system.

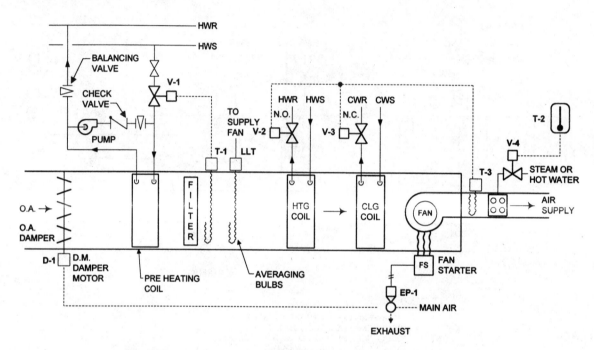

LEGEND:

T-1 = MODULATING CAPILLARY THERMOSTAT
T-2 = MODULATING ROOM THERMOSTAT
T-3 = MODULATING CAPILLARY DUCT THERMOSTAT
V-1 = MODULATING HOT WATER VALVE
V-2 = N.O. MODULATING 2-WAY HEATING COIL VALVE
V-3 = N.C. MODULATING 2-WAY COOLING COIL VALVE
V-4 = MODULATING STEAM/HOT WATER REHEAT COIL VALVE
D-1 = TWO POSITION OA DAMPER MOTOR
EP-1 = SOLENOID AIR VALVE
LLT = LOW LIMIT ELECTRIC FREEZE PROTECTION

100% O.A. Unit, Heating and Cooling, with Steam Preheat Coil, Hot Water Heating, Chilled Water Cooling, and steam Humidifier from Room Humidistat

Whenever the fan runs, the EP-1 is energized and O.A. dampers are opened. If the system is very large, provisions should be made to be sure the dampers are open before the fan starts. When the fan stops, the O.A. damper closes.

Room thermostat T-2 controls valves V-2 and V-3 in sequence to maintain space conditions. Thermostat T-1 controls the steam preheat coil based upon the O.A. temperature to prevent freezing. The V-1 is two position and stays open as long as the O.A. temperature is below freezing. Room humidistat H-1, through high-limit humidistat H-2, controls the humidifier valve V-4.

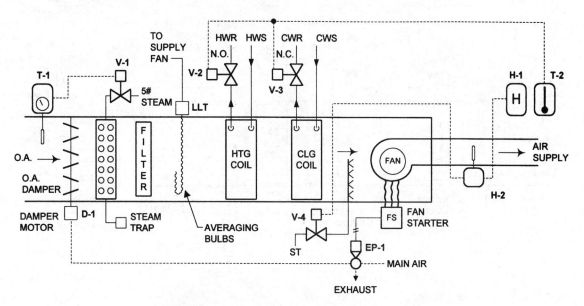

LEGEND:

T-1	=	TWO POSITION CAPILLARY THERMOSTAT
T-2	=	MODULATING ROOM THERMOSTAT
H-1	=	MODULATING ROOM HUMIDISTAT
H-2	=	HIGH LIMIT DUCT HUMIDISTAT
V-1	=	TWO POSITION PREHEAT STEAM COIL VALVE
V-2	=	N.O. MODULATING 2-WAY HEATING COIL VALVE
V-3	=	N.C. MODULATING 2-WAY COOLING COIL VALVE
V-4	=	N.C. MODULATING HUMIDIFIER STEAM VALVE
D-1	=	TWO POSITION OA DAMPER MOTOR
EP-1	=	SOLENOID AIR VALVE
LLT	=	LOW LIMIT ELECTRIC FREEZE PROTECTION THERMOSTAT

100% O.A. Unit, Heating and Cooling, with Steam Preheat Coil, Hot Water Heating, DX Cooling, and Steam Humidifier from Room Humidistat

Whenever the fan runs, the EP-1 is energized and the O.A. damper opens. If the unit is large, arrangements need to be made to be sure the damper is open before the fan runs.

Thermostat T-1 controls valve V-1 on the steam preheat coil. Thermostat LLT stops the fan on freezing conditions at that location. Room thermostat T-2 controls valve V-2 and PE switch PE-1 in sequence to satisfy room conditions. Room humidistat H-1 controls humidifier valve V-3 through high-limit duct humidistat H-2.

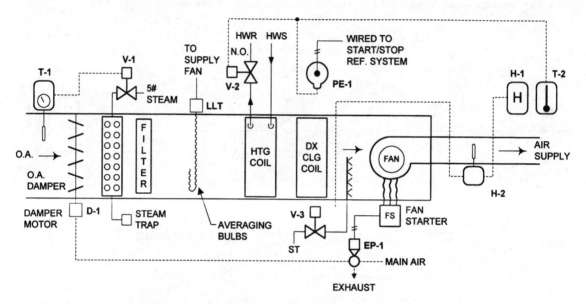

LEGEND:

T-1 = TWO POSITION CAPILLARY THERMOSTAT
T-2 = MODULATING ROOM THERMOSTAT
H-1 = MODULATING ROOM HUMIDISTAT
H-2 = HIGH LIMIT DUCT HUMIDISTAT
V-1 = TWO POSITION PREHEAT STEAM COIL VALVE
V-2 = N.O. MODULATING 2-WAY HEATING COIL VALVE
V-3 = N.C. MODULATING HUMIDIFIER STEAM VALVE
D-1 = TWO POSITION OA DAMPER MOTOR
EP-1 = SOLENOID AIR VALVE
PE-1 = PRESSURE/ELECTRIC SWITCH
LLT = LOW LIMIT ELECTRIC FREEZE PROTECTION THERMOSTAT

100% O.A. Unit, Ventilation Heating and Cooling, with Steam Preheat Coil, Hot Water Heating, Chilled Water Cooling, and Steam Humidifier from Room Humidistat

Whenever the fan runs, the EP-1 is energized and the O.A. damper opens. If the unit is large, arrangements need to be made to be sure the damper is open before the fan runs.

Thermostat T-1 controls valve V-1 on the steam preheat coil. Thermostat LLT stops the fan on freezing conditions at that location. Duct thermostat T-2 controls preheat coil face and bypass dampers. Room thermostat controls valves V-2 and V-3 in sequence to maintain space conditions. Humidistat H-1 controls humidifier valve V-4 through high-limit duct humidistat H-2.

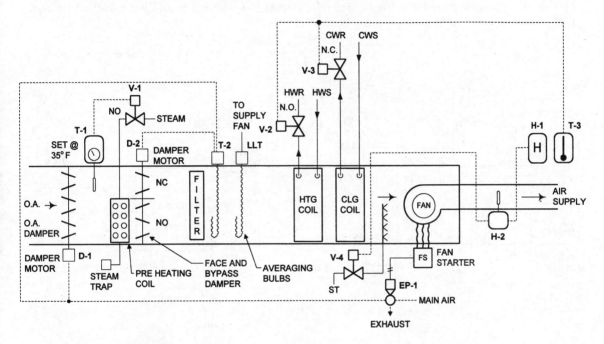

LEGEND:

T-1 = TWO POSITION CAPILLARY THERMOSTAT
T-2 = MODULATING CAPILLARY THERMOSTAT
T-3 = MODULATING ROOM THERMOSTAT
H-1 = MODULATING ROOM HUMIDISTAT
H-2 = HIGH LIMIT DUCT HUMIDISTAT
V-1 = TWO POSITION PREHEAT STEAM COIL VALVE
V-2 = N.O. MODULATING 2-WAY HEATING COIL VALVE
V-3 = N.C. MODULATING 2-WAY COOLING COIL VALVE
V-4 = N.C. MODULATING HUMIDIFIER STEAM VALVE
D-1 = TWO POSITION OA DAMPER MOTOR
D-2 = FACE & BYPASS DAMPER MOTOR
EP-1 = SOLENOID AIR VALVE
LLT = LOW LIMIT ELECTRIC FREEZE PROTECTION THERMOSTAT

100% O.A. Unit, Heating and Cooling, with Steam Preheat Coil, Hot Water Heating, and DX Cooling using Face and Bypass Dampers

Whenever the fan runs, the EP-1 is energized and the O.A. damper opens. If the unit is large, arrangements need to be made to be sure the damper is open before the fan runs.

Thermostat T-1 controls valve V-1 on the steam preheat coil. Thermostat LLT stops the fan on freezing conditions at that location. Duct thermostat T-2 controls preheat coil face and bypass dampers. The room thermostat controls valve V-2 and DX cooling coil face and bypass dampers in sequence with PE switch PE-1 to maintain space conditions. When cooling is no longer needed and the face damper is closed, PE switch PE-1 shuts off the cooling. Humidistat H-1 controls humidifier valve V-3 through high-limit duct humidistat H-2.

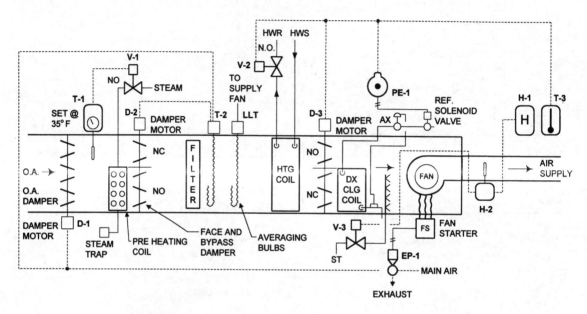

LEGEND:

T-1 = TWO POSITION CAPILLARY THERMOSTAT
T-2 = MODULATING CAPILLARY THERMOSTAT
T-3 = MODULATING ROOM THERMOSTAT
H-1 = MODULATING ROOM HUMIDISTAT
H-2 = HIGH LIMIT DUCT HUMIDISTAT
V-1 = TWO POSITION PREHEAT STEAM COIL VALVE
V-2 = N.O. MODULATING 2-WAY HEATING COIL VALVE
V-3 = N.C. MODULATING HUMIDIFIER STEAM VALVE
D-1 = TWO POSITION OA DAMPER MOTOR
D-2 = FACE & BYPASS DAMPER MOTOR
D-3 = FACE & BYPASS DAMPER MOTOR
EP-1 = SOLENOID AIR VALVE
PE-1 = PRESSURE/ELECTRIC SWITCH
LLT = LOW LIMIT ELECTRIC FREEZE PROTECTION THERMOSTAT

100% O.A. Unit, Heating and Cooling, with Hot Water Preheat Coil, Hot Water Heating, DX Cooling, and Steam Humidifier from Room Humidistat

Whenever the fan runs, the EP-1 is energized and the O.A. damper opens. If the unit is large, arrangements need to be made to be sure the damper is open before the fan runs.

Thermostat T-1 controls valve V-1 on the hot water preheat coil. Thermostat LLT stops the fan on freezing conditions at that location. Room thermostat controls valve V-2 and DX cooling coil in sequence with PE switch PE-1 to maintain space conditions. When cooling is no longer needed, PE-1 shuts off the cooling. Humidistat H-1 controls humidifier valve V-3 through high-limit duct humidistat H-2.

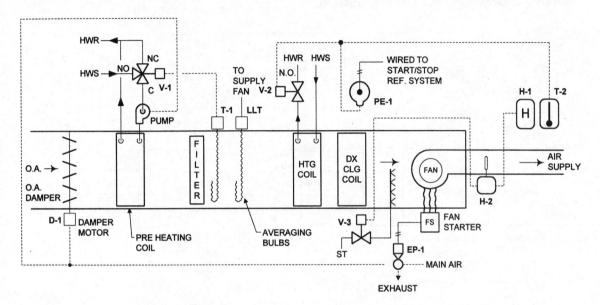

LEGEND:

T-1 = MODULATING CAPILLARY THERMOSTAT
T-2 = MODULATING ROOM THERMOSTAT
H-1 = MODULATING ROOM HUMIDISTAT
H-2 = HIGH LIMIT DUCT HUMIDISTAT
V-1 = MODULATING HOT WATER VALVE
V-2 = N.O. MODULATING 2-WAY HEATING COIL VALVE
V-3 = N.C. MODULATING HUMIDIFIER STEAM VALVE
D-1 = TWO POSITION OA DAMPER MOTOR
EP-1 = SOLENOID AIR VALVE
PE-1 = PRESSURE/ELECTRIC SWITCH
LLT = LOW LIMIT ELECTRIC FREEZE PROTECTION THERMOSTAT

100% O.A. Unit, Heating and Cooling, with Hot Water Preheat Coil, Hot Water Heating, DX Cooling, Face and Bypass Dampers, and Steam Humidifier from Room Humidistat

Whenever the fan runs, the EP-1 is energized and the O.A. damper opens. If the unit is large, arrangements need to be made to be sure the damper is open before the fan runs.

Thermostat T-1 controls valve V-1 on the hot water preheat coil. Thermostat LLT stops the fan on freezing conditions at that location. Room thermostat T-2 controls valve V-2 and damper motor D-2 on DX coil face and bypass dampers in sequence with PE switch PE-1 to maintain space conditions. When the face damper is closed, PE-1 shuts down the cooling. Humidistat H-1 controls humidifier valve V-3 through high-limit duct humidistat H-2.

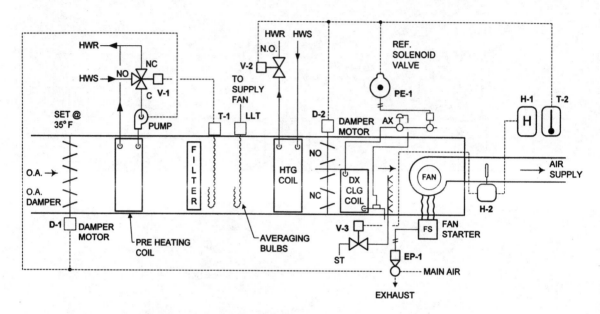

LEGEND:

T-1	=	MODULATING CAPILLARY THERMOSTAT
T-2	=	MODULATING ROOM THERMOSTAT
H-1	=	MODULATING ROOM HUMIDISTAT
H-2	=	HIGH LIMIT DUCT HUMIDISTAT
V-1	=	MODULATING HOT WATER VALVE
V-2	=	N.O. MODULATING 2-WAY HEATING COIL VALVE
V-3	=	N.C. MODULATING HUMIDIFIER STEAM VALVE
D-1	=	TWO POSITION OA DAMPER MOTOR
D-2	=	FACE & BYPASS DAMPER MOTOR
EP-1	=	SOLENOID AIR VALVE
PE-1	=	PRESSURE/ELECTRIC SWITCH
LLT	=	LOW LIMIT ELECTRIC FREEZE PROTECTION THERMOSTAT

100% O.A. Unit, Heating and Cooling, with Hot Water Preheat Coil, Hot Water Heating, DX Cooling, and Steam Humidifier from Room Humidistat

Whenever the fan runs, the EP-1 is energized and the O.A. damper opens. If the unit is large, arrangements need to be made to be sure the damper is open before the fan runs.

Thermostat T-1 controls valve V-1 on the hot water preheat coil. Thermostat LLT stops the fan on freezing conditions at that location. Room thermostat T-2 controls valve V-2 and PE switch PE-1 to maintain space conditions. When the space conditions are satisfied, PE-1 shuts down the cooling. Humidistat H-1 controls humidifier valve V-3 through high-limit duct humidistat H-2.

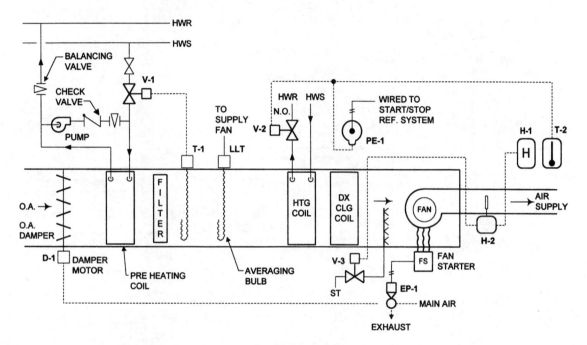

LEGEND:

T-1 = MODULATING CAPILLARY THERMOSTAT
T-2 = MODULATING ROOM THERMOSTAT
H-1 = MODULATING ROOM HUMIDISTAT
H-2 = HIGH LIMIT DUCT HUMIDISTAT
V-1 = MODULATING HOT WATER VALVE
V-2 = N.O. MODULATING 2-WAY HEATING COIL VALVE
V-3 = N.C. MODULATING HUMIDIFIER STEAM VALVE
D-1 = TWO POSITION OA DAMPER MOTOR
EP-1 = SOLENOID AIR VALVE
PE-1 = PRESSURE/ELECTRIC SWITCH
LLT = LOW LIMIT ELECTRIC FREEZE PROTECTION THERMOSTAT

100% O.A. Unit, Heating and Cooling, with Hot Water Preheat Coil, Hot Water Heating, Chilled Water Cooling, and Steam Humidifier from Room Humidistat

Whenever the fan runs, the EP-1 is energized and the O.A. damper opens. If the unit is large, arrangements need to be made to be sure the damper is open before the fan runs.

Thermostat T-1 controls valve V-1 on the hot water preheat coil. Thermostat LLT stops the fan on freezing conditions at that location. Room thermostat T-2 controls valve V-2 and V-3 on heating and cooling coils. Humidistat H-1 controls humidifier valve V-4 through high-limit duct humidistat H-2.

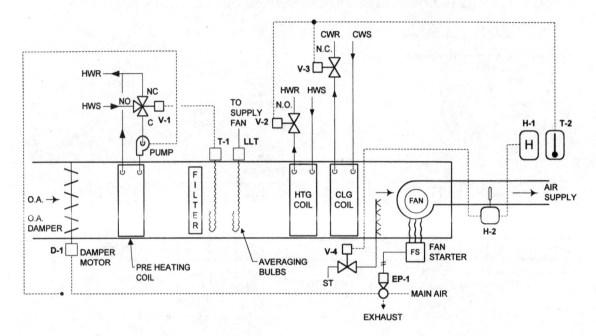

LEGEND:

T-1	= MODULATING CAPILLARY THERMOSTAT
T-2	= MODULATING ROOM THERMOSTAT
H-1	= MODULATING ROOM HUMIDISTAT
H-2	= HIGH LIMIT DUCT HUMIDISTAT
V-1	= MODULATING HOT WATER VALVE
V-2	= N.O. MODULATING 2-WAY HEATING COIL VALVE
V-3	= N.C. MODULATING 2-WAY COOLING COIL VALVE
V-4	= MODULATING STEAM/HOT WATER REHEAT COIL VALVE
D-1	= TWO POSITION OA DAMPER MOTOR
EP-1	= SOLENOID AIR VALVE
LLT	= LOW LIMIT ELECTRIC FREEZE PROTECTION

100% O.A. Unit, Heating and Cooling, with Hot Water Preheat Coil, Hot Water Heating, Chilled Water Cooling, and Steam Humidifier from Room Humidistat

Whenever the fan runs, the EP-1 is energized and the O.A. damper opens. If the unit is large, arrangements need to be made to be sure the damper is open before the fan runs.

Thermostat T-1 controls valve V-1 on the hot water preheat coil. Thermostat LLT stops the fan on freezing conditions at that location. Room thermostat T-2 controls valve V-2 and V-3 on heating and cooling coils. Humidistat H-1 controls humidifier valve V-4 through high-limit duct humidistat H-2.

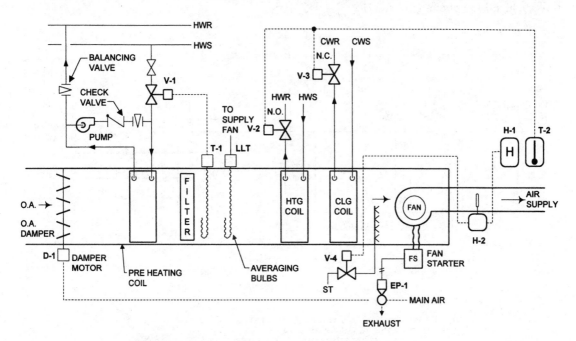

LEGEND:

T-1 = MODULATING CAPILLARY THERMOSTAT
T-2 = MODULATING ROOM THERMOSTAT
H-1 = MODULATING ROOM HUMIDISTAT
H-2 = HIGH LIMIT DUCT HUMIDISTAT
V-1 = MODULATING HOT WATER VALVE
V-2 = N.O. MODULATING 2-WAY HEATING COIL VALVE
V-3 = N.C. MODULATING 2-WAY COOLING COIL VALVE
V-4 = N.C. MODULATING HUMIDIFIER STEAM VALVE
D-1 = TWO POSITION OA DAMPER MOTOR
EP-1 = SOLENOID AIR VALVE
LLT = LOW LIMIT ELECTRIC FREEZE PROTECTION THERMOSTAT

100% O.A. Unit, Heating and Cooling, with Steam Preheat Coil, Chilled Water Cooling, Steam Humidifier from Room Humidistat, and Reheat Coil from Room Thermostat

Whenever the fan runs, the EP-1 is energized and the O.A. damper opens. If the unit is large, arrangements need to be made to be sure the damper is open before the fan runs.

Thermostat T-1 controls valve V-1 on the steam preheat coil. Thermostat LLT stops the fan on freezing conditions at that location. Duct thermostat T-3 controls valve V-3 on the cooling coil. Room thermostat(s) T-2 controls valve V-2 on reheat coil(s). Humidistat H-1 controls humidifier valve V-4 through high-limit duct humidistat H-2.

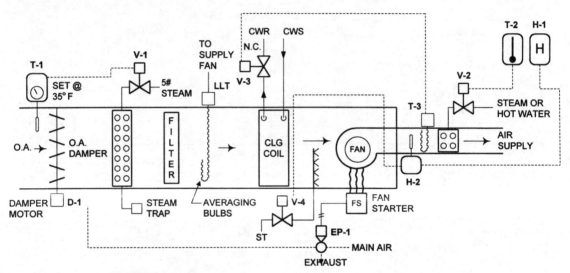

LEGEND:

T-1 = TWO POSITION CAPILLARY THERMOSTAT
T-2 = MODULATING ROOM THERMOSTAT
T-3 = MODULATING CAPILLARY DUCT THERMOSTAT
H-1 = MODULATING ROOM HUMIDISTAT
H-2 = HIGH LIMIT DUCT HUMIDISTAT
V-1 = TWO POSITION PREHEAT STEAM COIL VALVE
V-2 = MODULATING STEAM/HOT WATER REHEAT COIL VALVE
V-3 = N.C. MODULATING 2-WAY COOLING COIL VALVE
V-4 = N.C. MODULATING HUMIDIFIER STEAM VALVE
D-1 = TWO POSITION OA DAMPER MOTOR
EP-1 = SOLENOID AIR VALVE
LLT = LOW LIMIT ELECTRIC FREEZE PROTECTION THERMOSTAT

100% O.A. Unit, Heating and Cooling, with Steam Preheat Coil, DX Cooling, Steam Humidifier from Room Humidistat, and Reheat Coil from Room Thermostat

Whenever the fan runs, the EP-1 is energized and the O.A. damper opens. If the unit is large, arrangements need to be made to be sure the damper is open before the fan runs.

Thermostat T-1 controls valve V-1 on the steam preheat coil. Thermostat LLT stops the fan on freezing conditions at that location. Duct thermostat T-3 controls PE switch PE-1 on the DX cooling coil. Room thermostat(s) T-2 controls valve(s) V-2 on reheat coil(s). Humidistat H-1 controls humidifier valve V-3 through high-limit duct humidistat H-2. There may be more than one reheat coil.

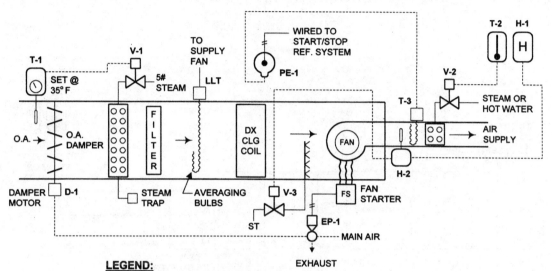

LEGEND:

T-1 = TWO POSITION CAPILLARY THERMOSTAT
T-2 = MODULATING ROOM THERMOSTAT
T-3 = MODULATING CAPILLARY DUCT THERMOSTAT
H-1 = MODULATING ROOM HUMIDISTAT
H-2 = HIGH LIMIT DUCT HUMIDISTAT
V-1 = TWO POSITION PREHEAT STEAM COIL VALVE
V-2 = MODULATING STEAM/HOT WATER REHEAT COIL VALVE
V-3 = N.C. MODULATING HUMIDIFIER STEAM VALVE
D-1 = TWO POSITION OA DAMPER MOTOR
EP-1 = SOLENOID AIR VALVE
PE-1 = PRESSURE/ELECTRIC SWITCH
LLT = LOW LIMIT ELECTRIC FREEZE PROTECTION THERMOSTAT

100% O.A. Unit, Heating and Cooling, with Steam Preheat Coil, Hot Water Heating, Chilled Water Cooling, Steam Humidifier from Room Humidistat, and Reheat Coil from Room Thermostat

Whenever the fan runs, the EP-1 is energized and the O.A. damper opens. If the unit is large, arrangements need to be made to be sure the damper is open before the fan runs.

Thermostat T-1 controls valve V-1 on the steam preheat coil. Thermostat LLT stops the fan on freezing conditions at that location. Duct thermostat T-3 controls valves V-2 and V-3 on heating and cooling coils. Room thermostat(s) T-2 control valve(s) V-4 on reheat coil(s). Humidistat H-1 controls humidifier valve V-5 through high-limit duct humidistat H-2. There may be more than one reheat coil.

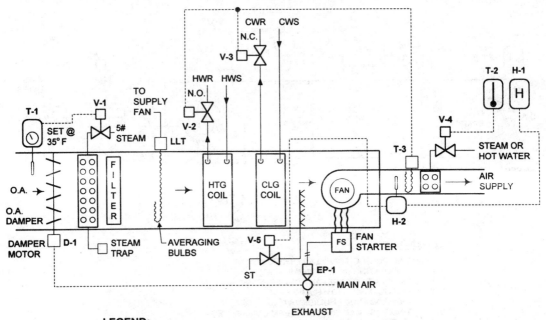

LEGEND:

T-1	=	TWO POSITION CAPILLARY THERMOSTAT
T-2	=	MODULATING ROOM THERMOSTAT
T-3	=	MODULATING CAPILLARY DUCT THERMOSTAT
H-1	=	MODULATING ROOM HUMIDISTAT
H-2	=	HIGH LIMIT DUCT HUMIDISTAT
V-1	=	TWO POSITION PREHEAT STEAM COIL VALVE
V-2	=	N.O. MODULATING 2-WAY HEATING COIL VALVE
V-3	=	N.C. MODULATING 2-WAY COOLING COIL VALVE
V-4	=	MODULATING STEAM/HOT WATER REHEAT COIL VALVE
V-5	=	N.C. MODULATING HUMIDIFIER STEAM VALVE
D-1	=	TWO POSITION OA DAMPER MOTOR
EP-1	=	SOLENOID AIR VALVE
LLT	=	LOW LIMIT ELECTRIC FREEZE PROTECTION THERMOSTAT

100% O.A, Unit, Heating and Cooling, with Steam Preheat Coil, Hot Water Heating, DX Cooling, Steam Humidifier from Room Humidistat, and Reheat Coil from Room Thermostat

Whenever the fan runs, the EP-1 is energized and the O.A. damper opens. If the unit is large, arrangements need to be made to be sure the damper is open before the fan runs.

Thermostat T-1 controls valve V-1 on the steam preheat coil. Thermostat LLT stops the fan on freezing conditions at that location. Duct thermostat T-3 controls valve V-2 and PE switch PE-1 on the hot water heating and DX cooling coil. Room thermostat(s) T-2 control valve(s) V-3 on the reheat coil(s). Humidistat H-1 controls humidifier valve V-4 through high-limit duct humidistat H-2. There may be more than one reheat coil.

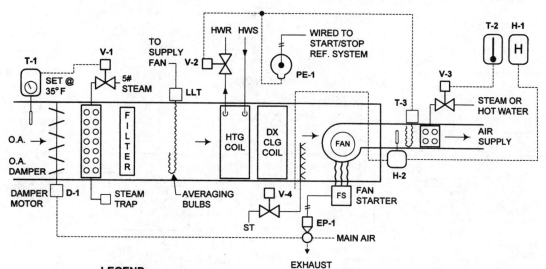

LEGEND:

T-1	=	TWO POSITION CAPILLARY THERMOSTAT
T-2	=	MODULATING ROOM THERMOSTAT
T-3	=	MODULATING CAPILLARY DUCT THERMOSTAT
H-1	=	MODULATING ROOM HUMIDISTAT
H-2	=	HIGH LIMIT DUCT HUMIDISTAT
V-1	=	TWO POSITION PREHEAT STEAM COIL VALVE
V-2	=	N.O. MODULATING 2-WAY HEATING COIL VALVE
V-3	=	MODULATING STEAM/HOT WATER REHEAT COIL VALVE
V-4	=	N.C. MODULATING HUMIDIFIER STEAM VALVE
D-1	=	TWO POSITION OA DAMPER MOTOR
EP-1	=	SOLENOID AIR VALVE
PE-1	=	PRESSURE/ELECTRIC SWITCH
LLT	=	LOW LIMIT ELECTRIC FREEZE PROTECTION THERMOSTAT

100% O.A. Unit, Heating and Cooling, with Steam Preheat Coil, Hot Water Heating Coil, Chilled Water Cooling Coil, Steam Humidifier from Room Humidistat, and Reheat Coil from Room Thermostat

Whenever the fan runs, the EP-1 is energized and the O.A. damper opens. If the unit is large, arrangements need to be made to be sure the damper is open before the fan runs.

Thermostat T-1 controls valve V-1 on the steam preheat coil. Thermostat T-2 controls face and bypass dampers on the preheat coil. LLT stops the fan on freezing conditions at that location. Duct thermostat T-4 controls valve V-2 and V-3 on heating and cooling coils. Room thermostat(s) T-3 control valve(s) V-4 on reheat coil(s). Humidistat H-1 controls humidifier valve V-5 through high-limit duct humidistat H-2. There may be more than one reheat coil.

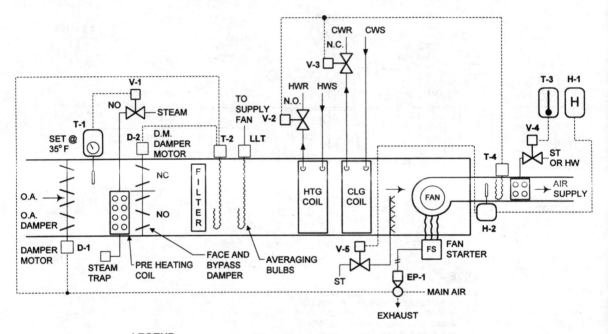

LEGEND:

T-1	= TWO POSITION CAPILLARY THERMOSTAT
T-2	= MODULATING CAPILLARY THERMOSTAT
T-3	= MODULATING ROOM THERMOSTAT
T-4	= MODULATING CAPILLARY DUCT THERMOSTAT
H-1	= MODULATING ROOM HUMIDISTAT
H-2	= HIGH LIMIT DUCT HUMIDISTAT
V-1	= TWO POSITION PREHEAT STEAM COIL VALVE
V-2	= N.O. MODULATING 2-WAY HEATING COIL VALVE
V-3	= N.C. MODULATING 2-WAY COOLING COIL VALVE
V-4	= MODULATING STEAM/HOT WATER REHEAT COIL VALVE
V-5	= N.C. MODULATING HUMIDIFIER STEAM VALVE
D-1	= TWO POSITION OA DAMPER MOTOR
D-2	= FACE & BYPASS DAMPER MOTOR
EP-1	= SOLENOID AIR VALVE
LLT	= LOW LIMIT ELECTRIC FREEZE PROTECTION THERMOSTAT

100% O.A. Unit, Heating and Cooling, with Steam Preheat Coil, Hot Water Heating, DX Cooling, Face and Bypass Dampers, Steam Humidifier from Room Humidistat, and Reheat Coil from Room Thermostat

Whenever the fan runs, the EP-1 is energized and O.A. dampers are opened. If the system is very large, provisions should be made to be sure the dampers are open before the fan starts. When the fan stops the O.A. damper closes.

Duct thermostat T-1 controls the two-position valve on the steam preheat coil. Thermostat T-2 controls the preheat coil face and bypass dampers to maintain downstream temperatures. Discharge duct thermostat T-4 controls the heating coil valve V-2 and the DX coil face and bypass dampers D-3 in sequence with PE switch PE-1. When the face damper is closed, the PE switch shuts down the DX cooling. Freeze protection thermostat LLT stops the fan on an indication of freezing conditions at that location. Room thermostat(s) T-3 control the valve V-3 on the reheat coil(s). There may be more than one reheat coil in the system. Room humidistat H-1, through high-limit humidistat H-2, controls the humidifier valve V-4.

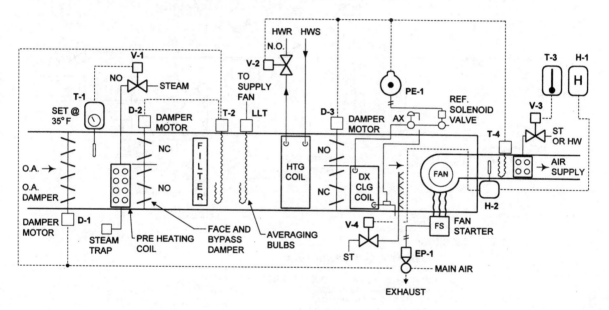

LEGEND:

T-1	= TWO POSITION CAPILLARY THERMOSTAT
T-2	= MODULATING CAPILLARY THERMOSTAT
T-3	= MODULATING ROOM THERMOSTAT
T-4	= MODULATING CAPILLARY DUCT THERMOSTAT
H-1	= MODULATING ROOM HUMIDISTAT
H-2	= HIGH LIMIT DUCT HUMIDISTAT
V-1	= TWO POSITION PREHEAT STEAM COIL VALVE
V-2	= N.O. MODULATING 2-WAY HEATING COIL VALVE
V-3	= MODULATING STEAM/HOT WATER REHEAT COIL VALVE
V-4	= N.C. MODULATING HUMIDIFIER STEAM VALVE
D-1	= TWO POSITION OA DAMPER MOTOR
D-2	= FACE & BYPASS DAMPER MOTOR
D-3	= FACE & BYPASS DAMPER MOTOR
EP-1	= SOLENOID AIR VALVE
PE-1	= PRESSURE/ELECTRIC SWITCH
LLT	= LOW LIMIT ELECTRIC FREEZE PROTECTION

100% O.A. Unit, Heating and Cooling, with Hot Water Preheat Coil, Hot Water Heating, DX Cooling, Steam Humidifier from Room Humidistat, and Reheat Coil from Room Thermostat

Whenever the fan runs, the EP-1 is energized and the O.A. damper opens. If the unit is large, arrangements need to be made to be sure the damper is open before the fan runs.

Thermostat T-1 controls valve V-1 on the hot water preheat coil. LLT stops the fan on freezing conditions at that location. Duct thermostat T-3 controls valve V-2 on the heating coil and PE switch PE-1 on the cooling coil. Room thermostat(s) T-2 control valve(s) V-3 on reheat coil(s). Humidistat H-1 controls humidifier valve V-4 through high-limit duct humidistat H-2. There may be more than one reheat coil.

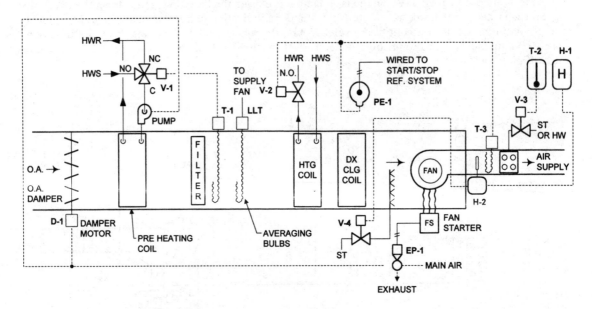

LEGEND:

T-1 = MODULATING CAPILLARY THERMOSTAT
T-2 = MODULATING ROOM THERMOSTAT
T-3 = MODULATING CAPILLARY DUCT THERMOSTAT
H-1 = MODULATING ROOM HUMIDISTAT
H-2 = HIGH LIMIT DUCT HUMIDISTAT
V-1 = MODULATING HOT WATER VALVE
V-2 = N.O. MODULATING 2-WAY HEATING COIL VALVE
V-3 = MODULATING STEAM/HOT WATER REHEAT COIL VALVE
V-4 = N.C. MODULATING HUMIDIFIER STEAM VALVE
D-1 = TWO POSITION OA DAMPER MOTOR
EP-1 = SOLENOID AIR VALVE
PE-1 = PRESSURE/ELECTRIC SWITCH
LLT = LOW LIMIT ELECTRIC FREEZE PROTECTION THERMOSTAT

100% O.A. Unit, Heating and Cooling, with Hot Water Preheat Coil, Hot Water Heating, DX Cooling, Face and Bypass Dampers, Steam Humidifier from Room Humidistat, and Reheat Coil from Room Thermostat

Whenever the fan runs, the EP-1 is energized and O.A. dampers are opened. If the system is very large, provisions should be made to be sure the dampers are open before the fan starts. When the fan stops, the O.A. damper closes.

Duct thermostat T-1 controls the hot water preheat coil. Discharge duct thermostat T-3 controls the heating coil valve V-2 and the DX coil face and bypass dampers D-2 in sequence with PE switch PE-1. When the face damper is closed, the PE switch shuts down the DX cooling. Freeze protection thermostat LLT stops the fan on an indication of freezing conditions at that location. Room thermostat(s) T-2 control the valve V-3 on the reheat coil(s). There may be more than one reheat coil in the system. Room humidistat H-1, through high-limit humidistat H-2, controls the humidifier valve V-4.

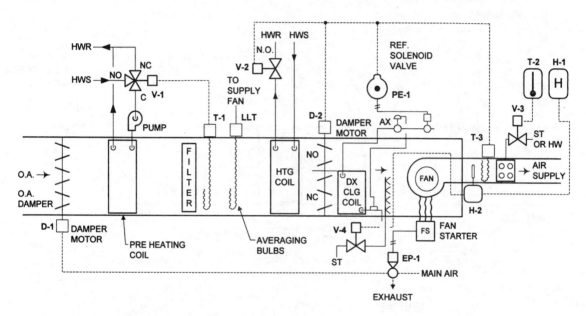

LEGEND:

T-1 = MODULATING CAPILLARY THERMOSTAT
T-2 = MODULATING ROOM THERMOSTAT
T-3 = MODULATING CAPILLARY DUCT THERMOSTAT
H-1 = MODULATING ROOM HUMIDISTAT
H-2 = HIGH LIMIT DUCT HUMIDISTAT
V-1 = MODULATING HOT WATER VALVE
V-2 = N.O. MODULATING 2-WAY HEATING COIL VALVE
V-3 = MODULATING STEAM/HOT WATER REHEAT COIL VALVE
V-4 = N.C. MODULATING HUMIDIFIER STEAM VALVE
D-1 = TWO POSITION OA DAMPER MOTOR
D-2 = FACE & BYPASS DAMPER
EP-1 = SOLENOID AIR VALVE
PE-1 = PRESSURE/ELECTRIC SWITCH
LLT = LOW LIMIT ELECTRIC FREEZE PROTECTION THERMOSTAT

100% O.A. Unit, Heating and Cooling, with Hot Water Preheat Coil, Hot Water Heating, DX Cooling, Steam Humidifier from Room Humidistat, and Reheat Coil from Room Thermostat

Whenever the fan runs, the EP-1 is energized and the O.A. damper opens. If the unit is large, arrangements need to be made to be sure the damper is open before the fan runs.

Thermostat T-1 controls valve V-1 on the hot water preheat coil. LLT stops the fan on freezing conditions at that location. Duct thermostat T-3 controls valve V-2 on the heating coil and PE switch PE-1 on the cooling coil. Room thermostat(s) T-2 control valve(s) V-3 on reheat coil(s). Humidistat H-1 controls humidifier valve V-4 through high-limit duct humidistat H-2. There may be more than one reheat coil.

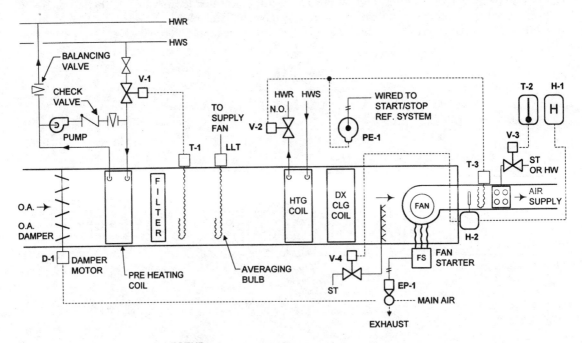

LEGEND:

T-1	= MODULATING CAPILLARY THERMOSTAT
T-2	= MODULATING ROOM THERMOSTAT
T-3	= MODULATING CAPILLARY DUCT THERMOSTAT
H-1	= MODULATING ROOM HUMIDISTAT
H-2	= HIGH LIMIT DUCT HUMIDISTAT
V-1	= MODULATING HOT WATER VALVE
V-2	= N.O. MODULATING 2-WAY HEATING COIL VALVE
V-3	= MODULATING STEAM/HOT WATER REHEAT COIL VALVE
V-4	= N.C. MODULATING HUMIDIFIER STEAM VALVE
D-1	= TWO POSITION OA DAMPER MOTOR
EP-1	= SOLENOID AIR VALVE
PE-1	= PRESSURE/ELECTRIC SWITCH
LLT	= LOW LIMIT ELECTRIC FREEZE PROTECTION THERMOSTAT

100% O.A. Unit, Heating and Cooling, with Hot Water Preheat Coil, Hot Water Heating, DX Cooling, Face and Bypass Dampers, Steam Humidifier from Room Humidistat, and Reheat Coil from Room Thermostat

Whenever the fan runs, the EP-1 is energized and the O.A. damper opens. If the unit is large, arrangements need to be made to be sure the damper is open before the fan runs.

Thermostat T-1 controls valve V-1 on the hot water preheat coil. LLT stops the fan on freezing conditions at that location. Duct thermostat T-3 controls valve V-2 on the heating coil and PE switch PE-1 along with damper motor D-2 on DX cooling coil face and bypass dampers. PE-1 shuts down the cooling when the face damper is closed. Room thermostat(s) T-2 control valve(s) V-3 on reheat coil(s). Humidistat H-1 controls humidifier valve V-4 through high-limit duct humidistat H-2. There may be more than one reheat coil.

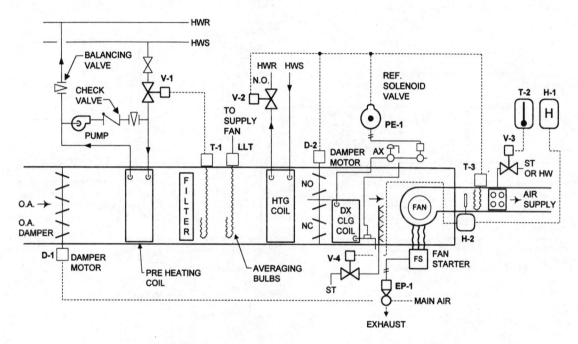

LEGEND:

T-1	=	MODULATING CAPILLARY THERMOSTAT
T-2	=	MODULATING ROOM THERMOSTAT
T-3	=	MODULATING CAPILLARY DUCT THERMOSTAT
H-1	=	MODULATING ROOM HUMIDISTAT
H-2	=	HIGH LIMIT DUCT HUMIDISTAT
V-1	=	MODULATING HOT WATER VALVE
V-2	=	N.O. MODULATING 2-WAY HEATING COIL VALVE
V-3	=	MODULATING STEAM/HOT WATER REHEAT COIL VALVE
V-4	=	N.C. MODULATING HUMIDIFIER STEAM VALVE
D-1	=	TWO POSITION OA DAMPER MOTOR
D-2	=	FACE & BYPASS DAMPER MOTOR
EP-1	=	SOLENOID AIR VALVE
PE-1	=	PRESSURE/ELECTRIC SWITCH
LLT	=	LOW LIMIT ELECTRIC FREEZE PROTECTION THERMOSTAT

100% O.A. Unit, Heating and Cooling, with Hot Water Preheat Coil, Hot Water Heating, Chilled Water Cooling, Steam Humidifier from Room Humidistat, and Reheat Coil from Room Thermostat

Whenever the fan runs, the EP-1 is energized and O.A. dampers are opened. If the system is very large, provisions should be made to be sure the dampers are open before the fan starts. When the fan stops, the O.A. damper closes.

Duct thermostat T-1 controls the hot water preheat coil. The pump on the preheat coil must run at all times. Discharge duct thermostat T-3 controls the heating coil valve V-2 and cooling coil V-3. Freeze protection thermostat LLT stops the fan on an indication of freezing conditions at that location. Room thermostat(s) T-2 control the valve V-4 on the reheat coil(s). There may be more than one reheat coil in the system. Room humidistat H-1, through high-limit humidistat H-2, controls the humidifier valve V-5.

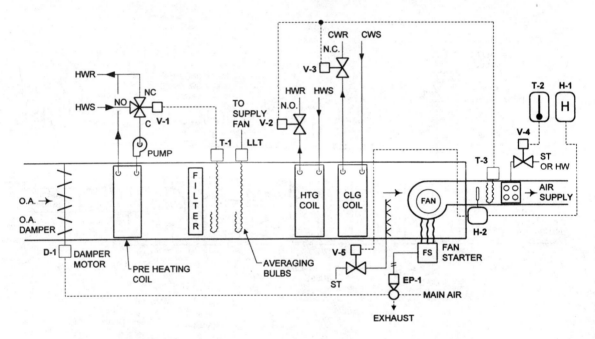

LEGEND:

T-1	=	MODULATING CAPILLARY THERMOSTAT
T-2	=	MODULATING ROOM THERMOSTAT
T-3	=	MODULATING CAPILLARY DUCT THERMOSTAT
H-1	=	MODULATING ROOM HUMIDISTAT
H-2	=	HIGH LIMIT DUCT HUMIDISTAT
V-1	=	MODULATING HOT WATER VALVE
V-2	=	N.O. MODULATING 2-WAY HEATING COIL VALVE
V-3	=	N.C. MODULATING 2-WAY COOLING COIL VALVE
V-4	=	MODULATING STEAM/HOT WATER REHEAT COIL VALVE
V-5	=	N.C. MODULATING HUMIDIFIER STEAM VALVE
D-1	=	TWO POSITION OA DAMPER MOTOR
EP-1	=	SOLENOID AIR VALVE
LLT	=	LOW LIMIT ELECTRIC FREEZE PROTECTION THERMOSTAT

100% O.A. Unit, Heating and Cooling, with Hot Water Preheat Coil, Hot water Heating, Chilled Water Cooling, Steam Humidifier from Room Humidistat, and Reheat Coil from Room Thermostat

Whenever the fan runs, the EP-1 is energized and O.A. dampers are opened. If the system is very large, provisions should be made to be sure the dampers are open before the fan starts. When the fan stops, the O.A. damper closes.

Duct thermostat T-1 controls the hot water preheat coil. The pump on the preheat coil must run at all times. Discharge duct thermostat T-3 controls the heating coil valve V-2 and cooling coil V-3. Freeze protection thermostat LLT stops the fan on an indication of freezing conditions at that location. Room thermostat(s) T-2 control the valve(s) V-4 on the reheat coil(s). There may be more than one reheat coil in the system. Room humidistat H-1, through high-limit humidistat H-2, controls the humidifier valve V-5.

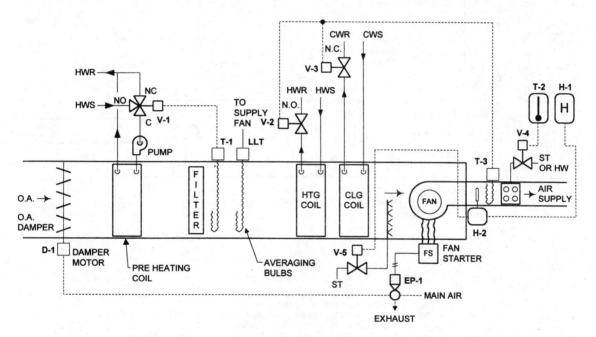

LEGEND:

T-1	=	MODULATING CAPILLARY THERMOSTAT
T-2	=	MODULATING ROOM THERMOSTAT
T-3	=	MODULATING CAPILLARY DUCT THERMOSTAT
H-1	=	MODULATING ROOM HUMIDISTAT
H-2	=	HIGH LIMIT DUCT HUMIDISTAT
V-1	=	MODULATING HOT WATER VALVE
V-2	=	N.O. MODULATING 2-WAY HEATING COIL VALVE
V-3	=	N.C. MODULATING 2-WAY COOLING COIL VALVE
V-4	=	MODULATING STEAM/HOT WATER REHEAT COIL VALVE
V-5	=	N.C. MODULATING HUMIDIFIER STEAM VALVE
D-1	=	TWO POSITION OA DAMPER MOTOR
EP-1	=	SOLENOID AIR VALVE
LLT	=	LOW LIMIT ELECTRIC FREEZE PROTECTION THERMOSTAT

Heating-Only Air-Handling Unit, Steam Heating Coil, with O.A., R.A., and REL Dampers and Mixed Air Controls

Whenever the fan runs, the EP-1 is energized and O.A., R.A., and REL dampers are placed under automatic controls. When the fan stops, all dampers return to their normal positions. Room thermostat T-2 controls valve V-1 through low-limit duct thermostat T-3. T-1 controls mixed air temperature. LLT stops the fan on falling temperature at that point to prevent freezing the coil.

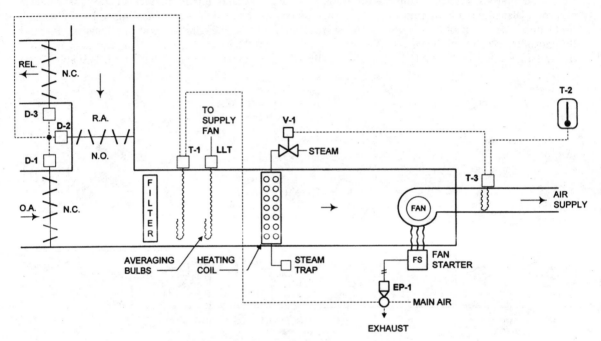

LEGEND:

T-1 = MODULATING CAPILLARY DUCT THERMOSTAT
T-2 = MODULATING ROOM THERMOSTAT
T-3 = MODULATING LOW LIMIT CAPILLARY DUCT THERMOSTAT
V-1 = N.O. MODULATING STEAM COIL VALVE
D-1 = OUTSIDE AIR DAMPER MOTOR
D-2 = RETURN AIR DAMPER MOTOR
D-3 = RELIEF AIR DAMPER MOTOR
EP-1 = SOLENOID AIR VALVE
LLT = LOW LIMIT ELECTRIC FREEZE PROTECTION

Heating-Only Air-Handling Unit, Hot Water Heating Coil, with O.A., R.A., and REL Dampers and Mixed Air Controls

Whenever the fan runs, the EP-1 is energized and O.A., R.A., and REL dampers are placed under automatic controls. When the fan stops, all dampers return to their normal positions. Room thermostat T-2 controls valve V-1 through low-limit duct thermostat T-3. T-1 controls mixed air temperature. LLT stops the fan on falling temperature at that point to prevent freezing the coil.

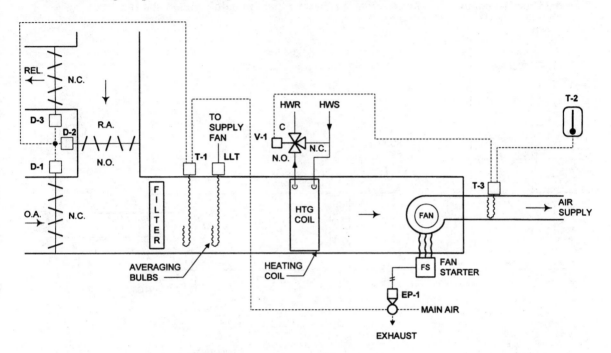

LEGEND:

T-1	=	MODULATING CAPILLARY DUCT THERMOSTAT
T-2	=	MODULATING ROOM THERMOSTAT
T-3	=	MODULATING LOW LIMIT CAPILLARY DUCT THERMOSTAT
V-1	=	N.O. MODULATING 3-WAY HW COIL VALVE
D-1	=	OUTSIDE AIR DAMPER MOTOR
D-2	=	RETURN AIR DAMPER MOTOR
D-3	=	RELIEF AIR DAMPER MOTOR
EP-1	=	SOLENOID AIR VALVE
LLT	=	LOW LIMIT ELECTRIC FREEZE PROTECTION

Heating-Only Air-Handling Unit, Steam Heating Coil, O.A., R.A., and REL Dampers, Mixed Air Controls, and Steam Humidifier from Room Humidistat

Whenever the fan runs, the EP-1 is energized and O.A., R.A., and REL dampers are placed under automatic controls. When the fan stops, all dampers return to their normal positions.

Room thermostat T-2 controls valve V-1 through low-limit duct thermostat T-3. T-1 controls mixed air temperature. LLT stops the fan on falling temperature at that point to prevent freezing the coil. Room humidistat H-1 controls humidifier valve V-2 through high-limit duct humidistat H-2.

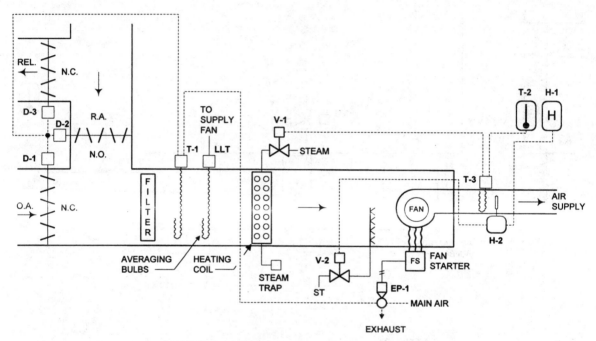

LEGEND:

T-1	=	MODULATING CAPILLARY DUCT THERMOSTAT
T-2	=	MODULATING ROOM THERMOSTAT
T-3	=	MODULATING LOW LIMIT CAPILLARY DUCT THERMOSTAT
H-1	=	MODULATING ROOM HUMIDISTAT
H-2	=	HIGH LIMIT DUCT HUMIDISTAT
V-1	=	N.O. MODULATING STEAM COIL VALVE
V-2	=	N.C. MODULATING HUMIDIFIER STEAM VALVE
D-1	=	OUTSIDE AIR DAMPER MOTOR
D-2	=	RETURN AIR DAMPER MOTOR
D-3	=	RELIEF AIR DAMPER MOTOR
EP-1	=	SOLENOID AIR VALVE
LLT	=	LOW LIMIT ELECTRIC FREEZE PROTECTION

Heating-Only Air-Handling Unit, Hot Water Heating Coil, O.A., R.A., and REL Dampers, Mixed Air Controls, and Steam Humidifier from Room Humidistat

Whenever the fan runs, the EP-1 is energized and O.A., R.A., and REL dampers are placed under automatic controls. When the fan stops, all dampers return to their normal positions.

Room thermostat T-2 controls valve V-1 through low-limit duct thermostat T-3. T-1 controls mixed air temperature. LLT stops the fan on falling temperature at that point to prevent freezing the coil. Room humidistat H-1 controls humidifier valve V-2 through high-limit duct humidistat H-2.

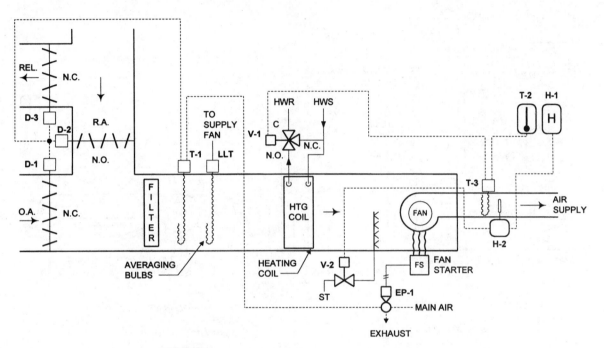

LEGEND:

T-1	=	MODULATING CAPILLARY DUCT THERMOSTAT
T-2	=	MODULATING ROOM THERMOSTAT
T-3	=	MODULATING LOW LIMIT CAPILLARY DUCT THERMOSTAT
H-1	=	MODULATING ROOM HUMIDISTAT
H-2	=	HIGH LIMIT DUCT HUMIDISTAT
V-1	=	N.O. MODULATING 3-WAY HW COIL VALVE
V-2	=	N.C. MODULATING HUMIDIFIER STEAM VALVE
D-1	=	OUTSIDE AIR DAMPER MOTOR
D-2	=	RETURN AIR DAMPER MOTOR
D-3	=	RELIEF AIR DAMPER MOTOR
EP-1	=	SOLENOID AIR VALVE
LLT	=	LOW LIMIT ELECTRIC FREEZE PROTECTION

Heating-Only Air-Handling Unit, Hot Water Heating Coil Using Face and Bypass Dampers, with O.A., R.A., and REL Dampers and Mixed Air Controls

Whenever the fan runs, the EP-1 is energized and O.A., R.A., and REL dampers are placed under automatic controls. When the fan stops, all dampers return to their normal positions.

Room thermostat T-2 controls valve V-1 and coil face and bypass dampers to maintain space conditions. LLT stops the fan on falling temperature at that point to prevent freezing the coil. Duct thermostat T-1 controls mixed air temperature by modulating O.A., R.A., and REL dampers.

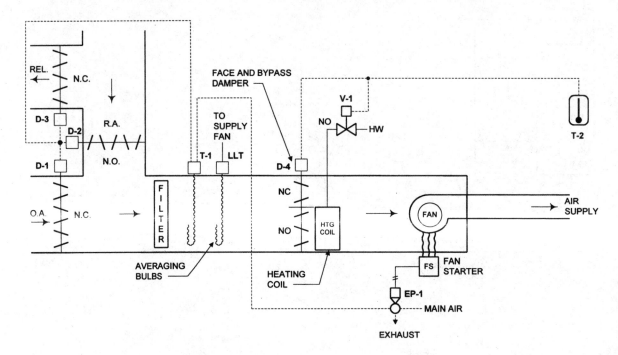

LEGEND:

T-1 = MODULATING CAPILLARY DUCT THERMOSTAT
T-2 = MODULATING ROOM THERMOSTAT
V-1 = N.O. MODULATING 2-WAY HW COIL VALVE
D-1 = OUTSIDE AIR DAMPER MOTOR
D-2 = RETURN AIR DAMPER MOTOR
D-3 = RELIEF AIR DAMPER MOTOR
D-4 = HEATING COIL FACE & BYPASS DAMPER MOTOR
EP-1 = SOLENOID AIR VALVE
LLT = LOW LIMIT ELECTRIC FREEZE PROTECTION

Heating-Only Air-Handling Unit, Hot Water Heating Coil Using Face and Bypass Dampers, with O.A., R.A., and REL Dampers Using Mixed Air Controls, with Steam Humidifier from Room Humidistat

Whenever the fan runs, the EP-1 is energized and O.A., R.A., and REL dampers are placed under automatic controls. When the fan stops, all dampers return to their normal positions.

Room thermostat T-2 controls valve V-1 and coil face and bypass dampers D-4 to maintain space conditions. LLT stops the fan on falling temperature at that point to prevent freezing the coil. Duct thermostat T-1 controls mixed air temperature by modulating O.A., R.A., and REL dampers. Room humidistat H-1 controls, through high-limit duct humidistat H-2, the humidifier valve V-2.

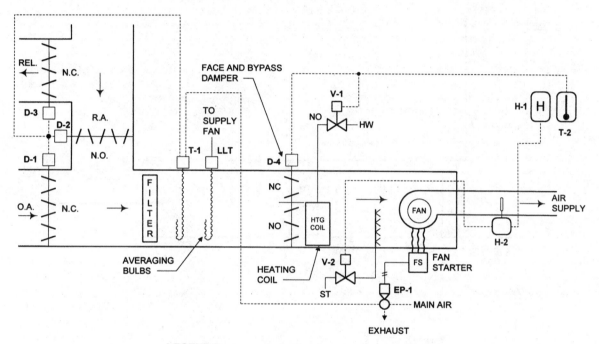

LEGEND:

T-1	=	MODULATING CAPILLARY DUCT THERMOSTAT
T-2	=	MODULATING ROOM THERMOSTAT
H-1	=	MODULATING ROOM HUMIDISTAT
H-2	=	HIGH LIMIT DUCT HUMIDISTAT
V-1	=	N.O. MODULATING 2-WAY HW COIL VALVE
V-2	=	N.C. MODULATING HUMIDIFIER STEAM VALVE
D-1	=	OUTSIDE AIR DAMPER MOTOR
D-2	=	RETURN AIR DAMPER MOTOR
D-3	=	RELIEF AIR DAMPER MOTOR
D-4	=	HEATING COIL FACE & BYPASS DAMPER MOTOR
EP-1	=	SOLENOID AIR VALVE
LLT	=	LOW LIMIT ELECTRIC FREEZE PROTECTION

Heating and Cooling Air-Handling Unit, Steam Heating and DX Cooling, with O.A., R.A., and REL Dampers Using Mixed Air Controls

Whenever the fan runs, the EP-1 is energized and O.A., R.A., and REL dampers are placed under automatic controls. When the fan stops, all dampers return to their normal positions.

Room thermostat T-2 controls valve V-1 on the heating coil in sequence with PE switch PE-1, which controls the DX cooling coil. LLT stops the fan on falling temperature at that point to prevent freezing the coil. Duct thermostat T-1 controls mixed air temperature by modulating O.A., R.A., and REL dampers. Minimum position S-1 can be set to maintain a minimum amount of O.A. as long as the fan is running.

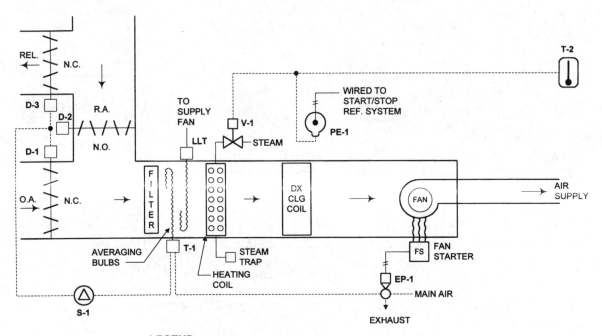

LEGEND:

T-1   = MODULATING CAPILLARY DUCT THERMOSTAT
T-2   = MODULATING ROOM THERMOSTAT
V-1   = N.O. MODULATING STEAM COIL VALVE
D-1   = OUTSIDE AIR DAMPER MOTOR
D-2   = RETURN AIR DAMPER MOTOR
D-3   = RELIEF AIR DAMPER MOTOR
EP-1 = SOLENOID AIR VALVE
PE-1 = PRESSURE/ELECTRIC SWITCH
S-1   = MINIMUM POSITION SWITCH
LLT   = LOW LIMIT ELECTRIC FREEZE PROTECTION

Heating and Cooling Air-Handling Unit, Steam Heating and DX Cooling Using Face and Bypass Dampers, with O.A., R.A., and REL Dampers Using Mixed Air Controls

Whenever the fan runs, the EP-1 is energized and O.A., R.A., and REL dampers are placed under automatic controls. When fan stops, the dampers return to their normal positions.

Duct thermostat T-1 controls the O.A., R.A., and REL dampers to maintain mix plenum temperatures. Minimum position switch S-1 can be set to maintain a minimum amount of O.A. as long as the fan is running, regardless of the actions of T-1. Room thermostat T-2 controls valve V-1 and DX cooling coil face and bypass damper D-4 in sequence with PE switch PE-1. When the face damper is closed, the PE switch shuts down the DX cooling.

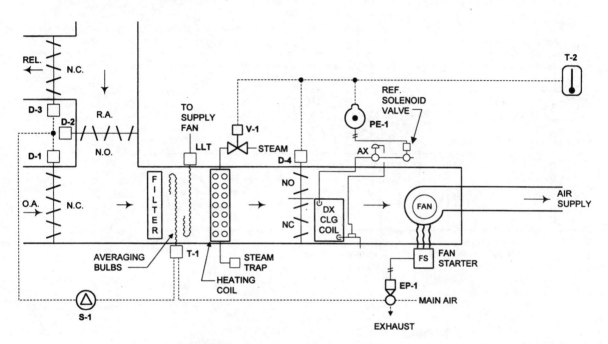

LEGEND:

T-1   = MODULATING CAPILLARY DUCT THERMOSTAT
T-2   = MODULATING ROOM THERMOSTAT
V-1   = N.O. MODULATING STEAM COIL VALVE
D-1   = OUTSIDE AIR DAMPER MOTOR
D-2   = RETURN AIR DAMPER MOTOR
D-3   = RELIEF AIR DAMPER MOTOR
D-4   = FACE & BYPASS DAMPER MOTOR
EP-1 = SOLENOID AIR VALVE
PE-1 = PRESSURE/ELECTRIC SWITCH
S-1   = MINIMUM POSITION SWITCH
LLT   = LOW LIMIT ELECTRIC FREEZE PROTECTION

Heating and Cooling Air-Handling Unit, Steam Heating and Chilled Water Cooling, with O.A., R.A., and REL Dampers Using Mixed Air Controls

Whenever the fan runs, the EP-1 is energized and O.A., R.A., and REL dampers are placed under automatic controls. When the fan stops, all dampers return to their normal positions.

Room thermostat T-2 controls valve V-1 on the heating coil in sequence with valve V-2 on the cooling coil. LLT stops the fan on falling temperature at that point to prevent freezing the coil. Duct thermostat T-1 controls mixed air temperature by modulating O.A., R.A., and REL dampers. Minimum position S-1 can be set to maintain a minimum amount of O.A. as long as the fan is running.

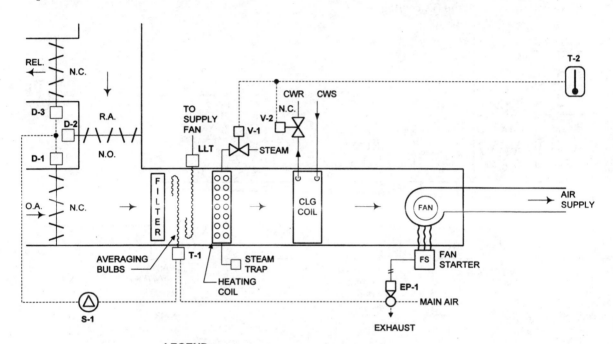

LEGEND:

T-1	=	MODULATING CAPILLARY DUCT THERMOSTAT
T-2	=	MODULATING ROOM THERMOSTAT
V-1	=	N.O. MODULATING STEAM COIL VALVE
V-2	=	N.C. MODULATING 2-WAY COOLING COIL VALVE
D-1	=	OUTSIDE AIR DAMPER MOTOR
D-2	=	RETURN AIR DAMPER MOTOR
D-3	=	RELIEF AIR DAMPER MOTOR
EP-1	=	SOLENOID AIR VALVE
S-1	=	MINIMUM POSITION SWITCH
LLT	=	LOW LIMIT ELECTRIC FREEZE PROTECTION

Heating and Cooling Air-Handling Unit, Hot Water Heating and DX Cooling Using Face and Bypass Dampers, with O.A., R.A., and REL Dampers Using Mixed Air Controls

Whenever the fan runs, the EP-1 is energized and O.A., R.A., and REL dampers are placed under automatic controls. When the fan stops, the dampers return to their normal positions.

Duct thermostat T-1 controls the O.A., R.A., and REL dampers to maintain mix plenum temperatures. Minimum position switch S-1 can be set to maintain a minimum amount of O.A. as long as the fan is running, regardless of the actions of T-1. Room thermostat T-2 controls steam valve V-1 and DX cooling coil face and bypass damper D-4 in sequence with PE switch PE-1. When the face damper is closed, the PE switch shuts down the DX cooling.

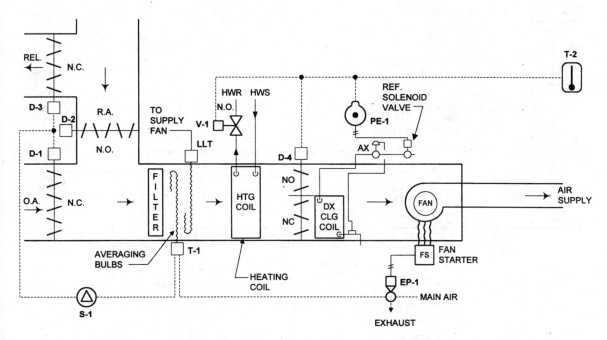

LEGEND:

T-1 = MODULATING CAPILLARY DUCT THERMOSTAT
T-2 = MODULATING ROOM THERMOSTAT
V-1 = N.O. MODULATING 2-WAY HEATING COIL VALVE
D-1 = OUTSIDE AIR DAMPER MOTOR
D-2 = RETURN AIR DAMPER MOTOR
D-3 = RELIEF AIR DAMPER MOTOR
D-4 = FACE & BYPASS DAMPER MOTOR
EP-1 = SOLENOID AIR VALVE
PE-1 = PRESSURE/ELECTRIC SWITCH
S-1 = MINIMUM POSITION SWITCH
LLT = LOW LIMIT ELECTRIC FREEZE PROTECTION

Heating and Cooling Air-Handling Unit, Hot Water Heating and DX Cooling, with O.A., R.A., and REL Dampers Using Mixed Air Controls

Whenever the fan runs, the EP-1 is energized and O.A., R.A., and REL dampers are placed under automatic controls. When the fan stops, all dampers return to their normal positions.

Room thermostat T-2 controls valve V-1 on the heating coil in sequence with valve PE switch PE-1 on the DX cooling coil. LLT stops the fan on falling temperature at that point to prevent freezing the coil. Duct thermostat T-1 controls mixed air temperature by modulating O.A., R.A., and REL dampers. Minimum position S-1 can be set to maintain a minimum amount of O.A. as long as the fan is running.

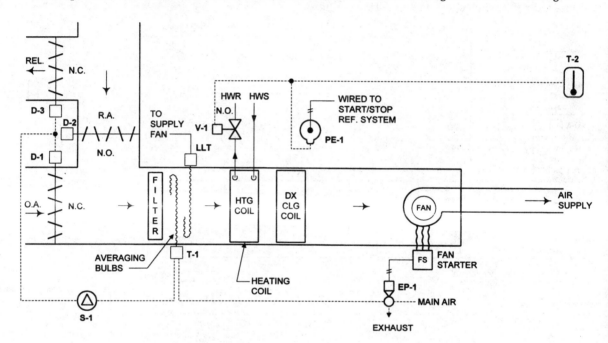

LEGEND:

T-1 = MODULATING CAPILLARY DUCT THERMOSTAT
T-2 = MODULATING ROOM THERMOSTAT
V-1 = N.O. MODULATING 2-WAY HEATING COIL VALVE
D-1 = OUTSIDE AIR DAMPER MOTOR
D-2 = RETURN AIR DAMPER MOTOR
D-3 = RELIEF AIR DAMPER MOTOR
EP-1 = SOLENOID AIR VALVE
PE-1 = PRESSURE/ELECTRIC SWITCH
S-1 = MINIMUM POSITION SWITCH
LLT = LOW LIMIT ELECTRIC FREEZE PROTECTION

Heating and Cooling Air-Handling Unit, Steam Heating and DX Cooling, with O.A., R.A., and REL Dampers Using Mixed Air Controls, and Steam Humidifier from Room Humidistat

Whenever the fan runs, the EP-1 is energized and O.A., R.A., and REL dampers are placed under automatic controls. When the fan stops, all dampers return to their normal positions.

Room thermostat T-2 controls valve V-1 on the heating coil in sequence with valve PE switch PE-1 on the DX cooling coil. LLT stops the fan on falling temperature at that point to prevent freezing the coil. Duct thermostat T-1 controls mixed air temperature by modulating O.A., R.A., and REL dampers. Minimum position S-1 can be set to maintain a minimum amount of O.A. as long as the fan is running. Room humidistat H-1 controls valve V-2 through high-limit duct humidistat H-2.

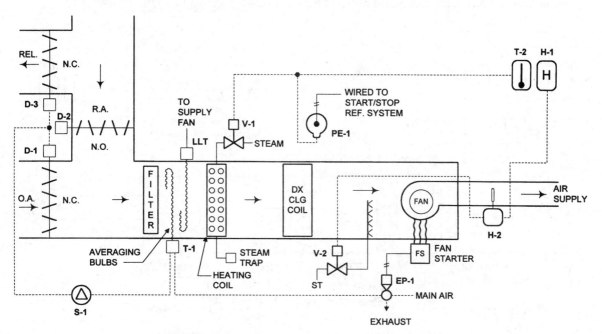

LEGEND:

T-1	=	MODULATING CAPILLARY DUCT THERMOSTAT
T-2	=	MODULATING ROOM THERMOSTAT
H-1	=	MODULATING ROOM HUMIDISTAT
H-2	=	HIGH LIMIT DUCT HUMIDISTAT
V-1	=	N.O. MODULATING STEAM COIL VALVE
V-2	=	N.C. MODULATING HUMIDIFIER STEAM VALVE
D-1	=	OUTSIDE AIR DAMPER MOTOR
D-2	=	RETURN AIR DAMPER MOTOR
D-3	=	RELIEF AIR DAMPER MOTOR
EP-1	=	SOLENOID AIR VALVE
PE-1	=	PRESSURE/ELECTRIC SWITCH
S-1	=	MINIMUM POSITION SWITCH
LLT	=	LOW LIMIT ELECTRIC FREEZE PROTECTION

Heating and Cooling Air-Handling Unit, Steam Heating and DX Cooling Using Face and Bypass Dampers, with O.A., R.A., and REL Dampers Using Mixed Air Controls, and Steam Humidifier from Room Humidistat

Whenever the fan runs, the EP-1 is energized and O.A., R.A., and REL dampers are placed under automatic controls. When the fan stops, all dampers return to their normal positions.

Room thermostat T-2 controls valve V-1 on the heating coil in sequence with face and bypass dampers and PE switch PE-1 on the DX cooling coil. When the face damper is closed, the PE switch shuts down the cooling. LLT stops the fan on falling temperature at that point to prevent freezing the coil. Duct thermostat T-1 controls mixed air temperature by modulating O.A., R.A., and REL dampers. Minimum position S-1 can be set to maintain a minimum amount of O.A. as long as the fan is running. Room humidistat H-1, through high-limit duct humidistat H-2, controls humidifier valve V-2.

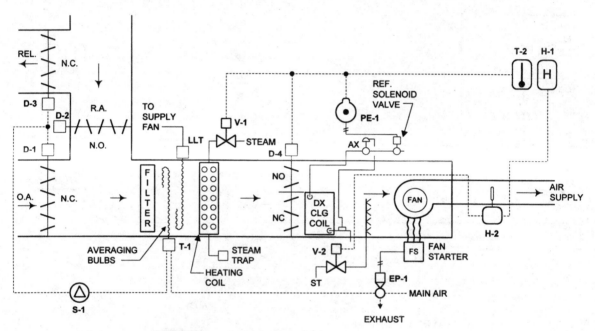

LEGEND:

T-1	= MODULATING CAPILLARY DUCT THERMOSTAT
T-2	= MODULATING ROOM THERMOSTAT
H-1	= MODULATING ROOM HUMIDISTAT
H-2	= HIGH LIMIT DUCT HUMIDISTAT
V-1	= N.O. MODULATING STEAM COIL VALVE
V-2	= N.C. MODULATING HUMIDIFIER STEAM VALVE
D-1	= OUTSIDE AIR DAMPER MOTOR
D-2	= RETURN AIR DAMPER MOTOR
D-3	= RELIEF AIR DAMPER MOTOR
D-4	= FACE & BYPASS DAMPER MOTOR
EP-1	= SOLENOID AIR VALVE
PE-1	= PRESSURE/ELECTRIC SWITCH
S-1	= MINIMUM POSITION SWITCH
LLT	= LOW LIMIT ELECTRIC FREEZE PROTECTION

Heating and Cooling Air-Handling Unit, Steam Heating and Chilled Water Cooling, with O.A., R.A., and REL Dampers Using Mixed Air Controls, and Steam Humidifier from Room Humidistat

Whenever the fan runs, the EP-1 is energized and O.A., R.A., and REL dampers are placed under automatic controls. When the fan stops, all dampers return to their normal positions.

Room thermostat T-2 controls valve V-1 on the heating coil in sequence with valve V-2 on the cooling coil. LLT stops the fan on falling temperature at that point to prevent freezing the coil. Duct thermostat T-1 controls mixed air temperature by modulating O.A., R.A., and REL dampers. Minimum position S-1 can be set to maintain a minimum amount of O.A. as long as the fan is running. Room humidistat H-1, through high-limit duct humidistat H-2, controls humidifier valve V-3.

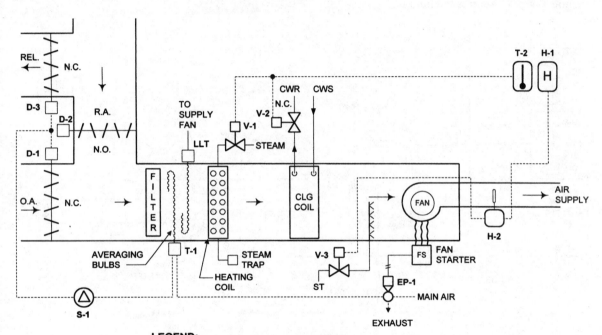

LEGEND:

T-1	= MODULATING CAPILLARY DUCT THERMOSTAT
T-2	= MODULATING ROOM THERMOSTAT
H-1	= MODULATING ROOM HUMIDISTAT
H-2	= HIGH LIMIT DUCT HUMIDISTAT
V-1	= N.O. MODULATING STEAM COIL VALVE
V-2	= N.C. MODULATING 2-WAY COOLING COIL VALVE
V-3	= N.C. MODULATING HUMIDIFIER STEAM VALVE
D-1	= OUTSIDE AIR DAMPER MOTOR
D-2	= RETURN AIR DAMPER MOTOR
D-3	= RELIEF AIR DAMPER MOTOR
EP-1	= SOLENOID AIR VALVE
S-1	= MINIMUM POSITION SWITCH
LLT	= LOW LIMIT ELECTRIC FREEZE PROTECTION

Heating and Cooling Air-Handling Unit, Hot Water Heating and DX Cooling Using Face and Bypass Dampers, with O.A., R.A., and REL Dampers Using Mixed Air Controls, and Steam Humidifier from Room Humidistat

Whenever the fan runs, the EP-1 is energized and O.A., R.A., and REL dampers are placed under automatic controls. When the fan stops, the dampers return to their normal positions.

Duct thermostat T-1 controls the O.A., R.A., and REL dampers to maintain mix plenum temperatures. Minimum position switch S-1 can be set to maintain a minimum amount of O.A. as long as the fan is running, regardless of the actions of T-1. Room thermostat T-2 controls steam valve V-1 and DX cooling coil face and bypass damper D-4 in sequence with PE switch PE-1. When the face damper is closed, the PE switch shuts down the DX cooling. Room humidistat H-1, through high-limit duct humidistat H-2, controls the humidifier valve V-2.

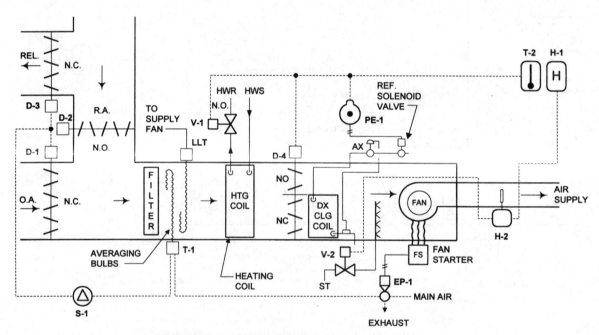

LEGEND:

T-1 = MODULATING CAPILLARY DUCT THERMOSTAT
T-2 = MODULATING ROOM THERMOSTAT
H-1 = MODULATING ROOM HUMIDISTAT
H-2 = HIGH LIMIT DUCT HUMIDISTAT
V-1 = N.O. MODULATING 2-WAY HEATING COIL VALVE
V-2 = N.C. MODULATING HUMIDIFIER STEAM VALVE
D-1 = OUTSIDE AIR DAMPER MOTOR
D-2 = RETURN AIR DAMPER MOTOR
D-3 = RELIEF AIR DAMPER MOTOR
D-4 = FACE & BYPASS DAMPER MOTOR
EP-1 = SOLENOID AIR VALVE
PE-1 = PRESSURE/ELECTRIC SWITCH
S-1 = MINIMUM POSITION SWITCH
LLT = LOW LIMIT ELECTRIC FREEZE PROTECTION

Heating and Cooling Air-Handling Unit, Hot Water Heating and DX Cooling, with O.A., R.A., and REL Dampers with Mixed Air Controls and Room Humidistat

Whenever the fan runs, the EP-1 is energized and O.A., R.A., and REL dampers are placed under automatic controls. When the fan stops, all dampers return to their normal positions.

Room thermostat T-2 controls valve V-1 on the heating coil in sequence with PE switch PE-1 on the DX cooling coil. LLT stops the fan on falling temperature at that point to prevent freezing the coil. Duct thermostat T-1 controls mixed air temperature by modulating O.A., R.A., and REL dampers. Minimum position S-1 can be set to maintain a minimum amount of O.A. as long as the fan is running. Room humidistat H-1, through high-limit duct humidistat H-2, controls humidifier valve V-2.

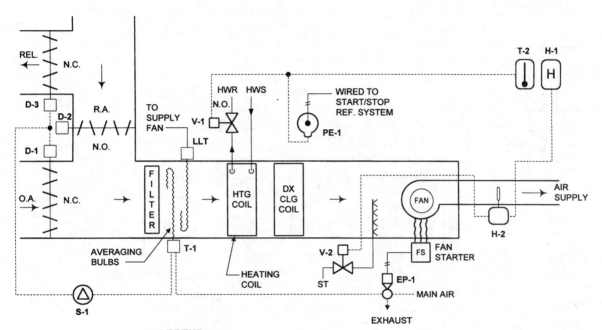

LEGEND:

T-1 = MODULATING CAPILLARY DUCT THERMOSTAT
T-2 = MODULATING ROOM THERMOSTAT
H-1 = MODULATING ROOM HUMIDISTAT
H-2 = HIGH LIMIT DUCT HUMIDISTAT
V-1 = N.O. MODULATING 2-WAY HEATING COIL VALVE
V-2 = N.C. MODULATING HUMIDIFIER STEAM VALVE
D-1 = OUTSIDE AIR DAMPER MOTOR
D-2 = RETURN AIR DAMPER MOTOR
D-3 = RELIEF AIR DAMPER MOTOR
EP-1 = SOLENOID AIR VALVE
PE-1 = PRESSURE/ELECTRIC SWITCH
S-1 = MINIMUM POSITION SWITCH
LLT = LOW LIMIT ELECTRIC FREEZE PROTECTION

Heating and Cooling Air-Handling Unit, Steam Heating Coil and DX Cooling Coil, with O.A., R.A., and REL Dampers Using Economizer Control Cycle

Whenever the fan runs, the EP-1 is energized and O.A., R.A., and REL dampers are placed under automatic controls. When the fan stops, all dampers return to their normal positions.

Room thermostat T-2 controls valve V-1 on the heating coil in sequence with PE switch PE-1 on the DX cooling coil. LLT stops the fan on falling temperature at that point to prevent freezing the coil. Duct thermostat T-1 controls mixed air temperature through O.A. limit duct thermostat T-3 by modulating O.A., R.A., and REL dampers. T-3 and T-1 act as economizer controls to allow for up to 100% O.A. when conditions permit. Minimum position relay S-1 can be set to maintain a minimum amount of O.A. as long as the fan is running.

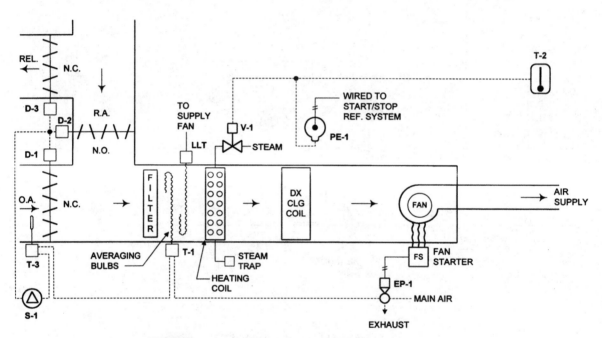

LEGEND:

T-1	= MODULATING CAPILLARY DUCT THERMOSTAT
T-2	= MODULATING ROOM THERMOSTAT
T-3	= MODULATING CAPILLARY THERMOSTAT
V-1	= N.O. MODULATING STEAM COIL VALVE
D-1	= OUTSIDE AIR DAMPER MOTOR
D-2	= RETURN AIR DAMPER MOTOR
D-3	= RELIEF AIR DAMPER MOTOR
EP-1	= SOLENOID AIR VALVE
PE-1	= PRESSURE/ELECTRIC SWITCH
S-1	= MINIMUM POSITION SWITCH
LLT	= LOW LIMIT ELECTRIC FREEZE PROTECTION

Heating and Cooling Air-Handling Unit, Steam Heating and DX Cooling Using Face and Bypass Dampers, with O.A., R.A., and REL Dampers Using Economizer Control Cycle

Whenever the fan runs, the EP-1 is energized and O.A., R.A., and REL dampers are placed under automatic controls. When the fan stops, all dampers return to their normal positions.

Room thermostat T-2 controls valve V-1 on the heating coil in sequence with face and bypass dampers D-4 on the DX cooling coil. PE switch PE-1 shuts down the cooling when the face damper is closed. LLT stops the fan on falling temperature at that point to prevent freezing the coil. Duct thermostat T-1 controls mixed air temperature through O.A. limit duct thermostat T-3 by modulating O.A., R.A., and REL dampers. T-3 and T-1 act as economizer controls to allow for up to 100% O.A. when conditions permit. Minimum position relay S-1 can be set to maintain a minimum amount of O.A. as long as the fan is running.

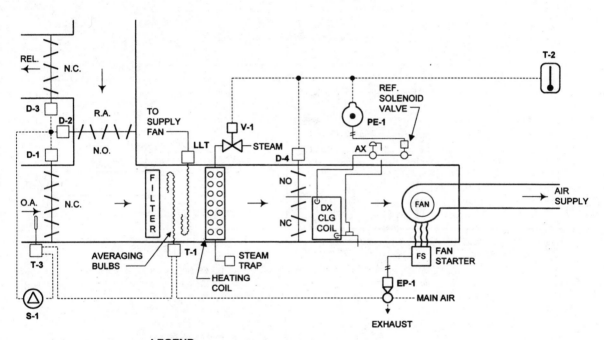

LEGEND:

T-1 = MODULATING CAPILLARY DUCT THERMOSTAT
T-2 = MODULATING ROOM THERMOSTAT
T-3 = MODULATING CAPILLARY THERMOSTAT
V-1 = N.O. MODULATING STEAM COIL VALVE
D-1 = OUTSIDE AIR DAMPER MOTOR
D-2 = RETURN AIR DAMPER MOTOR
D-3 = RELIEF AIR DAMPER MOTOR
D-4 = FACE & BYPASS DAMPER MOTOR
EP-1 = SOLENOID AIR VALVE
PE-1 = PRESSURE/ELECTRIC SWITCH
S-1 = MINIMUM POSITION SWITCH
LLT = LOW LIMIT ELECTRIC FREEZE PROTECTION

Heating and Cooling Air-Handling Unit, Steam Heating and Chilled Water Cooling, with O.A., R.A., and REL Dampers Using Economizer Control Cycle

Whenever the fan runs, the EP-1 is energized and O.A., R.A., and REL dampers are placed under automatic controls. When the fan stops, all dampers return to their normal positions.

Room thermostat T-2 controls valve V-1 on the heating coil in sequence with valve V-2 on the cooling coil. LLT stops the fan on falling temperature at that point to prevent freezing the coil. Duct thermostat T-1 controls mixed air temperature through O.A. limit duct thermostat T-3 by modulating O.A., R.A., and REL dampers. T-3 and T-1 act as economizer controls to allow for up to 100% O.A. when conditions permit. Minimum position relay S-1 can be set to maintain a minimum amount of O.A. as long as the fan is running.

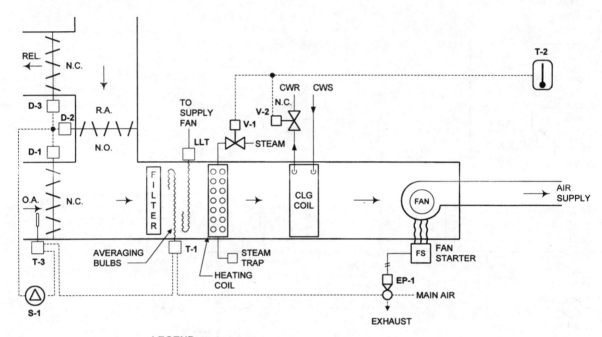

LEGEND:

T-1	= MODULATING CAPILLARY DUCT THERMOSTAT
T-2	= MODULATING ROOM THERMOSTAT
T-3	= MODULATING CAPILLARY THERMOSTAT
V-1	= N.O. MODULATING STEAM COIL VALVE
V-2	= N.C. MODULATING 2-WAY COOLING COIL VALVE
D-1	= OUTSIDE AIR DAMPER MOTOR
D-2	= RETURN AIR DAMPER MOTOR
D-3	= RELIEF AIR DAMPER MOTOR
EP-1	= SOLENOID AIR VALVE
S-1	= MINIMUM POSITION SWITCH
LLT	= LOW LIMIT ELECTRIC FREEZE PROTECTION

Heating and Cooling Air-Handling Unit, Hot Water Heating and DX Cooling Using Face and Bypass Dampers, with O.A., R.A., and REL Dampers Using Economizer Control Cycle

Whenever the fan runs, the EP-1 is energized and O.A., R.A., and REL dampers are placed under automatic controls. When the fan stops, all dampers return to their normal positions.

Room thermostat T-2 controls valve V-1 on the heating coil in sequence with face and bypass dampers on the DX cooling coil. PE switch PE-1 shuts down the cooling when the face damper is closed. LLT stops the fan on falling temperature at that point to prevent freezing the coil. Duct thermostat T-1 controls mixed air temperature through O.A. limit duct thermostat T-3 by modulating O.A., R.A., and REL dampers. T-3 and T-1 act as economizer controls to allow for up to 100% O.A. when conditions permit. Minimum position relay S-1 can be set to maintain a minimum amount of O.A. as long as the fan is running.

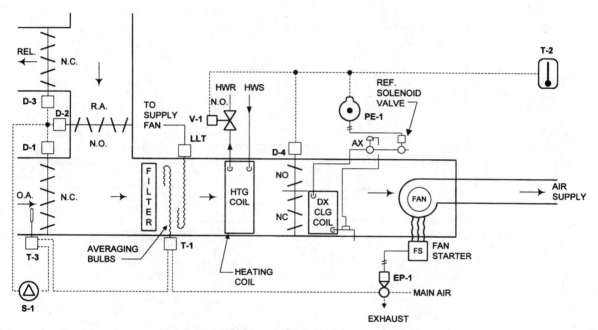

LEGEND:

T-1 = MODULATING CAPILLARY DUCT THERMOSTAT
T-2 = MODULATING ROOM THERMOSTAT
T-3 = MODULATING CAPILLARY THERMOSTAT
V-1 = N.O. MODULATING 2-WAY HEATING COIL VALVE
D-1 = OUTSIDE AIR DAMPER MOTOR
D-2 = RETURN AIR DAMPER MOTOR
D-3 = RELIEF AIR DAMPER MOTOR
D-4 = FACE & BYPASS DAMPER MOTOR
EP-1 = SOLENOID AIR VALVE
PE-1 = PRESSURE/ELECTRIC SWITCH
S-1 = MINIMUM POSITION SWITCH
LLT = LOW LIMIT ELECTRIC FREEZE PROTECTION

Heating and Cooling Air-Handling Unit, Hot Water Heating and DX Cooling, with O.A., R.A., and REL Dampers Using Economizer Control Cycle

Whenever the fan runs, the EP-1 is energized and O.A., R.A., and REL dampers are placed under automatic controls. When the fan stops, all dampers return to their normal positions.

Room thermostat T-2 controls valve V-1 on the heating coil in sequence with PE switch PE-1 on the DX cooling coil. LLT stops the fan on falling temperature at that point to prevent freezing the coil. Duct thermostat T-1 controls mixed air temperature through O.A. limit duct thermostat T-3 by modulating O.A., R.A., and REL dampers. T-3 and T-1 act as economizer controls to allow for up to 100% O.A. when conditions permit. Minimum position relay S-1 can be set to maintain a minimum amount of O.A. as long as the fan is running.

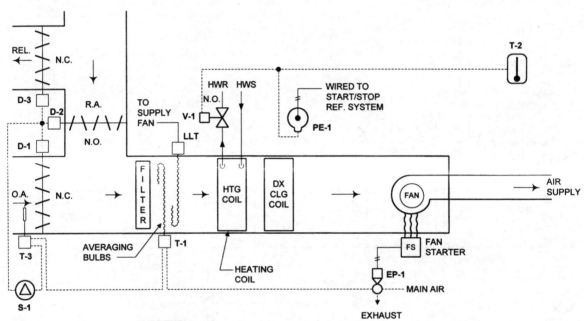

LEGEND:

T-1	=	MODULATING CAPILLARY DUCT THERMOSTAT
T-2	=	MODULATING ROOM THERMOSTAT
T-3	=	MODULATING CAPILLARY THERMOSTAT
V-1	=	N.O. MODULATING 2-WAY HEATING COIL VALVE
D-1	=	OUTSIDE AIR DAMPER MOTOR
D-2	=	RETURN AIR DAMPER MOTOR
D-3	=	RELIEF AIR DAMPER MOTOR
EP-1	=	SOLENOID AIR VALVE
PE-1	=	PRESSURE/ELECTRIC SWITCH
S-1	=	MINIMUM POSITION SWITCH
LLT	=	LOW LIMIT ELECTRIC FREEZE PROTECTION

Heating and Cooling Air-Handling Unit, Steam Heating and DX Cooling, with O.A., R.A., and REL Dampers Using Economizer Control Cycle with Steam Humidifier from Room Humidistat

Whenever the fan runs, the EP-1 is energized and O.A., R.A., and REL dampers are placed under automatic controls. When fan stops, the dampers return to their normal positions.

Duct thermostat T-1 controls the O.A., R.A., and REL dampers through the O.A. high-limit duct thermostat T-3. T-3 and T-1 act as economizer controls. They can allow up to 100% O.A. when conditions are favorable. Minimum position switch S-1 can be set to maintain a minimum amount of O.A. as long as the fan is running, regardless of the actions of T-1 and T-3. Room thermostat T-2 controls steam valve V-1 and DX cooling coil through PE switch PE-1. Room humidistat H-1, through high-limit duct humidistat H-2, controls the humidifier valve V-2.

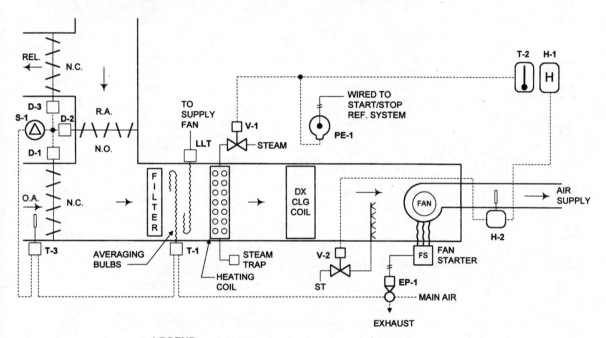

LEGEND:

T-1 = MODULATING CAPILLARY DUCT THERMOSTAT
T-2 = MODULATING ROOM THERMOSTAT
T-3 = MODULATING CAPILLARY THERMOSTAT
H-1 = MODULATING ROOM HUMIDISTAT
H-2 = HIGH LIMIT DUCT HUMIDISTAT
V-1 = N.O. MODULATING STEAM COIL VALVE
V-2 = N.C. MODULATING HUMIDIFIER STEAM VALVE
D-1 = OUTSIDE AIR DAMPER MOTOR
D-2 = RETURN AIR DAMPER MOTOR
D-3 = RELIEF AIR DAMPER MOTOR
EP-1 = SOLENOID AIR VALVE
PE-1 = PRESSURE/ELECTRIC SWITCH
S-1 = MINIMUM POSITION SWITCH
LLT = LOW LIMIT ELECTRIC FREEZE PROTECTION

Heating and Cooling Air-Handling Unit, Steam Heating and DX Cooling Using Face and Bypass Dampers, with O.A., R.A., and REL Dampers Using Economizer Control Cycle with Steam Humidifier from Room Humidistat

Whenever the fan runs, the EP-1 is energized and O.A., R.A., and REL dampers are placed under automatic controls. When the fan stops, all dampers return to their normal positions.

Room thermostat T-2 controls valve V-1 on the heating coil in sequence with face and bypass dampers D-4 on the DX cooling coil. PE switch PE-1 shuts down the cooling when the face damper is closed. LLT stops the fan on falling temperature at that point to prevent freezing the coil. Duct thermostat T-1 controls mixed air temperature through O.A. limit duct thermostat T-3 by modulating O.A., R.A., and REL dampers. T-3 and T-1 act as economizer controls to allow for up to 100% O.A. when conditions permit. Minimum position relay S-1 can be set to maintain a minimum amount of O.A. as long as the fan is running. Room humidistat H-1 controls, through high-limit duct humidistat H-2, the humidifier valve V-2.

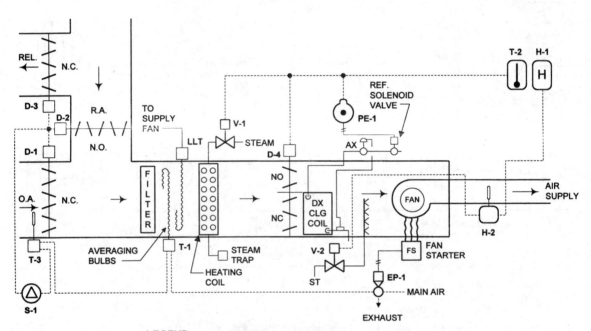

LEGEND:

T-1	=	MODULATING CAPILLARY DUCT THERMOSTAT
T-2	=	MODULATING ROOM THERMOSTAT
T-3	=	MODULATING CAPILLARY THERMOSTAT
H-1	=	MODULATING ROOM HUMIDISTAT
H-2	=	HIGH LIMIT DUCT HUMIDISTAT
V-1	=	N.O. MODULATING STEAM COIL VALVE
V-2	=	N.C. MODULATING HUMIDIFIER STEAM VALVE
D-1	=	OUTSIDE AIR DAMPER MOTOR
D-2	=	RETURN AIR DAMPER MOTOR
D-3	=	RELIEF AIR DAMPER MOTOR
D-4	=	FACE & BYPASS DAMPER MOTOR
EP-1	=	SOLENOID AIR VALVE
PE-1	=	PRESSURE/ELECTRIC SWITCH
S-1	=	MINIMUM POSITION SWITCH
LLT	=	LOW LIMIT ELECTRIC FREEZE PROTECTION

Heating/Cooling Air-Handling Unit, Steam Heating, Chilled Water Cooling, with O.A., R.A., and REL Dampers Using Economizer Control Cycle and Steam Humidifier from Room Humidistat

Whenever the fan runs, the EP-1 is energized and O.A., R.A., and REL dampers are placed under automatic controls. When fan stops, the dampers return to their normal positions.

Duct thermostat T-1 controls the O.A., R.A., and REL dampers through the O.A. high-limit duct thermostat T-3. T-3 and T-1 act as economizer controls. They can allow up to 100% O.A. when conditions are favorable. Minimum position switch S-1 can be set to maintain a minimum amount of O.A. as long as the fan is running, regardless of the actions of T-1 and T-3. Room thermostat T-2 controls steam valve V-1 and valve V-2 on the chilled water coil. Room humidistat H-1, through high-limit duct humidistat H-2, controls the humidifier valve V-3.

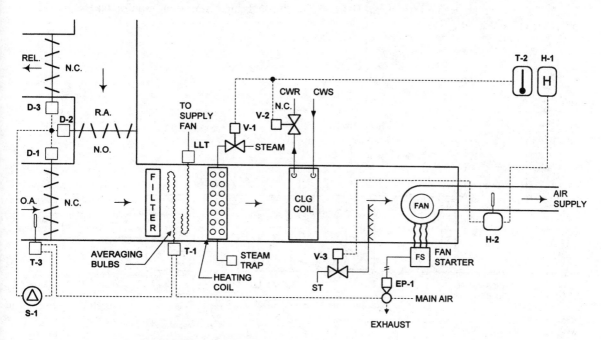

LEGEND:

T-1 = MODULATING CAPILLARY DUCT THERMOSTAT
T-2 = MODULATING ROOM THERMOSTAT
T-3 = MODULATING CAPILLARY THERMOSTAT
H-1 = MODULATING ROOM HUMIDISTAT
H-2 = HIGH LIMIT DUCT HUMIDISTAT
V-1 = N.O. MODULATING STEAM COIL VALVE
V-2 = N.C. MODULATING 2-WAY COOLING COIL VALVE
V-3 = N.C. MODULATING HUMIDIFIER STEAM VALVE
D-1 = OUTSIDE AIR DAMPER MOTOR
D-2 = RETURN AIR DAMPER MOTOR
D-3 = RELIEF AIR DAMPER MOTOR
EP-1 = SOLENOID AIR VALVE
S-1 = MINIMUM POSITION SWITCH
LLT = LOW LIMIT ELECTRIC FREEZE PROTECTION

Heating and Cooling Air-Handling Unit, Hot Water Heating and DX Cooling Using Face and Bypass Dampers, with O.A., R.A., and REL Dampers Using Economizer Control Cycle with Steam Humidifier from Room Humidistat

Whenever the fan runs, the EP-1 is energized and O.A., R.A., and REL dampers are placed under automatic controls. When the fan stops, all dampers return to their normal positions.

Room thermostat T-2 controls valve V-1 on the heating coil in sequence with face and bypass dampers on the DX cooling coil. PE switch PE-1 shuts down the cooling coil when the face damper is closed. LLT stops the fan on falling temperature at that point to prevent freezing the coil. Duct thermostat T-1 controls mixed air temperature through O.A. limit duct thermostat T-3 by modulating O.A., R.A., and REL dampers. T-3 and T-1 act as economizer controls to allow for up to 100% O.A. when conditions permit. Minimum position relay S-1 can be set to maintain a minimum amount of O.A. as long as the fan is running. Room humidistat H-1 controls, through high-limit duct humidistat H-2, the humidifier valve V-2.

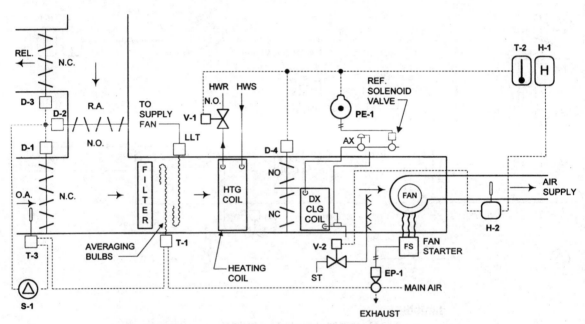

LEGEND:

T-1	= MODULATING CAPILLARY DUCT THERMOSTAT
T-2	= MODULATING ROOM THERMOSTAT
T-3	= MODULATING CAPILLARY THERMOSTAT
H-1	= MODULATING ROOM HUMIDISTAT
H-2	= HIGH LIMIT DUCT HUMIDISTAT
V-1	= N.O. MODULATING 2-WAY HEATING COIL VALVE
V-2	= N.C. MODULATING HUMIDIFIER STEAM VALVE
D-1	= OUTSIDE AIR DAMPER MOTOR
D-2	= RETURN AIR DAMPER MOTOR
D-3	= RELIEF AIR DAMPER MOTOR
D-4	= FACE & BYPASS DAMPER MOTOR
EP-1	= SOLENOID AIR VALVE
PE-1	= PRESSURE/ELECTRIC SWITCH
S-1	= MINIMUM POSITION SWITCH
LLT	= LOW LIMIT ELECTRIC FREEZE PROTECTION

Heating and Cooling Air-Handling Unit, Hot Water Heating DX cooling, O.A., R.A., and REL Dampers Using Economizer Control Cycle with Steam Humidifier from Room Humidistat

Whenever the fan runs, the EP-1 is energized and O.A., R.A., and REL dampers are placed under automatic controls. When the fan stops, all dampers return to their normal positions.

Room thermostat T-2 controls valve V-1 on the heating coil in sequence with PE switch PE-1 on the DX coding coil. LLT stops the fan on falling temperature at that point to prevent freezing the coil. Duct thermostat T-1 controls mixed air temperature through O.A. limit duct thermostat T-3 by modulating O.A., R.A., and REL dampers. T-3 and T-1 act as economizer controls to allow for up to 100% O.A. when conditions permit. Minimum position relay S-1 can be set to maintain a minimum amount of O.A. as long as the fan is running. Room humidistat H-1 controls through high-limit duct humidistat H-2 the humidifier valve V-2.

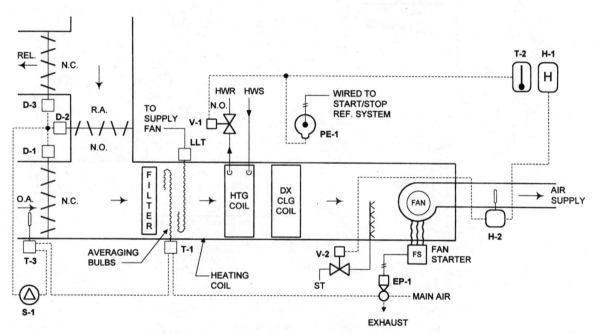

LEGEND:

T-1	= MODULATING CAPILLARY DUCT THERMOSTAT
T-2	= MODULATING ROOM THERMOSTAT
T-3	= MODULATING CAPILLARY THERMOSTAT
H-1	= MODULATING ROOM HUMIDISTAT
H-2	= HIGH LIMIT DUCT HUMIDISTAT
V-1	= N.O. MODULATING 2-WAY HEATING COIL VALVE
V-2	= N.C. MODULATING HUMIDIFIER STEAM VALVE
D-1	= OUTSIDE AIR DAMPER MOTOR
D-2	= RETURN AIR DAMPER MOTOR
D-3	= RELIEF AIR DAMPER MOTOR
EP-1	= SOLENOID AIR VALVE
PE-1	= PRESSURE/ELECTRIC SWITCH
S-1	= MINIMUM POSITION SWITCH
LLT	= LOW LIMIT ELECTRIC FREEZE PROTECTION

Heating-Only Air-Handling Unit, R.A. Fan, Steam Heating Coil, O.A., R.A., and REL Dampers under Control of Room Thermostat

Whenever the fan runs, the EP switch EP-1 is energized placing the dampers under automatic control. When the fan stops, the dampers return to their normal positions—O.A. and REL closed and R.A. open.

Room thermostat T-3 through the low-limit thermostat T-2 controls the O.A., R.A., and REL dampers through the mixed air thermostat T-1. T-2 also controls valve V-1 on the heating coil. Minimum position switch S-1 through relay R-1 sets a minimum amount of O.A. as long as the fan is running.

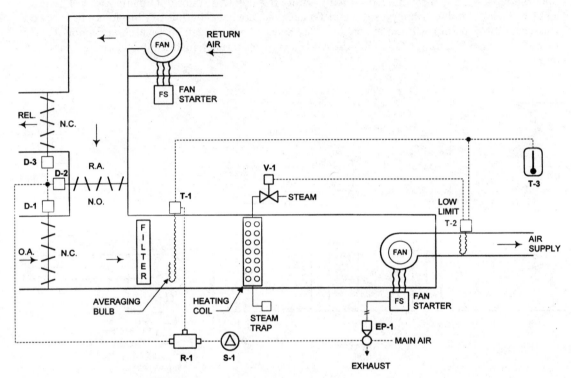

LEGEND:

T-1	=	MODULATING CAPILLARY DUCT THERMOSTAT
T-2	=	MODULATING CAPILLARY DUCT THERMOSTAT
T-3	=	MODULATING ROOM THERMOSTAT
V-1	=	N.O. MODULATING STEAM COIL VALVE
D-1	=	OUTSIDE AIR DAMPER MOTOR
D-2	=	RETURN AIR DAMPER MOTOR
D-3	=	RELIEF AIR DAMPER MOTOR
EP-1	=	SOLENOID AIR VALVE
R-1	=	HIGH SIGNAL SELECTOR
S-1	=	MINIMUM POSITION SWITCH

Heating-Only Air-Handling Unit, with R.A. Fan, Steam Heating Coil, O.A., R.A. and REL Dampers Using Mixed Air Control

Whenever the fan runs, the EP switch EP-1 is energized placing the dampers under automatic control. When the fan stops, the dampers return to their normal positions—O.A. and REL closed and R.A. open.

Room thermostat T-3 through the low-limit thermostat T-2 controls the valve V-1 on the heating coil. The O.A., R.A., and REL dampers are controlled by mixed air thermostat T-1. Minimum position switch S-1 through relay R-1 sets a minimum amount of O.A. as long as the fan is running.

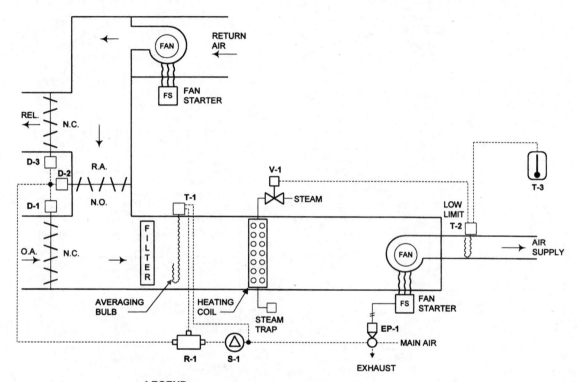

LEGEND:

T-1	= MODULATING CAPILLARY DUCT THERMOSTAT
T-2	= MODULATING CAPILLARY DUCT THERMOSTAT
T-3	= MODULATING ROOM THERMOSTAT
V-1	= N.O. MODULATING STEAM COIL VALVE
D-1	= OUTSIDE AIR DAMPER MOTOR
D-2	= RETURN AIR DAMPER MOTOR
D-3	= RELIEF AIR DAMPER MOTOR
EP-1	= SOLENOID AIR VALVE
R-1	= HIGH SIGNAL SELECTOR
S-1	= MINIMUM POSITION SWITCH

Heating-Only Air-Handling Unit, with R.A. Fan, Hot Water Heating Coil, O.A., R.A., and REL Dampers under Control of Room Thermostat

Whenever the fan runs, the EP switch EP-1 is energized placing the dampers under automatic control. When the fan stops, the dampers return to their normal positions—O.A. and REL closed and R.A. open.

Room thermostat T-3, through the low-limit thermostat T-2, controls valve V-1 on the heating coil as well as the O.A., R.A., and REL dampers. Minimum position switch S-1 through relay R-1 sets a minimum amount of O.A. as long as the fan is running.

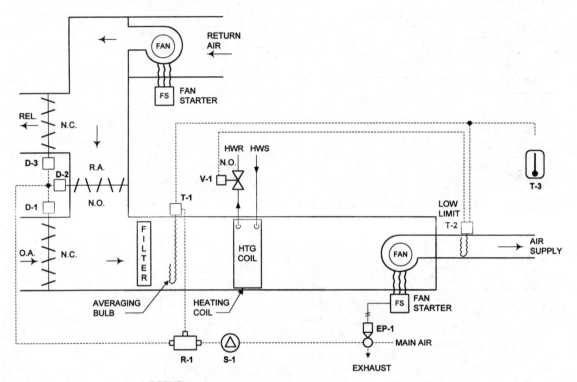

LEGEND:

T-1 = MODULATING CAPILLARY DUCT THERMOSTAT
T-2 = MODULATING CAPILLARY DUCT THERMOSTAT
T-3 = MODULATING ROOM THERMOSTAT
V-1 = N.O. MODULATING 2-WAY HEATING COIL VALVE
D-1 = OUTSIDE AIR DAMPER MOTOR
D-2 = RETURN AIR DAMPER MOTOR
D-3 = RELIEF AIR DAMPER MOTOR
EP-1 = SOLENOID AIR VALVE
R-1 = HIGH SIGNAL SELECTOR
S-1 = MINIMUM POSITION SWITCH

Heating-Only Air-Handling Unit, with R.A. Fan, Hot Water Heating Coil, O.A., R.A., and REL Dampers Using Mixed Air Controls

Whenever the fan runs, the EP switch EP-1 is energized placing the dampers under automatic control. When the fan stops, the dampers return to their normal positions—O.A. and REL closed and R.A. open.

Room thermostat T-3, through the low-limit thermostat T-2, controls valve V-1 on the heating coil. The O.A., R.A., and REL dampers are controlled by mixed air thermostat T-1. Minimum position switch S-1 through relay R-1 sets a minimum amount of O.A. as long as the fan is running.

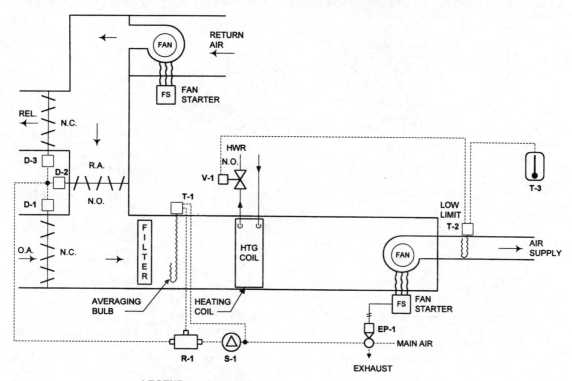

LEGEND:

T-1	= MODULATING CAPILLARY DUCT THERMOSTAT
T-2	= MODULATING CAPILLARY DUCT THERMOSTAT
T-3	= MODULATING ROOM THERMOSTAT
V-1	= N.O. MODULATING 2-WAY HEATING COIL VALVE
D-1	= OUTSIDE AIR DAMPER MOTOR
D-2	= RETURN AIR DAMPER MOTOR
D-3	= RELIEF AIR DAMPER MOTOR
EP-1	= SOLENOID AIR VALVE
R-1	= HIGH SIGNAL SELECTOR
S-1	= MINIMUM POSITION SWITCH

Heating-Only Air-Handling Unit, Steam Heating, with R.A. FAN. O.A., R.A., and REL Dampers under Control of Room Thermostat with Economizer Control Cycle Also

Whenever the fan runs, the EP switch EP-1 is energized placing the dampers under automatic control. When the fan stops, the dampers return to their normal positions. O.A. and REL closed and R.A. open.

Room thermostat T-3, through the low-limit thermostat T-2, controls valve V-1 on the heating coil. The O.A., R.A., and REL dampers are controlled by mixed air thermostat T-1 through low-limit O.A. thermostat T-4. T-1 and T-4 act as economizer thermostats. They can allow up to 100% O.A. when conditions are favorable. Minimum position switch S-1 through relay R-1 sets a minimum amount of O.A. as long as the fan is running.

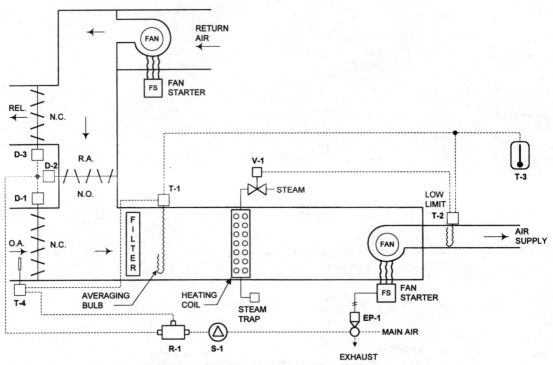

LEGEND:

T-1	=	MODULATING CAPILLARY DUCT THERMOSTAT
T-2	=	MODULATING CAPILLARY DUCT THERMOSTAT
T-3	=	MODULATING ROOM THERMOSTAT
T-4	=	MODULATING CAPILLARY THERMOSTAT
V-1	=	N.O. MODULATING STEAM COIL VALVE
D-1	=	OUTSIDE AIR DAMPER MOTOR
D-2	=	RETURN AIR DAMPER MOTOR
D-3	=	RELIEF AIR DAMPER MOTOR
EP-1	=	SOLENOID AIR VALVE
R-1	=	HIGH SIGNAL SELECTOR
S-1	=	MINIMUM POSITION SWITCH

Heating-Only Air-Handling Unit, with R.A. Fan, Steam Heating Coil, O.A., R.A., and REL Dampers Using Economizer Control Cycle

Whenever the fan runs, the EP-1 is energized and O.A., R.A., and REL dampers are placed under automatic controls. When the fan stops, all dampers return to their normal positions.

Room thermostat T-3 controls valve V-1 on the heating coil through low-limit duct thermostat T-2. Duct thermostat T-1 controls mixed air temperature through O.A. limit duct thermostat T-4 by modulating O.A., R.A., and REL dampers. Minimum position relay S-1 can be set to maintain a minimum amount of O.A. through high signal selector R-1 as long as the fan is running.

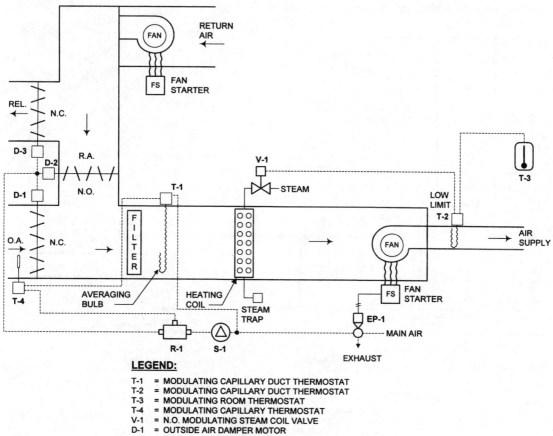

LEGEND:

T-1	=	MODULATING CAPILLARY DUCT THERMOSTAT
T-2	=	MODULATING CAPILLARY DUCT THERMOSTAT
T-3	=	MODULATING ROOM THERMOSTAT
T-4	=	MODULATING CAPILLARY THERMOSTAT
V-1	=	N.O. MODULATING STEAM COIL VALVE
D-1	=	OUTSIDE AIR DAMPER MOTOR
D-2	=	RETURN AIR DAMPER MOTOR
D-3	=	RELIEF AIR DAMPER MOTOR
EP-1	=	SOLENOID AIR VALVE
R-1	=	HIGH SIGNAL SELECTOR
S-1	=	MINIMUM POSITION SWITCH

Heating-Only Air-Handling Unit, with R.A. Fan, Hot Water Heating Coil, O.A., R.A., and REL Dampers under Control of Room Thermostat

Whenever the fan runs, the EP-1 is energized and O.A., R.A., and REL dampers are placed under automatic controls. When the fan stops, all dampers return to their normal positions.

Room thermostat T-3 controls valve V-1 on the heating coil through low-limit duct thermostat T-2. It also controls the O.A, R.A, and REL dampers through duct thermostat T-1 and high-limit O.A. duct thermostat T-4. T-1 and T-4 act as economizer controls which allow up to 100% O.A. when conditions permit. Minimum position relay S-1 can be set to maintain a minimum amount of O.A. through high signal selector R-1 as long as the fan is running. The room thermostat on a call for heating sequences both the coil valve and dampers to heat up the space.

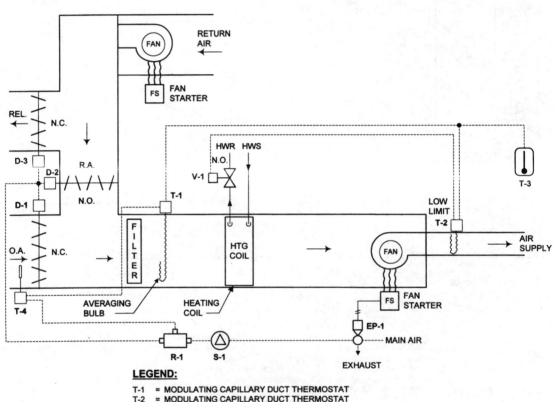

LEGEND:

T-1	=	MODULATING CAPILLARY DUCT THERMOSTAT
T-2	=	MODULATING CAPILLARY DUCT THERMOSTAT
T-3	=	MODULATING ROOM THERMOSTAT
T-4	=	MODULATING CAPILLARY THERMOSTAT
V-1	=	N.O. MODULATING 2-WAY HEATING COIL VALVE
D-1	=	OUTSIDE AIR DAMPER MOTOR
D-2	=	RETURN AIR DAMPER MOTOR
D-3	=	RELIEF AIR DAMPER MOTOR
EP-1	=	SOLENOID AIR VALVE
R-1	=	HIGH SIGNAL SELECTOR
S-1	=	MINIMUM POSITION SWITCH

Heating-only Air-Handling Unit, with R.A. Fan, Hot Water Heating Coil., O.A., R.A., and REL Dampers Using Economizer Control Cycle

Whenever the fan runs, the EP-1 is energized and O.A., R.A., and REL dampers are placed under automatic controls. When the fan stops, all dampers return to their normal positions.

Room thermostat T-3 controls valve V-1 on the heating coil through low-limit duct thermostat T-2. Duct thermostat T-1, through high-limit O.A. duct thermostat T-4, controls the O.A., R.A., and REL dampers. T-1 and T-4 act as economizer controls which allow up to 100% O.A. when conditions permit. Minimum position relay S-1 can be set to maintain a minimum amount of O.A. through high signal selector R-1 as long as the fan is running.

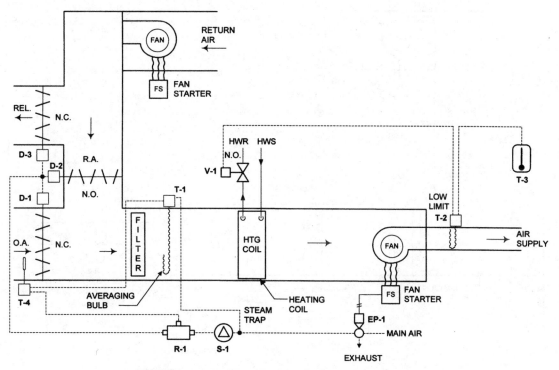

LEGEND:

T-1	=	MODULATING CAPILLARY DUCT THERMOSTAT
T-2	=	MODULATING CAPILLARY DUCT THERMOSTAT
T-3	=	MODULATING ROOM THERMOSTAT
T-4	=	MODULATING CAPILLARY THERMOSTAT
V-1	=	N.O. MODULATING 2-WAY HEATING COIL VALVE
D-1	=	OUTSIDE AIR DAMPER MOTOR
D-2	=	RETURN AIR DAMPER MOTOR
D-3	=	RELIEF AIR DAMPER MOTOR
EP-1	=	SOLENOID AIR VALVE
R-1	=	HIGH SIGNAL SELECTOR
S-1	=	MINIMUM POSITION SWITCH

Heating-Only Air-Handling Unit, with R.A. Fan, Hot Water Heating Coil, O.A., R.A., and REL Dampers Using Economizer Control Cycle

Whenever the fan runs, the EP-1 is energized and O.A., R.A., and REL dampers are placed under automatic controls. When the fan stops, all dampers return to their normal positions.

Room thermostat T-3 controls valve V-1 on the heating coil through low-limit duct thermostat T-2. It also controls the O.A., R.A., and REL dampers through duct thermostat T-1. Minimum position relay S-1 can be set to maintain a minimum amount of O.A. through high signal selector R-1 as long as the fan is running. Room humidistat H-1, through high-limit duct humidistat H-2, controls the humidifier valve V-2.

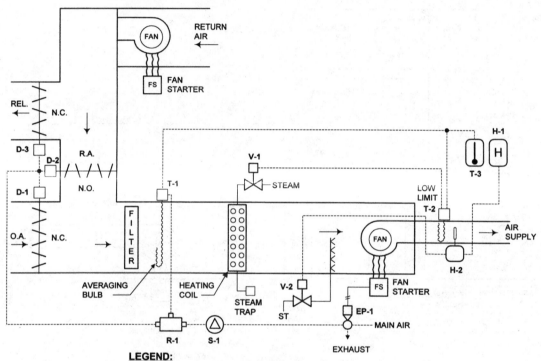

Heating-Only Air-Handling Unit, with R.A. Fan, Steam Heating Coil, O.A., R.A., and REL Dampers Using Mixed Air Controls, and Steam Humidifier from Room Humidistat with Economizer Controls

Whenever the fan runs, the EP switch EP-1 is energized, placing the dampers under automatic control. When the fan stops, the dampers return to their normal positions—O.A. and REL closed and R.A. open.

Room thermostat T-3, through the low-limit thermostat T-2, controls valve V-1 on the heating coil. The O.A., R.A., and REL dampers are controlled by mixed air thermostat T-1 through low-limit O.A. thermostat T-4. T-1 and T-4 act as economizer thermostats. They can allow up to 100% O.A. when conditions are favorable. Minimum position switch S-1 through relay R-1 sets a minimum amount of O.A. as long as the fan is running. Room humidistat H-1, through high-limit dust humidistat H-2, controls the humidifier.

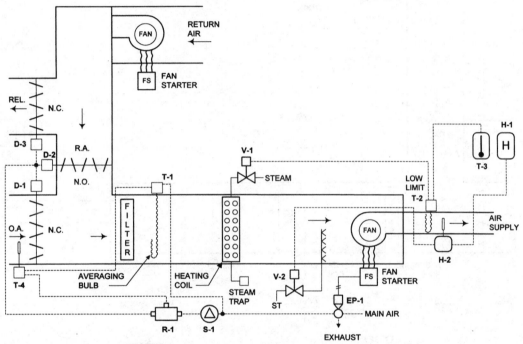

LEGEND:

T-1	=	MODULATING CAPILLARY DUCT THERMOSTAT
T-2	=	MODULATING CAPILLARY DUCT THERMOSTAT
T-3	=	MODULATING ROOM THERMOSTAT
T-4	=	MODULATING CAPILLARY THERMOSTAT
H-1	=	MODULATING ROOM HUMIDISTAT
H-2	=	HIGH LIMIT DUCT HUMIDISTAT
V-1	=	N.O. MODULATING STEAM COIL VALVE
V-2	=	N.C. MODULATING HUMIDIFIER STEAM VALVE
D-1	=	OUTSIDE AIR DAMPER MOTOR
D-2	=	RETURN AIR DAMPER MOTOR
D-3	=	RELIEF AIR DAMPER MOTOR
EP-1	=	SOLENOID AIR VALVE
R-1	=	HIGH SIGNAL SELECTOR
S-1	=	MINIMUM POSITION SWITCH

Heating-Only Air-Handling Unit, with R.A. Fan, Hot Water Heating, O.A., R.A., and REL Dampers under Control of Room Thermostat with Steam Humidifier from Room Humidistat

Whenever the fan runs, the EP-1 is energized and O.A., R.A., and REL dampers are placed under automatic controls. When the fan stops, all dampers return to their normal positions.

Room thermostat T-3 controls valve V-1 on the heating coil through low-limit duct thermostat T-2. It also controls the O.A., R.A., and REL dampers through duct thermostat T-1. Minimum position relay S-1 can be set to maintain a minimum amount of O.A. through high signal selector R-1 as long as the fan is running. Room humidistat H-1, through high-limit duct humidistat H-2, controls the humidifier valve V-2.

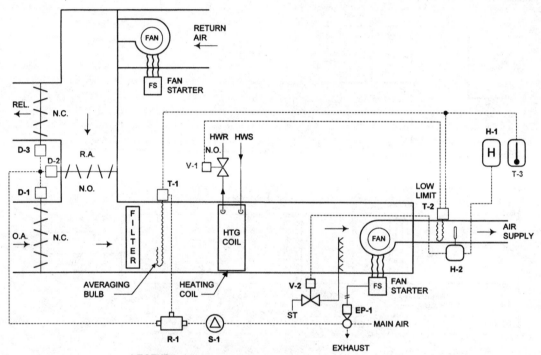

LEGEND:

T-1	=	MODULATING CAPILLARY DUCT THERMOSTAT
T-2	=	MODULATING CAPILLARY DUCT THERMOSTAT
T-3	=	MODULATING ROOM THERMOSTAT
H-1	=	MODULATING ROOM HUMIDISTAT
H-2	=	HIGH LIMIT DUCT HUMIDISTAT
V-1	=	N.O. MODULATING 2-WAY HEATING COIL VALVE
V-2	=	N.C. MODULATING HUMIDIFIER STEAM VALVE
D-1	=	OUTSIDE AIR DAMPER MOTOR
D-2	=	RETURN AIR DAMPER MOTOR
D-3	=	RELIEF AIR DAMPER MOTOR
EP-1	=	SOLENOID AIR VALVE
R-1	=	HIGH SIGNAL SELECTOR
S-1	=	MINIMUM POSITION SWITCH

Heating-Only Air-Handling Unit, with R.A. Fan, Hot Water Heating Coil, O.A., R.A., and REL Dampers Using Mixed Air Controls, and Steam Humidifier from Room Humidistat

Whenever the fan runs, the EP-1 is energized and O.A., R.A., and REL dampers are placed under automatic controls. When the fan stops, all dampers return to their normal positions.

Room thermostat T-3 controls valve V-1 on the heating coil through low-limit duct thermostat T-2. Thermostat T-1 controls the O.A., R.A., and REL dampers. Minimum position relay S-1 can be set to maintain a minimum amount of O.A. through high signal selector R-1 as long as the fan is running. Room humidistat H-1, through high-limit duct humidistat H-2, controls the humidifier valve V-2.

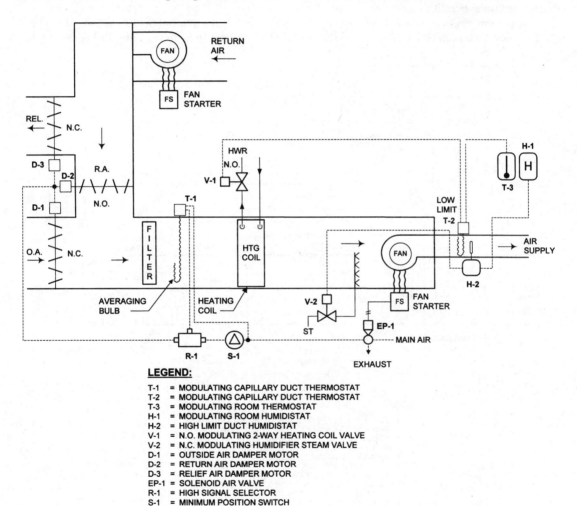

LEGEND:

T-1 = MODULATING CAPILLARY DUCT THERMOSTAT
T-2 = MODULATING CAPILLARY DUCT THERMOSTAT
T-3 = MODULATING ROOM THERMOSTAT
H-1 = MODULATING ROOM HUMIDISTAT
H-2 = HIGH LIMIT DUCT HUMIDISTAT
V-1 = N.O. MODULATING 2-WAY HEATING COIL VALVE
V-2 = N.C. MODULATING HUMIDIFIER STEAM VALVE
D-1 = OUTSIDE AIR DAMPER MOTOR
D-2 = RETURN AIR DAMPER MOTOR
D-3 = RELIEF AIR DAMPER MOTOR
EP-1 = SOLENOID AIR VALVE
R-1 = HIGH SIGNAL SELECTOR
S-1 = MINIMUM POSITION SWITCH

Heating-Only Air-Handling Unit, with R.A. Fan, Steam Heating Coil, O.A., R.A., and REL Dampers Using Economizer Control Cycle under Room Thermostat Control, and Steam Humidifier under Control of Room Humidistat

Whenever the fan runs, EP-1 is energized and O.A., R.A., and REL dampers are placed under automatic control. When the fan stops, the dampers return to their normal positions.

Room thermostat T-3 controls valve V-1 on the heating coil through low-limit duct thermostat T-2. It also controls the O.A., R.A., and REL dampers in sequence through duct thermostat T-1 and O.A. duct thermostat T-4. T-1 and T-4 act as economizer thermostats. T-1 and T-4 can admit up to 100% O.A. when conditions permit.

Minimum position relay S-1 can be set to maintain a minimum amount of O.A. as long as the fan is running. Room humidistat H-1, through high-limit duct humidistat H-2, controls humidifier valve V-2.

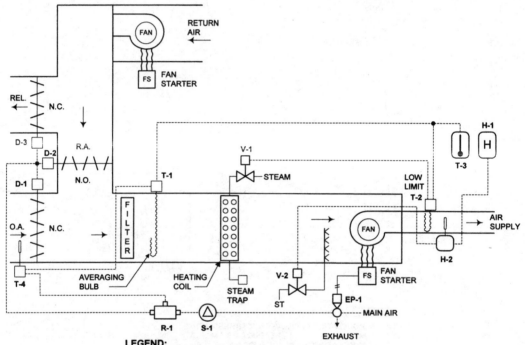

LEGEND:

T-1	=	MODULATING CAPILLARY DUCT THERMOSTAT
T-2	=	MODULATING CAPILLARY DUCT THERMOSTAT
T-3	=	MODULATING ROOM THERMOSTAT
T-4	=	MODULATING CAPILLARY THERMOSTAT
H-1	=	MODULATING ROOM HUMIDISTAT
H-2	=	HIGH LIMIT DUCT HUMIDISTAT
V-1	=	N.O. MODULATING STEAM COIL VALVE
V-2	=	N.C. MODULATING HUMIDIFIER STEAM VALVE
D-1	=	OUTSIDE AIR DAMPER MOTOR
D-2	=	RETURN AIR DAMPER MOTOR
D-3	=	RELIEF AIR DAMPER MOTOR
EP-1	=	SOLENOID AIR VALVE
R-1	=	HIGH SIGNAL SELECTOR
S-1	=	MINIMUM POSITION SWITCH

Heating-Only Air-Handling unit, with R.A. Fan, Steam Heating Coil, O.A., R.A., and REL Dampers Using Economizer Control Cycle with Steam Humidifier from Room Humidistat

Whenever the fan runs, the EP-1 is energized and O.A., R.A., and REL dampers are placed under automatic controls. When the fan stops, all dampers return to their normal positions.

Room thermostat T-3 controls valve V-1 on the heating coil through low-limit duct thermostat T-2. Duct thermostat T-1 controls the O.A., R.A., and REL dampers through O.A. high-limit duct thermostat T-4. T-1 and T-4 act as economizer controls allowing up to 100% O.A. when conditions are right. Minimum position relay S-1 can be set to maintain a minimum amount of O.A. through high signal selector R-1 as long as the fan is running. Room humidistat H-1, through high-limit duct humidistat H-2, controls the humidifier valve V-2.

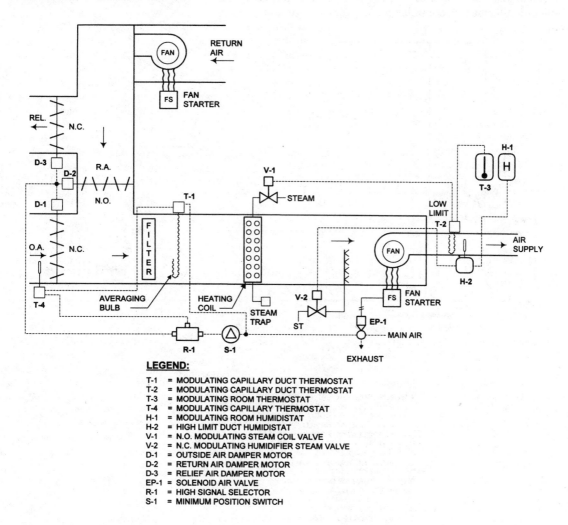

LEGEND:

T-1	=	MODULATING CAPILLARY DUCT THERMOSTAT
T-2	=	MODULATING CAPILLARY DUCT THERMOSTAT
T-3	=	MODULATING ROOM THERMOSTAT
T-4	=	MODULATING CAPILLARY THERMOSTAT
H-1	=	MODULATING ROOM HUMIDISTAT
H-2	=	HIGH LIMIT DUCT HUMIDISTAT
V-1	=	N.O. MODULATING STEAM COIL VALVE
V-2	=	N.C. MODULATING HUMIDIFIER STEAM VALVE
D-1	=	OUTSIDE AIR DAMPER MOTOR
D-2	=	RETURN AIR DAMPER MOTOR
D-3	=	RELIEF AIR DAMPER MOTOR
EP-1	=	SOLENOID AIR VALVE
R-1	=	HIGH SIGNAL SELECTOR
S-1	=	MINIMUM POSITION SWITCH

Heating-Only Air-Handling Unit, with R.A. Fan, Hot Water Heating, O.A., R.A., and REL Dampers Using Economizer Control Cycle under Room Thermostat with Steam Humidifier from Room Humidistat

Whenever the fan runs, the EP-1 is energized and O.A., R.A., and REL dampers are placed under automatic controls. When the fan stops, all dampers return to their normal positions.

Room thermostat T-3 controls valve V-1 on the heating coil through low-limit duct thermostat T-2. It also controls, through duct thermostat T-1, the O.A., R.A., and REL dampers through O.A. high-limit duct thermostat T-4. T-1 and T-4 act as economizer controls allowing up to 100% O.A. when conditions are right. Minimum position relay S-1 can be set to maintain a minimum amount of O.A. through high signal selector R-1 as long as the fan is running. Room humidistat H-1, through high-limit duct humidistat H-2, controls the humidifier valve V-2.

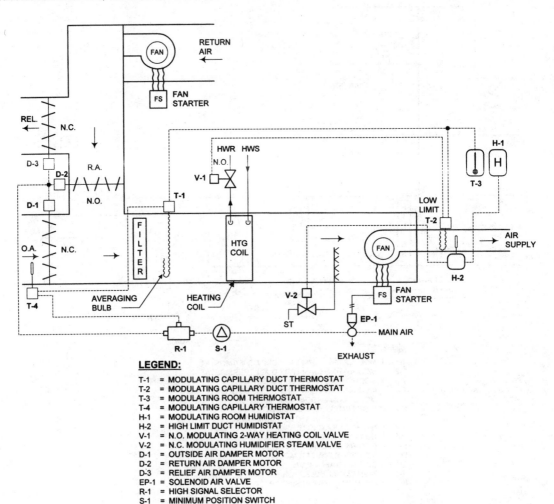

LEGEND:

T-1 = MODULATING CAPILLARY DUCT THERMOSTAT
T-2 = MODULATING CAPILLARY DUCT THERMOSTAT
T-3 = MODULATING ROOM THERMOSTAT
T-4 = MODULATING CAPILLARY THERMOSTAT
H-1 = MODULATING ROOM HUMIDISTAT
H-2 = HIGH LIMIT DUCT HUMIDISTAT
V-1 = N.O. MODULATING 2-WAY HEATING COIL VALVE
V-2 = N.C. MODULATING HUMIDIFIER STEAM VALVE
D-1 = OUTSIDE AIR DAMPER MOTOR
D-2 = RETURN AIR DAMPER MOTOR
D-3 = RELIEF AIR DAMPER MOTOR
EP-1 = SOLENOID AIR VALVE
R-1 = HIGH SIGNAL SELECTOR
S-1 = MINIMUM POSITION SWITCH

Heating-Only Air-Handling Unit, with R.A. Fan, Hot Water Heating Coil, O.A., R.A., and REL Dampers Using Economizer Control Cycle with Steam Humidifier from Room Humidistat

Whenever the fan runs, the EP-1 is energized and O.A., R.A., and REL dampers are placed under automatic controls. When the fan stops, all dampers return to their normal positions.

Room thermostat T-3 controls valve V-1 on the heating coil through low-limit duct thermostat T-2. Duct thermostat T-1 controls the O.A., R.A., and REL dampers through O.A. high-limit duct thermostat T-4. T-1 and T-4 act as economizer controls allowing up to 100% O.A. when conditions are right. Minimum position relay S-1 can be set to maintain a minimum amount of O.A. through high signal selector R-1 as long as the fan is running. Room humidistat H-1, through high-limit duct humidistat H-2, controls the humidifier valve V-2.

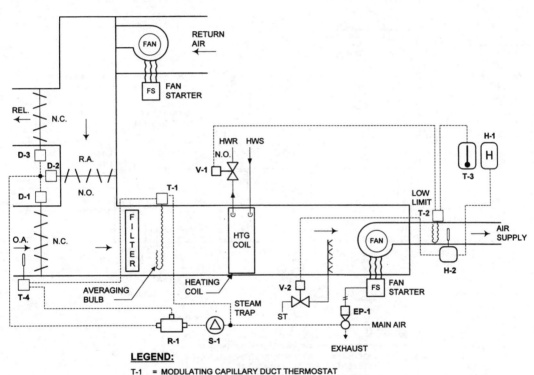

LEGEND:

T-1	= MODULATING CAPILLARY DUCT THERMOSTAT
T-2	= MODULATING CAPILLARY DUCT THERMOSTAT
T-3	= MODULATING ROOM THERMOSTAT
T-4	= MODULATING CAPILLARY THERMOSTAT
H-1	= MODULATING ROOM HUMIDISTAT
H-2	= HIGH LIMIT DUCT HUMIDISTAT
V-1	= N.O. MODULATING 2-WAY HEATING COIL VALVE
V-2	= N.C. MODULATING HUMIDIFIER STEAM VALVE
D-1	= OUTSIDE AIR DAMPER MOTOR
D-2	= RETURN AIR DAMPER MOTOR
D-3	= RELIEF AIR DAMPER MOTOR
EP-1	= SOLENOID AIR VALVE
R-1	= HIGH SIGNAL SELECTOR
S-1	= MINIMUM POSITION SWITCH

Heating and Cooling Air-Handling Unit, with R.A. Fan, Steam Heating and DX Cooling, O.A., R.A., and REL Dampers Using Mixed Air Controls

Whenever the fan runs, the EP-1 is energized and O.A., R.A., and REL dampers are placed under automatic controls. When the fan stops, all dampers return to their normal positions.

Room thermostat T-3 controls valve V-1 on the heating coil through low-limit duct thermostat T-2. It also controls PE switch PE-1 on the DX cooling coil. The O.A., R.A., and REL dampers are also controlled by the room thermostat in sequence. Minimum position relay S-1 can be set to maintain a minimum amount of O.A. through high signal selector R-1 as long as the fan is running.

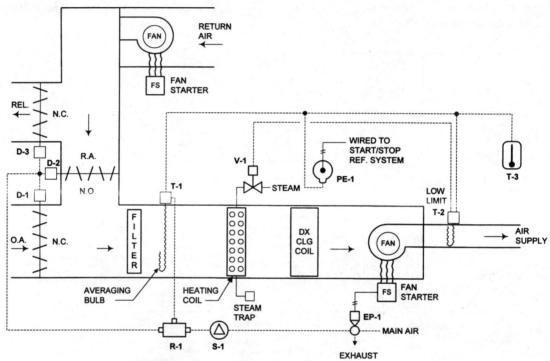

LEGEND:

T-1 = MODULATING CAPILLARY DUCT THERMOSTAT
T-2 = MODULATING CAPILLARY DUCT THERMOSTAT
T-3 = MODULATING ROOM THERMOSTAT
V-1 = N.O. MODULATING STEAM COIL VALVE
D-1 = OUTSIDE AIR DAMPER MOTOR
D-2 = RETURN AIR DAMPER MOTOR
D-3 = RELIEF AIR DAMPER MOTOR
EP-1 = SOLENOID AIR VALVE
PE-1 = PRESSURE/ELECTRIC SWITCH
R-1 = HIGH SIGNAL SELECTOR
S-1 = MINIMUM POSITION SWITCH

Heating and Cooling Air-Handling Unit, with R.A. Fan, Steam Heating, DX Cooling, O.A., R.A., and REL Dampers Using Economizer Cycle

Whenever the fan runs, the EP-1 is energized and O.A., R.A., and REL dampers are placed under automatic controls. When the fan stops, all dampers return to their normal positions.

Room thermostat T-3 controls valve V-1 on the heating coil through low-limit duct thermostat T-2. It also controls PE switch PE-1 on the DX cooling coil. The O.A., R.A., and REL dampers are controlled by thermostat T-1 through O.A. high-limit thermostat T-4. T-1 and T-4 act as economizer thermostats to allow for up to 100% O.A. when conditions are correct. Minimum position relay S-1 can be set to maintain a minimum amount of O.A. through high signal selector R-1 as long as the fan is running.

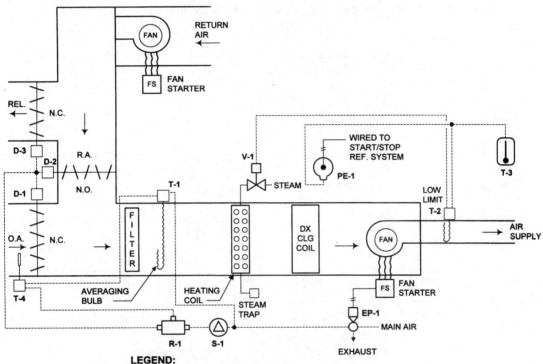

LEGEND:

T-1	= MODULATING CAPILLARY DUCT THERMOSTAT
T-2	= MODULATING CAPILLARY DUCT THERMOSTAT
T-3	= MODULATING ROOM THERMOSTAT
T-4	= MODULATING CAPILLARY THERMOSTAT
V-1	= N.O. MODULATING STEAM COIL VALVE
D-1	= OUTSIDE AIR DAMPER MOTOR
D-2	= RETURN AIR DAMPER MOTOR
D-3	= RELIEF AIR DAMPER MOTOR
EP-1	= SOLENOID AIR VALVE
PE-1	= PRESSURE/ELECTRIC SWITCH
R-1	= HIGH SIGNAL SELECTOR
S-1	= MINIMUM POSITION SWITCH

Heating and Cooling Air-Handling Unit, with R.A. Fan, Steam Heating Coil and DX Cooling Coil Using Face and Bypass Dampers, O.A., R.A., and REL Dampers Using Mixed Air Controls

Whenever the fan runs, the EP-1 is energized and O.A., R.A., and REL dampers are placed under automatic controls. When the fan stops, all dampers return to their normal positions.

Room thermostat T-3 controls valve V-1 on the heating coil through low-limit duct thermostat T-2. It also controls PE switch PE-1 and face and bypass dampers D-4 on the DX cooling coil. The PE switch shuts off the cooling when the face damper is closed. The O.A., R.A., and REL dampers are controlled by thermostat T-1 through O.A. high-limit thermostat T-4. T-1 and T-4 act as economizer thermostats to allow for up to 100% O.A. when conditions are correct. Minimum position relay S-1 can be set to maintain a minimum amount of O.A. through high signal selector R-1 as long as the fan is running.

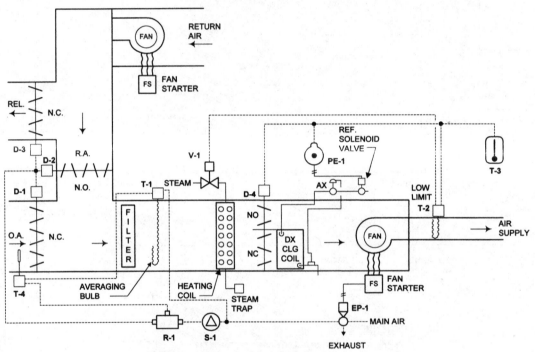

LEGEND:

T-1 = MODULATING CAPILLARY DUCT THERMOSTAT
T-2 = MODULATING CAPILLARY DUCT THERMOSTAT
T-3 = MODULATING ROOM THERMOSTAT
T-4 = MODULATING CAPILLARY THERMOSTAT
V-1 = N.O. MODULATING STEAM COIL VALVE
D-1 = OUTSIDE AIR DAMPER MOTOR
D-2 = RETURN AIR DAMPER MOTOR
D-3 = RELIEF AIR DAMPER MOTOR
D-4 = FACE & BYPASS DAMPER MOTOR
EP-1 = SOLENOID AIR VALVE
PE-1 = PRESSURE/ELECTRIC SWITCH
R-1 = HIGH SIGNAL SELECTOR
S-1 = MINIMUM POSITION SWITCH

Heating and Cooling Air-Handling Unit, with R.A. Fan, Steam Heating and Chilled Water Cooling, O.A., R.A., and REL Dampers Using Economizer Control Cycle

Whenever the fan runs, the EP-1 is energized and O.A., R.A., and REL dampers are placed under automatic controls. When the fan stops, all dampers return to their normal positions.

Room thermostat T-3 controls valve V-1 on the heating coil and V-2 on the cooling coil through low-limit duct thermostat T-2. The chilled water and steam must be manually operated summer and winter, so that they are not on at the same time. The O.A., R.A., and REL dampers are controlled by thermostat T-1 through O.A. high-limit thermostat T-4. T-1 and T-4 act as economizer thermostats to allow for up to 100% O.A. when conditions are correct. Minimum position relay S-1 can be set to maintain a minimum amount of O.A. through high signal selector R-1 as long as the fan is running.

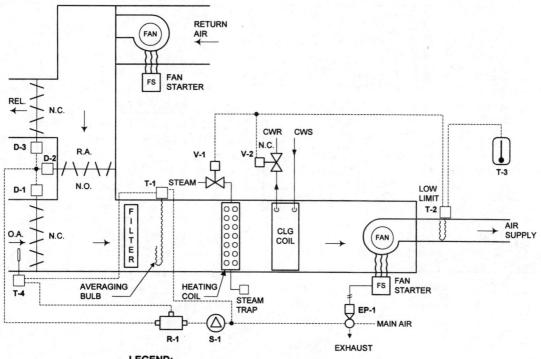

LEGEND:

T-1 = MODULATING CAPILLARY DUCT THERMOSTAT
T-2 = MODULATING CAPILLARY DUCT THERMOSTAT
T-3 = MODULATING ROOM THERMOSTAT
T-4 = MODULATING CAPILLARY THERMOSTAT
V-1 = N.O. MODULATING STEAM COIL VALVE
V-2 = N.C. MODULATING 2-WAY COOLING COIL VALVE
D-1 = OUTSIDE AIR DAMPER MOTOR
D-2 = RETURN AIR DAMPER MOTOR
D-3 = RELIEF AIR DAMPER MOTOR
EP-1 = SOLENOID AIR VALVE
R-1 = HIGH SIGNAL SELECTOR
S-1 = MINIMUM POSITION SWITCH

Heating and Cooling Air-Handling Unit, with R.A. Fan, Hot Water Heating and Chilled Water Cooling, O.A., R.A., and REL Dampers Using Economizer Control Cycle

Whenever the fan runs, the EP-1 is energized and O.A., R.A., and REL dampers are placed under automatic controls. When the fan stops, all dampers return to their normal positions.

   Room thermostat T-3 controls valve V-1 on the heating coil and V-2 on the cooling coil through low-limit duct thermostat T-2. The chilled water and steam must be manually operated summer and winter, so that they are not on at the same time. The O.A., R.A., and REL dampers are controlled by thermostat T-1 through O.A. high-limit thermostat T-4. T-1 and T-4 act as economizer thermostats to allow for up to 100% O.A. when conditions are correct. Minimum position relay S-1 can be set to maintain a minimum amount of O.A. through high signal selector R-1 as long as the fan is running.

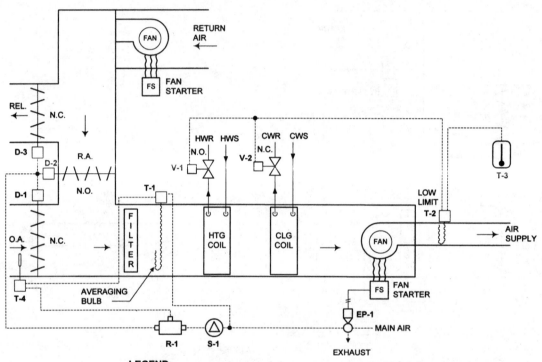

LEGEND:

T-1	= MODULATING CAPILLARY DUCT THERMOSTAT
T-2	= MODULATING CAPILLARY DUCT THERMOSTAT
T-3	= MODULATING ROOM THERMOSTAT
T-4	= MODULATING CAPILLARY THERMOSTAT
V-1	= N.O. MODULATING 2-WAY HEATING COIL VALVE
V-2	= N.C. MODULATING 2-WAY COOLING COIL VALVE
D-1	= OUTSIDE AIR DAMPER MOTOR
D-2	= RETURN AIR DAMPER MOTOR
D-3	= RELIEF AIR DAMPER MOTOR
EP-1	= SOLENOID AIR VALVE
R-1	= HIGH SIGNAL SELECTOR
S-1	= MINIMUM POSITION SWITCH

Heating and Cooling Air-Handling Unit, with R.A. Fan, Hot Water Heating and DX Cooling Using Face and Bypass Dampers, O.A., R.A., and REL Dampers Using Economizer Cycle

Whenever the fan runs, the EP-1 is energized and O.A., R.A., and REL dampers are placed under automatic controls. When the fan stops, all dampers return to their normal positions.

Room thermostat T-3 controls valve V-1 through the low-limit thermostat T-2 on the heating coil and face and bypass dampers D-4 on the DX cooling coil. PE switch PE-1 shuts down the DX coil when the face damper is closed. The DX coil and hot water coil must be manually operated summer and winter, so that they are not on at the same time. The O.A., R.A., and REL dampers are controlled by thermostat T-1 through O.A. high-limit thermostat T-4. T-1 and T-4 act as economizer thermostats to allow for up to 100% O.A. when conditions are correct. Minimum position relay S-1 can be set to maintain a minimum amount of O.A. through high signal selector R-1 as long as the fan is running.

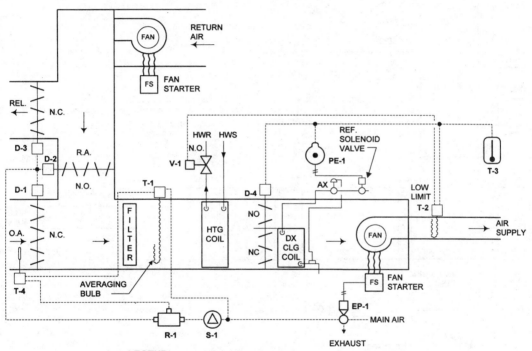

LEGEND:

T-1	= MODULATING CAPILLARY DUCT THERMOSTAT
T-2	= MODULATING CAPILLARY DUCT THERMOSTAT
T-3	= MODULATING ROOM THERMOSTAT
T-4	= MODULATING CAPILLARY THERMOSTAT
V-1	= N.O. MODULATING 2-WAY HEATING COIL VALVE
D-1	= OUTSIDE AIR DAMPER MOTOR
D-2	= RETURN AIR DAMPER MOTOR
D-3	= RELIEF AIR DAMPER MOTOR
D-4	= FACE & BYPASS DMAPER MOTOR
EP-1	= SOLENOID AIR VALVE
PE-1	= PRESSURE/ELECTRIC SWITCH
R-1	= HIGH SIGNAL SELECTOR
S-1	= MINIMUM POSITION SWITCH

Heating/Cooling Air-Handling Unit, with R.A. Fan, Hot Water Heating and DX Cooling, O.A., R.A., and REL Dampers Using Economizer Control Cycle

Whenever the fan runs, the EP-1 is energized and O.A., R.A., and REL dampers are placed under automatic controls. When the fan stops, all dampers return to their normal positions.

Room thermostat T-3 controls valve V-1 through the low-limit thermostat T-2 on the heating coil and the PE switch PE-1 on the DX cooling coil. DX coil and hot water coil must be manually operated summer and winter, so that they are not on at the same time. The O.A., R.A., and REL dampers are controlled by thermostat T-1 through O.A. high-limit thermostat T-4. T-1 and T-4 act as economizer thermostats to allow for up to 100% O.A. when conditions are correct. Minimum position relay S-1 can be set to maintain a minimum amount of O.A. through high signal selector R-1 as long as the fan is running.

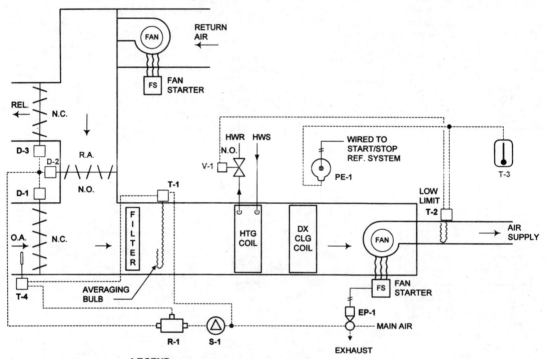

LEGEND:

T-1	= MODULATING CAPILLARY DUCT THERMOSTAT
T-2	= MODULATING CAPILLARY DUCT THERMOSTAT
T-3	= MODULATING ROOM THERMOSTAT
T-4	= MODULATING CAPILLARY THERMOSTAT
V-1	= N.O. MODULATING 2-WAY HEATING COIL VALVE
D-1	= OUTSIDE AIR DAMPER MOTOR
D-2	= RETURN AIR DAMPER MOTOR
D-3	= RELIEF AIR DAMPER MOTOR
EP-1	= SOLENOID AIR VALVE
PE-1	= PRESSURE/ELECTRIC SWITCH
R-1	= HIGH SIGNAL SELECTOR
S-1	= MINIMUM POSITION SWITCH

Heating and Cooling Air-Handling Unit, with R.A. Fan, Steam Heating Coil and DX Cooling, O.A., R.A., and REL Dampers Using Economizer Control Cycle

Whenever the fan runs, the EP-1 is energized and O.A., R.A., and REL dampers are placed under automatic controls. When the fan stops, all dampers return to their normal positions.

Room thermostat T-3 controls valve V-1 through the low-limit thermostat T-2 on the heating coil and the PE switch PE-1 on the DX cooling coil. DX coil and steam coil must be manually operated summer and winter, so that they are not on at the same time. The O.A., R.A., and REL dampers are controlled by thermostat T-1 through O.A. high-limit thermostat T-4. T-1 and T-4 act as economizer thermostats to allow for up to 100% O.A. when conditions are correct. Minimum position relay S-1 can be set to maintain a minimum amount of O.A. through high signal selector R-1 as long as the fan is running.

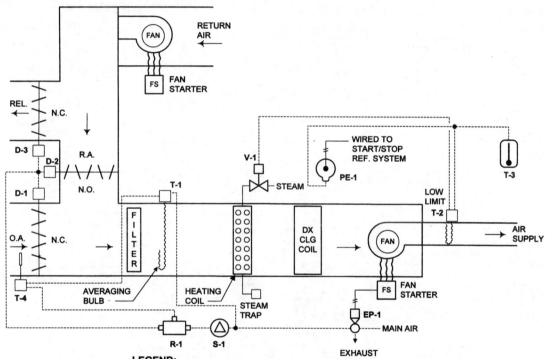

LEGEND:

T-1	=	MODULATING CAPILLARY DUCT THERMOSTAT
T-2	=	MODULATING CAPILLARY DUCT THERMOSTAT
T-3	=	MODULATING ROOM THERMOSTAT
T-4	=	MODULATING CAPILLARY THERMOSTAT
V-1	=	N.O. MODULATING STEAM COIL VALVE
D-1	=	OUTSIDE AIR DAMPER MOTOR
D-2	=	RETURN AIR DAMPER MOTOR
D-3	=	RELIEF AIR DAMPER MOTOR
EP-1	=	SOLENOID AIR VALVE
PE-1	=	PRESSURE/ELECTRIC SWITCH
R-1	=	HIGH SIGNAL SELECTOR
S-1	=	MINIMUM POSITION SWITCH

Heating and Cooling Air-Handling Unit, with R.A. Fan, Steam Heating and DX Cooling Using Face and Bypass Dampers, O.A., R.A., and REL Dampers Using Economizer Control Cycle

Whenever the fan runs, the EP-1 is energized and O.A., R.A., and REL dampers are placed under automatic controls. When the fan stops, all dampers return to their normal positions.

Room thermostat T-3 controls valve V-1 through the low-limit thermostat T-2 on the heating coil and the face and bypass dampers D-4, as well as PE switch PE-1 on the DX cooling coil. The DX cooling is shut down when the face damper is closed. DX coil and steam coil must be manually operated summer and winter, so that they are not on at the same time. The O.A., R.A., and REL dampers are controlled by thermostat T-1 through O.A. high-limit thermostat T-4. T-1 and T-4 act as economizer thermostats to allow for up to 100% O.A. when conditions are correct. Minimum position relay S-1 can be set to maintain a minimum amount of O.A. through high signal selector R-1 as long as the fan is running.

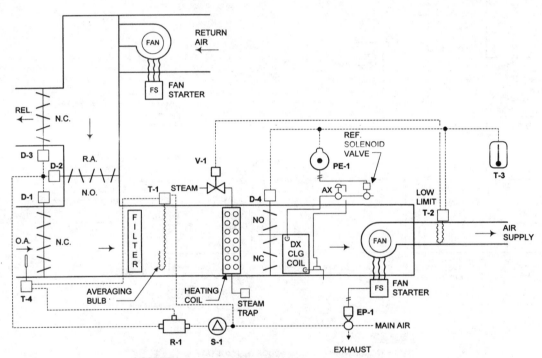

LEGEND:

T-1	=	MODULATING CAPILLARY DUCT THERMOSTAT
T-2	=	MODULATING CAPILLARY DUCT THERMOSTAT
T-3	=	MODULATING ROOM THERMOSTAT
T-4	=	MODULATING CAPILLARY THERMOSTAT
V-1	=	N.O. MODULATING STEAM COIL VALVE
D-1	=	OUTSIDE AIR DAMPER MOTOR
D-2	=	RETURN AIR DAMPER MOTOR
D-3	=	RELIEF AIR DAMPER MOTOR
D-4	=	FACE & BYPASS DAMPER MOTOR
EP-1	=	SOLENOID AIR VALVE
PE-1	=	PRESSURE/ELECTRIC SWITCH
R-1	=	HIGH SIGNAL SELECTOR
S-1	=	MINIMUM POSITION SWITCH

Heating and Cooling Air-Handling Unit, with R.A. Fan, Steam Heating and Chilled Water Cooling, O.A., R.A., and REL Dampers Using Economizer Control Cycle

Whenever the fan runs, the EP-1 is energized, and the O.A., R.A., and REL dampers are placed under automatic controls. When the fan stops, all dampers return to their normal positions.

Room thermostat T-3 controls valve V-1 through the low-limit thermostat T-2 on the heating coil and valve V-2 on the cooling coil. Cooling coil and steam coil must be manually operated summer and winter, so that they are not on at the same time. The O.A., R.A., and REL dampers are controlled by thermostat T-1 through O.A. high-limit thermostat T-4. T-1 and T-4 act as economizer thermostats to allow for up to 100% O.A. when conditions are correct. Minimum position relay S-1 can be set to maintain a minimum amount of O.A. through high signal selector R-1 as long as the fan is running.

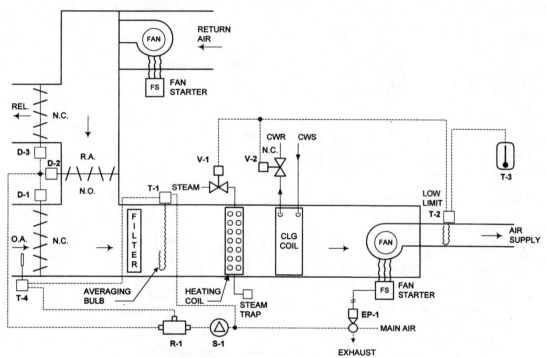

LEGEND:

T-1	=	MODULATING CAPILLARY DUCT THERMOSTAT
T-2	=	MODULATING CAPILLARY DUCT THERMOSTAT
T-3	=	MODULATING ROOM THERMOSTAT
T-4	=	MODULATING CAPILLARY THERMOSTAT
V-1	=	N.O. MODULATING STEAM COIL VALVE
V-2	=	N.C. MODULATING 2-WAY COOLING COIL VALVE
D-1	=	OUTSIDE AIR DAMPER MOTOR
D-2	=	RETURN AIR DAMPER MOTOR
D-3	=	RELIEF AIR DAMPER MOTOR
EP-1	=	SOLENOID AIR VALVE
R-1	=	HIGH SIGNAL SELECTOR
S-1	=	MINIMUM POSITION SWITCH

Heating and Cooling Air-Handling Unit, with R.A. Fan, Hot Water Heating and Chilled Water Cooling, O.A., R.A., and REL Dampers Using Economizer Control Cycle

Whenever the fan runs, the EP-1 is energized and O.A., R.A., and REL dampers are placed under automatic controls. When the fan stops, all dampers return to their normal positions.

Room thermostat T-3 controls valve V-1 through the low-limit thermostat T-2 on the heating coil and valve V-2 on the cooling coil. Cooling coil and hot water coil must be manually operated summer and winter, so that they are not on at the same time. The O.A., R.A., and REL dampers are controlled by thermostat T-1 through O.A. high-limit thermostat T-4. T-1 and T-4 act as economizer thermostats to allow for up to 100% O.A. when conditions are correct. Minimum position relay S-1 can be set to maintain a minimum amount of O.A. through high signal selector R-1 as long as the fan is running.

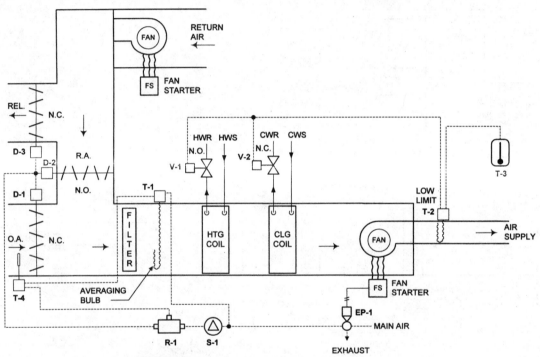

LEGEND:

T-1	=	MODULATING CAPILLARY DUCT THERMOSTAT
T-2	=	MODULATING CAPILLARY DUCT THERMOSTAT
T-3	=	MODULATING ROOM THERMOSTAT
T-4	=	MODULATING CAPILLARY THERMOSTAT
V-1	=	N.O. MODULATING 2-WAY HEATING COIL VALVE
V-2	=	N.C. MODULATING 2-WAY COOLING COIL VALVE
D-1	=	OUTSIDE AIR DAMPER MOTOR
D-2	=	RETURN AIR DAMPER MOTOR
D-3	=	RELIEF AIR DAMPER MOTOR
EP-1	=	SOLENOID AIR VALVE
R-1	=	HIGH SIGNAL SELECTOR
S-1	=	MINIMUM POSITION SWITCH

Heating and Cooling Air-Handling Unit, with R.A. Fan, Hot Water Heating and DX Cooling Using Face and Bypass Dampers, O.A., R.A., and REL Dampers Using Economizer Control Cycle

Whenever the fan runs, the EP-1 is energized and R.A. and REL dampers are placed under automatic controls. When the fan stops, all dampers return to their normal positions.

Room thermostat T-3 controls valve V-1 through the low-limit thermostat T-2 on the heating coil and face and bypass dampers on the DX coil. When the face damper is closed, PE switch PE-1 shuts down the DX cooling coil. DX cooling coil and hot water coil must be manually operated summer and winter, so that they are not on at the same tine. The O.A., R.A., and REL dampers are controlled by thermostat T-1 through O.A. high-limit thermostat T-4. T-1 and T-4 act as economizer thermostats to allow for up to 100% O.A. when conditions are correct. Minimum position relay S-1 can be set to maintain a minimum amount of O.A. through high signal selector R-1 as long as the fan is running.

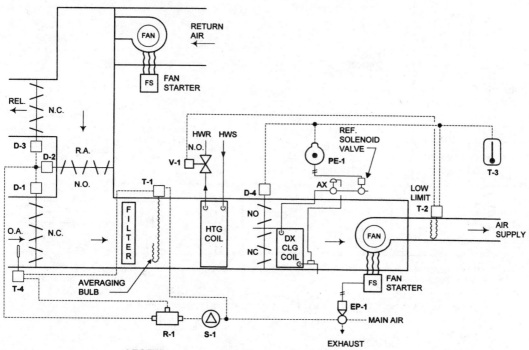

LEGEND:

T-1	= MODULATING CAPILLARY DUCT THERMOSTAT
T-2	= MODULATING CAPILLARY DUCT THERMOSTAT
T-3	= MODULATING ROOM THERMOSTAT
T-4	= MODULATING CAPILLARY THERMOSTAT
V-1	= N.O. MODULATING 2-WAY HEATING COIL VALVE
D-1	= OUTSIDE AIR DAMPER MOTOR
D-2	= RETURN AIR DAMPER MOTOR
D-3	= RELIEF AIR DAMPER MOTOR
D-4	= FACE & BYPASS DMAPER MOTOR
EP-1	= SOLENOID AIR VALVE
PE-1	= PRESSURE/ELECTRIC SWITCH
R-1	= HIGH SIGNAL SELECTOR
S-1	= MINIMUM POSITION SWITCH

Heating and Cooling Air-Handling Unit, with R.A. Fan, Hot Water Heating and DX Cooling, O.A., R.A., and REL Dampers Using Economizer Control Cycle

Whenever the fan runs, the EP-1 is energized and O.A., R.A., and REL dampers are placed under automatic controls. When the fan stops, all dampers return to their normal positions.

Room thermostat T-3 controls valve V-1 through the low-limit thermostat T-2 on the heating coil and DX cooling coil through PE switch PE-1. DX cooling coil and hot water coil must be manually operated summer and winter, so that they are not on at the same time. The O.A., R.A., and REL dampers are controlled by thermostat T-1 through O.A. high-limit thermostat T-4. T-1 and T-4 act as economizer thermostats to allow for up to 100% O.A. when conditions are correct. Minimum position relay S-1 can be set to maintain a minimum amount of O.A. through high signal selector R-1 as long as the fan is running. Room humidistat H-1 controls through high-limit duct humidistat H-2 valve V-2.

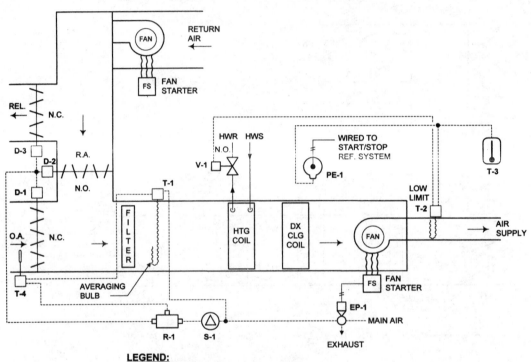

LEGEND:

T-1	=	MODULATING CAPILLARY DUCT THERMOSTAT
T-2	=	MODULATING CAPILLARY DUCT THERMOSTAT
T-3	=	MODULATING ROOM THERMOSTAT
T-4	=	MODULATING CAPILLARY THERMOSTAT
V-1	=	N.O. MODULATING 2-WAY HEATING COIL VALVE
D-1	=	OUTSIDE AIR DAMPER MOTOR
D-2	=	RETURN AIR DAMPER MOTOR
D-3	=	RELIEF AIR DAMPER MOTOR
EP-1	=	SOLENOID AIR VALVE
PE-1	=	PRESSURE/ELECTRIC SWITCH
R-1	=	HIGH SIGNAL SELECTOR
S-1	=	MINIMUM POSITION SWITCH

Heating and Cooling Air-Handling Unit, with R.A. Fan, Steam Heating, DX Cooling, O.A., R.A., and REL Dampers Using Mixed Air Control from Room Thermostat, Steam Humidifier from Room Humidistat

Whenever the fan runs, the EP-1 is energized and O.A., R.A., and REL dampers are placed under automatic controls. When the fan stops, all dampers return to their normal positions.

Room thermostat T-3 controls valve V-1 through the low-limit thermostat T-2 on the heating coil and DX cooling coil through PE switch PE-1. Thermostat T-3 also controls the O.A., R.A., and REL dampers in sequence. DX cooling coil and hot water coil must be manually operated summer and winter, so that they are not on at the same time. The O.A., R.A., and REL dampers are controlled by thermostat T-1 through O.A. high-limit thermostat T-4. T-1 and T-4 act as economizer thermostats to allow for up to 100% O.A. when conditions are correct. Minimum position relay S-1 can be set to maintain a minimum amount of O.A. through high signal selector R-1 as long as the fan is running. Room humidistat H-1, through duct high-limit humidistat H-2, controls humidifier valve V-2.

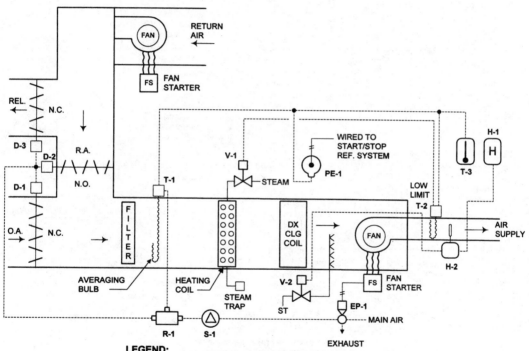

LEGEND:

T-1 = MODULATING CAPILLARY DUCT THERMOSTAT
T-2 = MODULATING CAPILLARY DUCT THERMOSTAT
T-3 = MODULATING ROOM THERMOSTAT
H-1 = MODULATING ROOM HUMIDISTAT
H-2 = HIGH LIMIT DUCT HUMIDISTAT
V-1 = N.O. MODULATING STEAM COIL VALVE
V-2 = N.C. MODULATING HUMIDIFIER STEAM VALVE
D-1 = OUTSIDE AIR DAMPER MOTOR
D-2 = RETURN AIR DAMPER MOTOR
D-3 = RELIEF AIR DAMPER MOTOR
EP-1 = SOLENOID AIR VALVE
PE-1 = PRESSURE/ELECTRIC SWITCH
R-1 = HIGH SIGNAL SELECTOR
S-1 = MINIMUM POSITION SWITCH

Heating and Cooling Air-Handling Unit, with R.A. Fan, Steam Heating Coil and DX Cooling, O.A., R.A., and REL Dampers Using Economizer Control Cycle with Steam Humidifier under Control of Room Humidistat

Whenever the fan runs, EP-1 is energized and O.A., R.A., and REL dampers are placed under automatic control. When the fan stops, the dampers return to their normal positions.

Room thermostat T-3 controls valve V-1 on the heating coil through low-limit duct thermostat T-2. It also controls the DX cooling through PE switch PE-1. The DX cooling and steam heating must be manually operated summer and winter so that they are not on at the same time. The O.A., R.A., and REL dampers are controlled by duct thermostat T-1 and O.A. limit thermostat T-4. T-1 and T-4 act as economizer thermostats to allow up to 100% O.A. when conditions permit. Minimum position relay S-1 can be set to maintain a minimum amount of O.A. as long as the fan is running. Room humidistat H-1, through high-limit duct humidistat H-2, controls humidifier valve V-2.

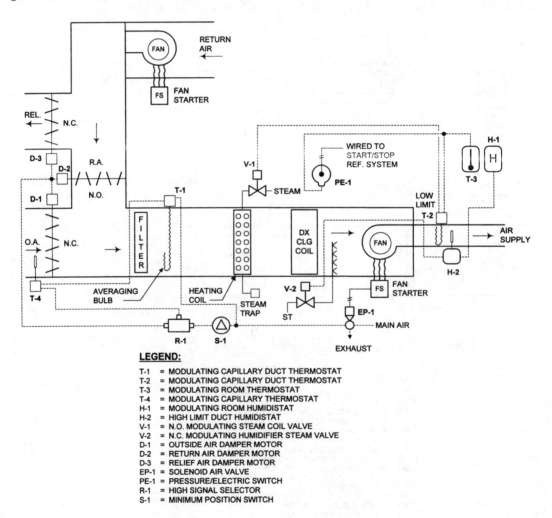

LEGEND:

T-1 = MODULATING CAPILLARY DUCT THERMOSTAT
T-2 = MODULATING CAPILLARY DUCT THERMOSTAT
T-3 = MODULATING ROOM THERMOSTAT
T-4 = MODULATING CAPILLARY THERMOSTAT
H-1 = MODULATING ROOM HUMIDISTAT
H-2 = HIGH LIMIT DUCT HUMIDISTAT
V-1 = N.O. MODULATING STEAM COIL VALVE
V-2 = N.C. MODULATING HUMIDIFIER STEAM VALVE
D-1 = OUTSIDE AIR DAMPER MOTOR
D-2 = RETURN AIR DAMPER MOTOR
D-3 = RELIEF AIR DAMPER MOTOR
EP-1 = SOLENOID AIR VALVE
PE-1 = PRESSURE/ELECTRIC SWITCH
R-1 = HIGH SIGNAL SELECTOR
S-1 = MINIMUM POSITION SWITCH

Heating and Cooling Air-Handling Unit, with R.A. Fan, Steam Heating and DX Cooling Using Face and Bypass Dampers, O.A., R.A., and REL Dampers Using Economizer Control Cycle with Steam Humidifier from Room Humidistat

Whenever the fan runs, the EP-1 is energized and O.A., R.A., and REL dampers are placed under automatic controls. When the fan stops, all dampers return to their normal positions.

Room thermostat T-3 controls valve V-1 through the low-limit thermostat T-2 on the heating coil and DX cooling coil through face and bypass dampers D-4 and PE switch PE-1. When the face damper is closed the PE switch PE-1 shuts down the cooling. DX cooling coil and steam coil must be manually operated summer and winter, so that they are not on at the same time. The O.A., R.A., and REL dampers are controlled by thermostat T-1 through O.A. high-limit thermostat T-4. T-1 and T-4 act as economizer thermostats. They allow for up to 100% O.A. when conditions are favorable. Minimum position relay S-1 can be set to maintain a minimum amount of O.A. through high signal selector R-1 as long as the fan is running. Room humidistat H-1, through duct high-limit humidistat H-2, controls humidifier valve V-2.

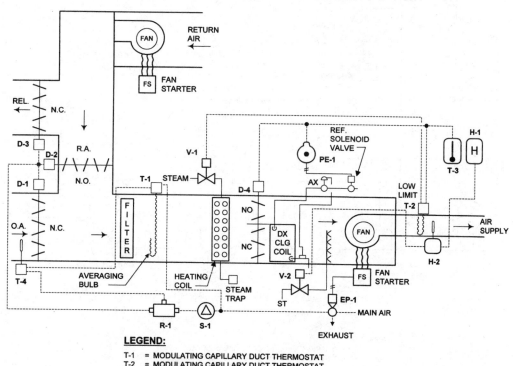

LEGEND:

T-1 = MODULATING CAPILLARY DUCT THERMOSTAT
T-2 = MODULATING CAPILLARY DUCT THERMOSTAT
T-3 = MODULATING ROOM THERMOSTAT
T-4 = MODULATING CAPILLARY THERMOSTAT
H-1 = MODULATING ROOM HUMIDISTAT
H-2 = HIGH LIMIT DUCT HUMIDISTAT
V-1 = N.O. MODULATING STEAM COIL VALVE
V-2 = N.C. MODULATING HUMIDIFIER STEAM VALVE
D-1 = OUTSIDE AIR DAMPER MOTOR
D-2 = RETURN AIR DAMPER MOTOR
D-3 = RELIEF AIR DAMPER MOTOR
D-4 = FACE & BYPASS DAMPER MOTOR
EP-1 = SOLENOID AIR VALVE
PE-1 = PRESSURE/ELECTRIC SWITCH
R-1 = HIGH SIGNAL SELECTOR
S-1 = MINIMUM POSITION SWITCH

Heating and Cooling Air-Handling Unit, with R.A. Fan, Steam Heating and Chilled Water Cooling, O.A., R.A., and REL Dampers Using Economizer Control Cycle with Steam Humidifier from Room Humidistat

Whenever the fan runs, the EP-1 is energized and O.A., R.A., and REL dampers are placed under automatic controls. When the fan stops, all dampers return to their normal positions.

Room thermostat T-3 controls valve V-1 through the low-limit thermostat T-2 on the heating coil and valve V-2 on the chilled water cooling coil. Chilled water coil and steam coil must be manually operated summer and winter, so that they are not on at the same time. The O.A., R.A., and REL dampers are controlled by thermostat T-1 through O.A. high-limit thermostat T-4. T-1 and T-4 act as economizer thermostats. They allow for up to 100% O.A. when conditions are favorable. Minimum position relay S-1 can be set to maintain a minimum amount of O.A. through high signal selector R-1 as long as the fan is running. Room humidistat H-1, through duct high-limit humidistat H-2, controls humidifier valve V-2.

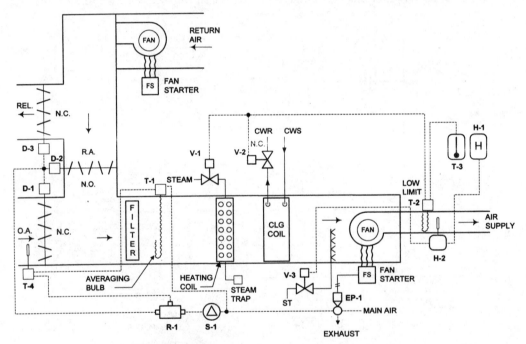

LEGEND:

T-1	=	MODULATING CAPILLARY DUCT THERMOSTAT
T-2	=	MODULATING CAPILLARY DUCT THERMOSTAT
T-3	=	MODULATING ROOM THERMOSTAT
T-4	=	MODULATING CAPILLARY THERMOSTAT
H-1	=	MODULATING ROOM HUMIDISTAT
H-2	=	HIGH LIMIT DUCT HUMIDISTAT
V-1	=	N.O. MODULATING STEAM COIL VALVE
V-2	=	N.C. MODULATING 2-WAY COOLING COIL VALVE
V-3	=	N.C. MODULATING HUMIDIFIER STEAM VALVE
D-1	=	OUTSIDE AIR DAMPER MOTOR
D-2	=	RETURN AIR DAMPER MOTOR
D-3	=	RELIEF AIR DAMPER MOTOR
EP-1	=	SOLENOID AIR VALVE
R-1	=	HIGH SIGNAL SELECTOR
S-1	=	MINIMUM POSITION SWITCH

Heating and Cooling Air-Handling Unit, with R.A. Fan, Hot Water Heating and Chilled Water Cooling, O.A., R.A., and REL Dampers Using Economizer Control Cycle with Steam Humidifier from Room Humidistat

Whenever the fan runs, the EP-1 is energized and O.A., R.A., and REL dampers are placed under automatic controls. When the fan stops, all dampers return to their normal positions.

Room thermostat T-3 controls valve V-1 through the low-limit thermostat T-2 on the heating coil and valve V-2 on the chilled water cooling coil. Chilled water coil and hot water coil must be manually operated summer and winter, so that they are not on at the same time. The O.A., R.A., and REL dampers are controlled by thermostat T-1 through O.A. high-limit thermostat T-4. T-1 and T-4 act as economizer thermostats. They allow for up to 100% O.A. when conditions are favorable. Minimum position relay S-1 can be set to maintain a minimum amount of O.A. through high signal selector R-1 as long as the fan is running. Room humidistat H-1, through duct high-limit humidistat H-2, controls humidifier valve V-2.

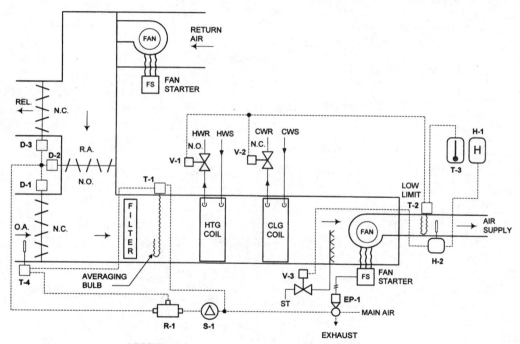

LEGEND:

T-1	=	MODULATING CAPILLARY DUCT THERMOSTAT
T-2	=	MODULATING CAPILLARY DUCT THERMOSTAT
T-3	=	MODULATING ROOM THERMOSTAT
T-4	=	MODULATING CAPILLARY THERMOSTAT
H-1	=	MODULATING ROOM HUMIDISTAT
H-2	=	HIGH LIMIT DUCT HUMIDISTAT
V-1	=	N.O. MODULATING 2-WAY HEATING COIL VALVE
V-2	=	N.C. MODULATING 2-WAY COOLING COIL VALVE
V-3	=	N.C. MODULATING HUMIDIFIER STEAM VALVE
D-1	=	OUTSIDE AIR DAMPER MOTOR
D-2	=	RETURN AIR DAMPER MOTOR
D-3	=	RELIEF AIR DAMPER MOTOR
EP-1	=	SOLENOID AIR VALVE
R-1	=	HIGH SIGNAL SELECTOR
S-1	=	MINIMUM POSITION SWITCH

Heating and Cooling Air-Handling Unit, with Return Air Fan, Hot Water Heating and DX Cooling Using Face and Bypass Dampers, O.A., R.A., and REL Dampers Using Economizer Control Cycle with Steam Humidifier from the Room Humidistat

Whenever the fan runs, the EP-1 is energized and O.A., R.A., and REL dampers are placed under automatic controls. When the fan stops, all dampers return to their normal positions.

Room thermostat T-3 controls valve V-1 through the low-limit thermostat T-2 on the heating coil and face and bypass dampers D-4 on the DX cooling coil. PE switch PE-1 shuts down the DX cooling coil when the face damper is closed. DX coil and hot water coil must be manually operated summer and winter, so that they are not on at the same time. The O.A., R.A., and REL dampers are controlled by thermostat T-1 through O.A. high-limit thermostat T-4. T-1 and T-4 act as economizer thermostats. They allow for up to 100% O.A. when conditions are favorable. Minimum position relay S-1 can be set to maintain a minimum amount of O.A. through high signal selector R-1 as long as the fan is running. Room humidistat H-1, through duct high-limit humidistat H-2, controls humidifier valve V-2.

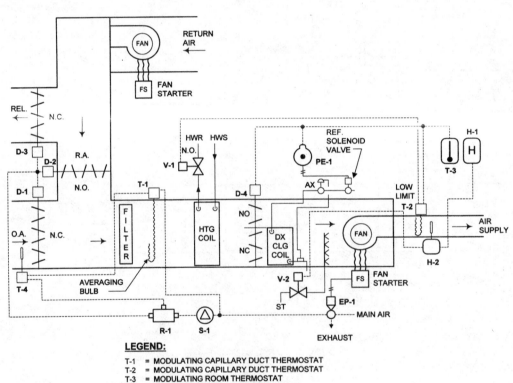

LEGEND:

T-1	= MODULATING CAPILLARY DUCT THERMOSTAT
T-2	= MODULATING CAPILLARY DUCT THERMOSTAT
T-3	= MODULATING ROOM THERMOSTAT
T-4	= MODULATING CAPILLARY THERMOSTAT
H-1	= MODULATING ROOM HUMIDISTAT
H-2	= HIGH LIMIT DUCT HUMIDISTAT
V-1	= N.O. MODULATING 2-WAY HEATING COIL VALVE
V-2	= N.C. MODULATING HUMIDIFIER STEAM VALVE
D-1	= OUTSIDE AIR DAMPER MOTOR
D-2	= RETURN AIR DAMPER MOTOR
D-3	= RELIEF AIR DAMPER MOTOR
D-4	= FACE & BYPASS DMAPER MOTOR
EP-1	= SOLENOID AIR VALVE
PE-1	= PRESSURE/ELECTRIC SWITCH
R-1	= HIGH SIGNAL SELECTOR
S-1	= MINIMUM POSITION SWITCH

Heating/Cooling Air-Handling Unit, with R.A. Fan, Hot Water Heating and DX Cooling, O.A., R.A., and REL Dampers Using Economizer Control Cycle with Steam Humidifier from Room Humidistat

Whenever the fan runs, the EP-1 is energized and O.A., R.A., and REL dampers are placed under automatic controls. When the fan stops, all dampers return to their normal positions.

Room thermostat T-3 controls valve V-1 through the low-limit thermostat T-2 on the heating coil and PE switch PE-1 on the DX cooling coil. DX cooling coil and hot water coil must be manually operated summer and winter, so that they are not on at the same time. The O.A., R.A., and REL dampers are controlled by thermostat T-1 through O.A. high-limit thermostat T-4. T-1 and T-4 act as economizer thermostats. They allow for up to 100% O.A. when conditions are favorable. Minimum position relay S-1 can be set to maintain a minimum amount of O.A. through high signal selector R-1 as long as the fan is running. Room humidistat H-1, through duct high-limit humidistat H-2, controls humidifier valve V-2.

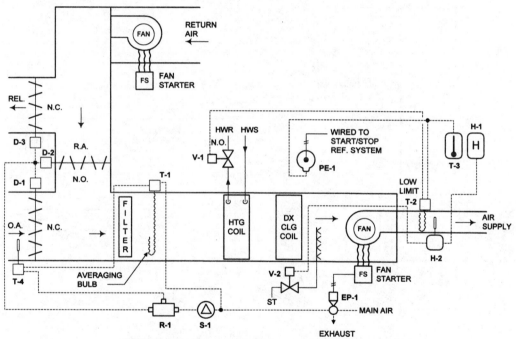

LEGEND:

T-1	=	MODULATING CAPILLARY DUCT THERMOSTAT
T-2	=	MODULATING CAPILLARY DUCT THERMOSTAT
T-3	=	MODULATING ROOM THERMOSTAT
T-4	=	MODULATING CAPILLARY THERMOSTAT
H-1	=	MODULATING ROOM HUMIDISTAT
H-2	=	HIGH LIMIT DUCT HUMIDISTAT
V-1	=	N.O. MODULATING 2-WAY HEATING COIL VALVE
V-2	=	N.C. MODULATING HUMIDIFIER STEAM VALVE
D-1	=	OUTSIDE AIR DAMPER MOTOR
D-2	=	RETURN AIR DAMPER MOTOR
D-3	=	RELIEF AIR DAMPER MOTOR
EP-1	=	SOLENOID AIR VALVE
PE-1	=	PRESSURE/ELECTRIC SWITCH
R-1	=	HIGH SIGNAL SELECTOR
S-1	=	MINIMUM POSITION SWITCH

Heating and Cooling Air-Handling Unit, with R.A. Fan, Steam Heating, DX Cooling, O.A., R.A., and REL Dampers Using Economizer Control Cycle under Control of Room Thermostat, with Steam Humidifier from Room Humidistat

Whenever the fan runs, the EP-1 is energized and O.A., R.A., and REL dampers are placed under automatic controls. When the fan stops, all dampers return to their normal positions.

Room thermostat T-3 controls valve V-1 through the low-limit thermostat T-2 on the heating coil and PE switch PE-1 on the DX cooling coil. DX cooling coil and steam coil must be manually operated summer and winter, so that they are not on at the same time. The O.A., R.A., and REL dampers are controlled by thermostat T-1 through O.A. high-limit thermostat T-4. T-1 and T-4 act as economizer thermostats. They allow for up to 100% O.A. when conditions are favorable. Minimum position relay S-1 can be set to maintain a minimum amount of O.A. through high signal selector R-1 as long as the fan is running. Room humidistat H-1, through duct high-limit humidistat H-2, controls humidifier valve V-2.

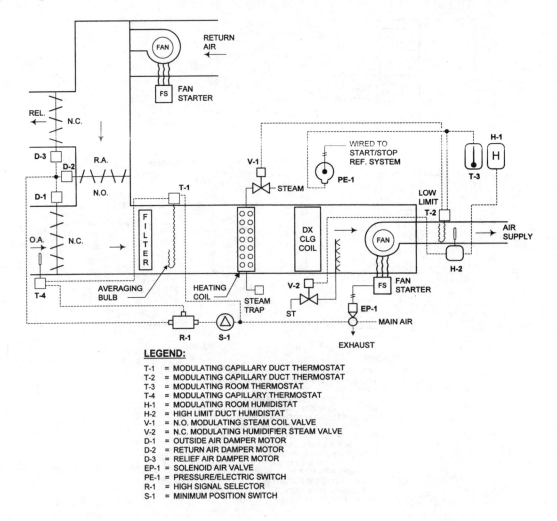

LEGEND:

T-1	= MODULATING CAPILLARY DUCT THERMOSTAT
T-2	= MODULATING CAPILLARY DUCT THERMOSTAT
T-3	= MODULATING ROOM THERMOSTAT
T-4	= MODULATING CAPILLARY THERMOSTAT
H-1	= MODULATING ROOM HUMIDISTAT
H-2	= HIGH LIMIT DUCT HUMIDISTAT
V-1	= N.O. MODULATING STEAM COIL VALVE
V-2	= N.C. MODULATING HUMIDIFIER STEAM VALVE
D-1	= OUTSIDE AIR DAMPER MOTOR
D-2	= RETURN AIR DAMPER MOTOR
D-3	= RELIEF AIR DAMPER MOTOR
EP-1	= SOLENOID AIR VALVE
PE-1	= PRESSURE/ELECTRIC SWITCH
R-1	= HIGH SIGNAL SELECTOR
S-1	= MINIMUM POSITION SWITCH

Heating and Cooling Air-Handling Unit, with R.A. Fan, Steam Heat and DX Cooling Using Face and Bypass Dampers, O.A., R.A., and REL Dampers Using Economizer Control Cycle, and Steam Humidifier from Room Humidistat

Whenever the fan runs, the EP-1 is energized and O.A., R.A., and REL dampers are placed under automatic controls. When the fan stops, all dampers return to their normal positions.

Room thermostat T-3 controls valve V-1 through the low-limit thermostat T-2 on the heating coil and face and bypass dampers D-4. PE switch PE-1 on the DX cooling shuts down the cooling when the face damper is closed. DX cooling coil and steam coil must be manually operated summer and winter, so that they are not on at the same time. The O.A., R.A., and REL dampers are controlled by thermostat T-1 through O.A. high-limit thermostat T-4. T-1 and T-4 act as economizer thermostats. They allow for up to 100% O.A. when conditions are favorable. Minimum position relay S-1 can be set to maintain a minimum amount of O.A. through high signal selector R-1 as long as the fan is running. Room humidistat H-1, through duct high-limit humidistat H-2, controls humidifier valve V-2.

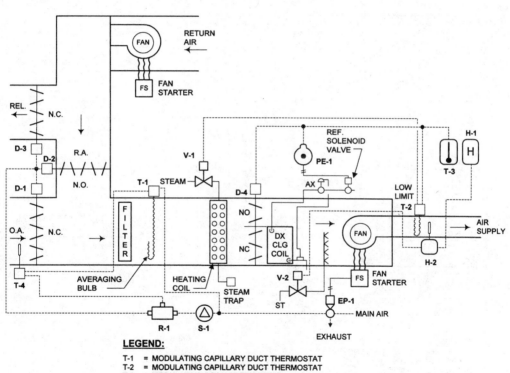

LEGEND:

T-1	=	MODULATING CAPILLARY DUCT THERMOSTAT
T-2	=	MODULATING CAPILLARY DUCT THERMOSTAT
T-3	=	MODULATING ROOM THERMOSTAT
T-4	=	MODULATING CAPILLARY THERMOSTAT
H-1	=	MODULATING ROOM HUMIDISTAT
H-2	=	HIGH LIMIT DUCT HUMIDISTAT
V-1	=	N.O. MODULATING STEAM COIL VALVE
V-2	=	N.C. MODULATING HUMIDIFIER STEAM VALVE
D-1	=	OUTSIDE AIR DAMPER MOTOR
D-2	=	RETURN AIR DAMPER MOTOR
D-3	=	RELIEF AIR DAMPER MOTOR
D-4	=	FACE & BYPASS DAMPER MOTOR
EP-1	=	SOLENOID AIR VALVE
PE-1	=	PRESSURE/ELECTRIC SWITCH
R-1	=	HIGH SIGNAL SELECTOR
S-1	=	MINIMUM POSITION SWITCH

Heating/Cooling Air-Handling Unit, with R.A. Fan, Steam Heat, and Chilled Water Cooling, O.A., R.A., and REL Dampers Using Economizer Control Cycle with Steam Humidifier from Room Humidistat

Whenever the fan runs, the EP-1 is energized and O.A., R.A., and REL dampers are placed under automatic controls. When the fan stops, all dampers return to their normal positions.

Room thermostat T-3 controls valve V-1 and valve V-2 in sequence through the low-limit thermostat T-2 on the heating coil and chilled water cooling coil. Chilled water cooling coil and steam coil must be manually operated summer and winter, so that they are not on at the same time. The O.A., R.A., and REL dampers are controlled by thermostat T-1 through O.A. high-limit thermostat T-4. T-1 and T-4 act as economizer thermostats. They allow for up to 100% O.A. when conditions are favorable. Minimum position relay S-1 can be set to maintain a minimum amount of O.A. through high signal selector R-1 as long as the fan is running. Room humidistat H-1, through duct high-limit humidistat H-2, controls humidifier valve V-2.

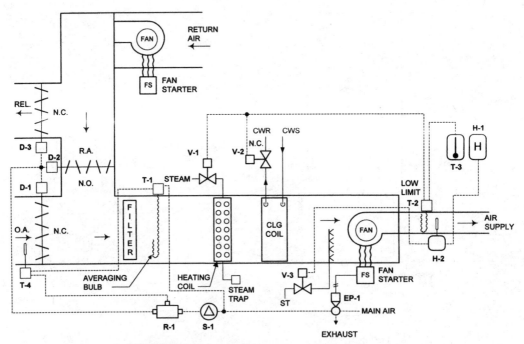

LEGEND:

T-1	= MODULATING CAPILLARY DUCT THERMOSTAT
T-2	= MODULATING CAPILLARY DUCT THERMOSTAT
T-3	= MODULATING ROOM THERMOSTAT
T-4	= MODULATING CAPILLARY THERMOSTAT
H-1	= MODULATING ROOM HUMIDISTAT
H-2	= HIGH LIMIT DUCT HUMIDISTAT
V-1	= N.O. MODULATING STEAM COIL VALVE
V-2	= N.C. MODULATING 2-WAY COOLING COIL VALVE
V-3	= N.C. MODULATING HUMIDIFIER STEAM VALVE
D-1	= OUTSIDE AIR DAMPER MOTOR
D-2	= RETURN AIR DAMPER MOTOR
D-3	= RELIEF AIR DAMPER MOTOR
EP-1	= SOLENOID AIR VALVE
R-1	= HIGH SIGNAL SELECTOR
S-1	= MINIMUM POSITION SWITCH

Heating and Cooling Air-Handling Unit, with R.A. Fan, Hot Water Heating and Chilled Water Cooling, O.A., R.A., and REL Dampers Using Economizer Control Cycle with Steam Humidifier from Room Humidistat

Whenever the fan runs, the EP-1 is energized and O.A., R.A., and REL dampers are placed under automatic controls. When the fan stops, all dampers return to their normal positions.

Room thermostat T-3 controls valve V-1 and valve V2 in sequence through the low-limit thermostat T-2 on the heating coil and chilled water cooling coil. Chilled water cooling coil and steam coil must be manually operated summer and winter, so that they are not on at the same time. The O.A., R.A., and REL dampers are controlled by thermostat T-1 through O.A. high-limit thermostat T-4. T-1 and T-4 act as economizer thermostats. They allow for up to 100% O.A. when conditions are favorable. Minimum position relay S-1 can be set to maintain a minimum amount of O.A. through high signal selector R-1 as long as the fan is running. Room humidistat H-1, through duct high-limit humidistat H-2, controls humidifier valve V-2.

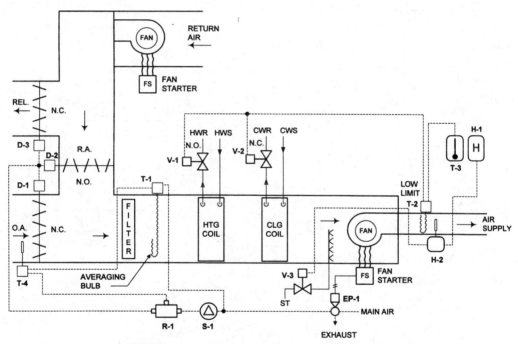

LEGEND:

T-1	=	MODULATING CAPILLARY DUCT THERMOSTAT
T-2	=	MODULATING CAPILLARY DUCT THERMOSTAT
T-3	=	MODULATING ROOM THERMOSTAT
T-4	=	MODULATING CAPILLARY THERMOSTAT
H-1	=	MODULATING ROOM HUMIDISTAT
H-2	=	HIGH LIMIT DUCT HUMIDISTAT
V-1	=	N.O. MODULATING 2-WAY HEATING COIL VALVE
V-2	=	N.C. MODULATING 2-WAY COOLING COIL VALVE
V-3	=	N.C. MODULATING HUMIDIFIER STEAM VALVE
D-1	=	OUTSIDE AIR DAMPER MOTOR
D-2	=	RETURN AIR DAMPER MOTOR
D-3	=	RELIEF AIR DAMPER MOTOR
EP-1	=	SOLENOID AIR VALVE
R-1	=	HIGH SIGNAL SELECTOR
S-1	=	MINIMUM POSITION SWITCH

Heating and Cooling Air-Handling Unit, with R.A. Fan, Hot Water Heating and DX Cooling, Face and Bypass Dampers, O.A., R.A., and REL Dampers Using Economizer Control Cycle with Steam Humidifier from Room Humidistat

Whenever the fan runs, the EP-1 is energized and O.A., R.A., and REL dampers are placed under automatic controls. When the fan stops, all dampers return to their normal positions.

Room thermostat T-3 controls valve V-1 through low-limit thermostat T-2 along with DX coil face and bypass dampers D-4 and PE switch PE-1. The PE switch PE-1 shuts down the cooling when the face damper is closed. DX cooling coil and heating coil must be manually operated summer and winter, so that they are not on at the same time. The O.A., R.A., and REL dampers are controlled by thermostat T-1 through O.A. high-limit thermostat T-4. T-1 and T-4 act as economizer thermostats. They allow for up to 100% O.A. when conditions are favorable. Minimum position relay S-1 can be set to maintain a minimum amount of O.A. through high signal selector R-1 as long as the fan is running. Room humidistat H-1, through duct high-limit humidistat H-2, controls humidifier valve V-2.

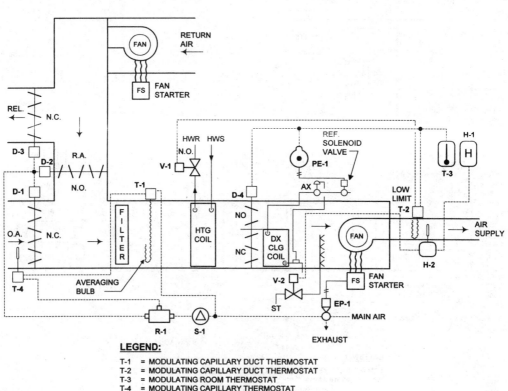

LEGEND:

T-1	=	MODULATING CAPILLARY DUCT THERMOSTAT
T-2	=	MODULATING CAPILLARY DUCT THERMOSTAT
T-3	=	MODULATING ROOM THERMOSTAT
T-4	=	MODULATING CAPILLARY THERMOSTAT
H-1	=	MODULATING ROOM HUMIDISTAT
H-2	=	HIGH LIMIT DUCT HUMIDISTAT
V-1	=	N.O. MODULATING 2-WAY HEATING COIL VALVE
V-2	=	N.C. MODULATING HUMIDIFIER STEAM VALVE
D-1	=	OUTSIDE AIR DAMPER MOTOR
D-2	=	RETURN AIR DAMPER MOTOR
D-3	=	RELIEF AIR DAMPER MOTOR
D-4	=	FACE & BYPASS DMAPER MOTOR
EP-1	=	SOLENOID AIR VALVE
PE-1	=	PRESSURE/ELECTRIC SWITCH
R-1	=	HIGH SIGNAL SELECTOR
S-1	=	MINIMUM POSITION SWITCH

Heating and Cooling Air-Handling Unit, with R.A. Fan, Hot Water Heating and DX Cooling, O.A., R.A., and REL Dampers Using Economizer Control Cycle with Steam Humidifier from Room Humidistat

Whenever the fan runs, the EP-1 is energized and O.A., R.A., and REL dampers are placed under automatic controls. When the fan stops, all dampers return to their normal positions.

Room thermostat T-3 controls valve V-1 through low-limit thermostat T-2 along with PE switch PE-1 on the DX cooling coil. DX cooling coil and heating coil must be manually operated summer and winter, so that they are not on at the same time. The O.A., R.A., and REL dampers are controlled by thermostat T-1 through O.A. high-limit thermostat T-4. T-1 and T-4 act as economizer thermostats. They allow for up to 100% O.A. when conditions are favorable. Minimum position relay S-1 can be set to maintain a minimum amount of O.A. through high signal selector R-1 as long as the fan is running. Room humidistat H-1, through duct high-limit humidistat H-2, controls humidifier valve V-2.

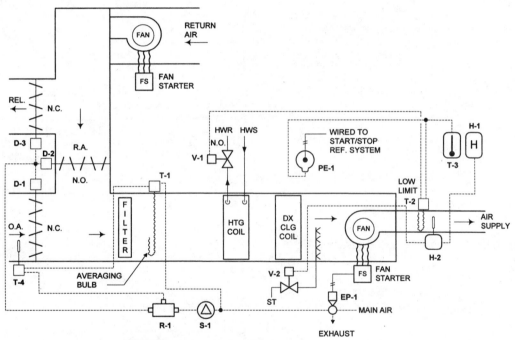

LEGEND:

T-1	=	MODULATING CAPILLARY DUCT THERMOSTAT
T-2	=	MODULATING CAPILLARY DUCT THERMOSTAT
T-3	=	MODULATING ROOM THERMOSTAT
T-4	=	MODULATING CAPILLARY THERMOSTAT
H-1	=	MODULATING ROOM HUMIDISTAT
H-2	=	HIGH LIMIT DUCT HUMIDISTAT
V-1	=	N.O. MODULATING 2-WAY HEATING COIL VALVE
V-2	=	N.C. MODULATING HUMIDIFIER STEAM VALVE
D-1	=	OUTSIDE AIR DAMPER MOTOR
D-2	=	RETURN AIR DAMPER MOTOR
D-3	=	RELIEF AIR DAMPER MOTOR
EP-1	=	SOLENOID AIR VALVE
PE-1	=	PRESSURE/ELECTRIC SWITCH
R-1	=	HIGH SIGNAL SELECTOR
S-1	=	MINIMUM POSITION SWITCH

CHAPTER 17

Dual-Path Units

Heating/Cooling Multizone Air-Handling Unit, Steam Heating and Chilled Water Cooling Coil, O.A., R.A., and REL Dampers Using Mixed Air Control, with up to 14 Zone Dampers

Whenever the fan runs, the EP-1 is energized and O.A., R.A., and REL dampers are placed under automatic controls. When the fan stops, all dampers return to their normal positions.

Room thermostat T-1 controls the multizone dampers to maintain space conditions. Submaster receiver controller RC-1, with sensors T-2 and T-3, maintain a variable hot deck temperature in accordance with O.A. temperature. Receiver controller RC-2 with sensor T-4 maintains a fixed cold deck temperature. Duct thermostat T-5 controls the O.A., R.A., and REL dampers to maintain a fixed temperature in the mix chamber. Minimum position switch S-1 allows for a minimum position of the O.A. damper as long as the fan is running.

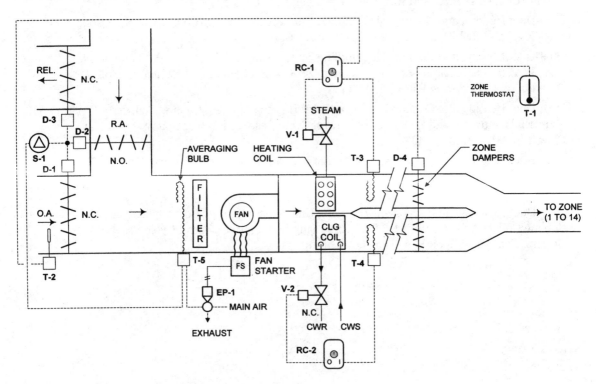

LEGEND:

T-1	=	MODULATING ROOM ZONE THERMOSTAT
T-2	=	OUTSIDE AIR MASTER TEMPERATURE SENSOR
T-3	=	HOT DECK TEMPERATURE SENSOR
T-4	=	COLD DECK TEMPERATURE SENSOR
T-5	=	MIXED AIR CONTROLLER
V-1	=	N.O. MODULATING STEAM COIL VALVE
V-2	=	N.C. MODULATING 2-WAY COOLING COIL VALVE
D-1	=	OUTSIDE AIR DAMPER MOTOR
D-2	=	RETURN AIR DAMPER MOTOR
D-3	=	RELIEF AIR DAMPER MOTOR
D-4	=	ZONE DAMPER MOTOR
EP-1	=	SOLENOID AIR VALVE
RC-1	=	RECEIVER CONTROLLER
RC-2	=	RECEIVER CONTROLLER
S-1	=	MINIMUM POSITION SWITCH

Heating and Cooling Multizone Air-Handling Unit, Steam Heating and DX Cooling, O.A., R.A., and REL Dampers Using Mixed Air Control, up to 14 Zone Dampers

Whenever the fan runs, the EP-1 is energized and O.A., R.A., and REL dampers are placed under automatic controls. When the fan stops, all dampers return to their normal positions.

Room thermostats T-1 control the multizone dampers to maintain space conditions. Submaster receiver controller RC-1, with sensors T-2 and T-3, maintain a variable hot deck temperature in accordance with O.A. temperature. Receiver controller RC-2 with sensor T-4 maintains a fixed cold deck temperature. The cooling coil is a DX coil and is controlled by PE switch PE-1. Duct thermostat T-5 controls the O.A., R.A., and REL dampers to maintain a fixed temperature in the mix chamber. Minimum position switch S-1 allows for a minimum position of the O.A. damper as long as the fan is running.

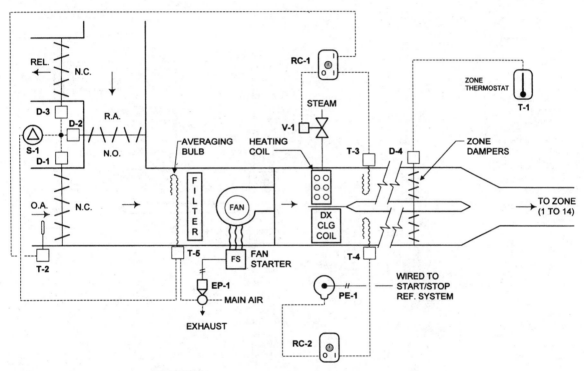

LEGEND:

T-1 = MODULATING ROOM ZONE THERMOSTAT
T-2 = OUTSIDE AIR MASTER TEMPERATURE SENSOR
T-3 = HOT DECK TEMPERATURE SENSOR
T-4 = COLD DECK TEMPERATURE SENSOR
T-5 = MIXED AIR CONTROLLER
V-1 = N.O. MODULATING STEAM COIL VALVE
D-1 = OUTSIDE AIR DAMPER MOTOR
D-2 = RETURN AIR DAMPER MOTOR
D-3 = RELIEF AIR DAMPER MOTOR
D-4 = ZONE DAMPER MOTOR
EP-1 = SOLENOID AIR VALVE
PE-1 = PRESSURE/ELECTRIC SWITCH
RC-1 = RECEIVER CONTROLLER
RC-2 = RECEIVER CONTROLLER
S-1 = MINIMUM POSITION SWITCH

Heating and Cooling Multizone Air-Handling Unit, Hot Water Heating and Chilled Water Cooling, O.A., R.A., and REL Dampers Using Mixed Air Control, up to 14 Zone Dampers

Whenever the fan runs, the EP-1 is energized and O.A., R.A., and REL dampers are placed under automatic controls. When the fan stops, all dampers return to their normal positions.

Room thermostats T-1 control the multizone dampers to maintain space conditions. Submaster receiver controller RC-1, with sensors T-2 and T-3, maintain a variable hot deck temperature in accordance with O.A. temperature. Receiver controller RC-2 with sensor T-4 maintains a fixed cold deck temperature by controlling valve V-2 on the cooling coil. Duct thermostat T-5 controls the O.A., R.A., and REL dampers to maintain a fixed temperature in the mix chamber. Minimum position switch S-1 allows for a minimum position of the O.A. damper as long as the fan is running.

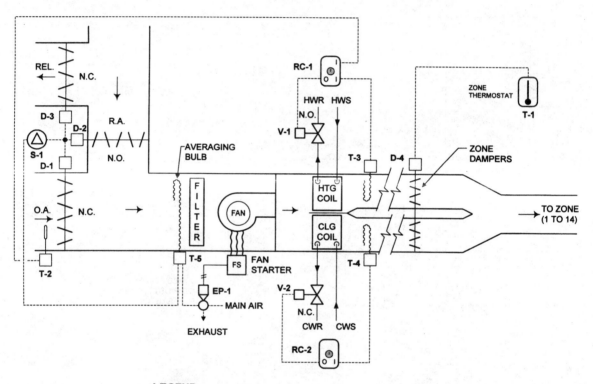

LEGEND:

T-1 = MODULATING ROOM ZONE THERMOSTAT
T-2 = OUTSIDE AIR MASTER TEMPERATURE SENSOR
T-3 = HOT DECK TEMPERATURE SENSOR
T-4 = COLD DECK TEMPERATURE SENSOR
T-5 = MIXED AIR CONTROLLER
V-1 = N.O. MODULATING 2-WAY HEATING COIL VALVE
V-2 = N.C. MODULATING 2-WAY COOLING COIL VALVE
D-1 = OUTSIDE AIR DAMPER MOTOR
D-2 = RETURN AIR DAMPER MOTOR
D-3 = RELIEF AIR DAMPER MOTOR
D-4 = ZONE DAMPER MOTOR
EP-1 = SOLENOID AIR VALVE
RC-1 = RECEIVER CONTROLLER
RC-2 = RECEIVER CONTROLLER
S-1 = MINIMUM POSITION SWITCH

Heating and Cooling Multizone Unit, Hot Water Heating and DX Cooling, O.A., R.A., and REL Dampers Using Mixed Air Control, up to 14 Zone Dampers

Whenever the fan runs, the EP-1 is energized and O.A., R.A., and REL dampers are placed under automatic controls. When the fan stops, all dampers return to their normal positions.

Room thermostats T-1 control the multizone dampers to maintain space conditions. Submaster receiver controller RC-1, with sensors T-2 and T-3, maintain a variable hot deck temperature in accordance with O.A. temperature. Receiver controller RC-2 with sensor T-4 maintains a fixed cold deck temperature by controlling PE switch PE-1 on the DX cooling coil. Duct thermostat T-5 controls the O.A., R.A., and REL dampers to maintain a fixed temperature in the mix chamber. Minimum position switch S-1 allows for a minimum position of the O.A. damper as long as the fan is running.

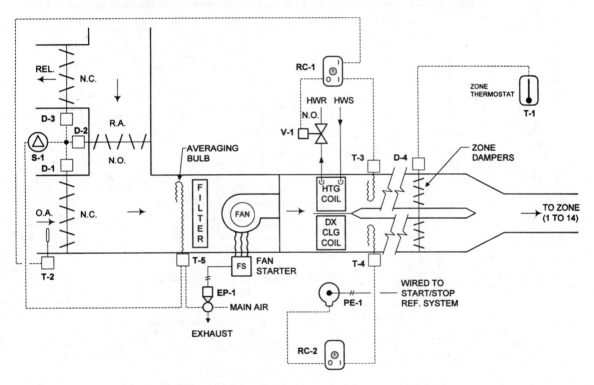

LEGEND:

T-1	= MODULATING ROOM ZONE THERMOSTAT
T-2	= OUTSIDE AIR MASTER TEMPERATURE SENSOR
T-3	= HOT DECK TEMPERATURE SENSOR
T-4	= COLD DECK TEMPERATURE SENSOR
T-5	= MIXED AIR CONTROLLER
V-1	= N.O. MODULATING 2-WAY HEATING COIL VALVE
D-1	= OUTSIDE AIR DAMPER MOTOR
D-2	= RETURN AIR DAMPER MOTOR
D-3	= RELIEF AIR DAMPER MOTOR
D-4	= ZONE DAMPER MOTOR
EP-1	= SOLENOID AIR VALVE
PE-1	= PRESSURE/ELECTRIC SWITCH
RC-1	= RECEIVER CONTROLLER
RC-2	= RECEIVER CONTROLLER
S-1	= MINIMUM POSITION SWITCH

Heating and Cooling Multizone Unit, Hot Water Heating and DX Cooling, O.A., R.A., and REL Dampers Using Economizer Control, up to 14 Zone Dampers

Whenever the fan runs, the EP-1 is energized and O.A., R.A., and REL dampers are placed under automatic controls. When the fan stops, all dampers return to their normal positions.

Room thermostats T-1 control the multizone dampers to maintain space conditions. Submaster receiver controller RC-1, with sensors T-2 and T-3, maintain a variable hot deck temperature in accordance with O.A. temperature.

Thermostat T-5 with sensing bulb in the cold deck controls PE switch PE-1 on the DX cooling coil. Duct thermostats T-2 and T-4 with sensing bulbs in the mix chamber control the O.A., R.A., and REL dampers to operate as economizer thermostats. T-2 and T-4 allow for up to 100% O.A. when conditions are favorable. Minimum position switch S-1 allows for a minimum position of the O.A. damper as long as the fan is running.

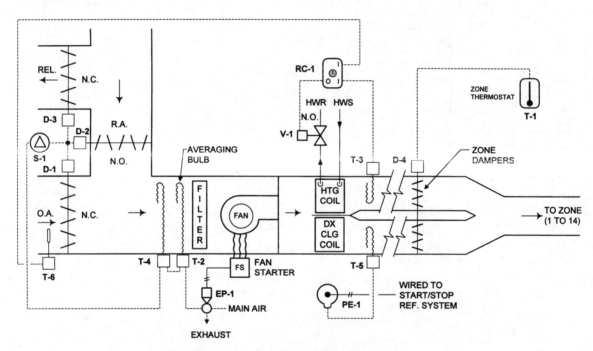

LEGEND:

T-1	= MODULATING ROOM ZONE THERMOSTAT
T-2	= MODULATING CAPILLARY DUCT THERMOSTAT
T-3	= MODULATING CAPILLARY DUCT SENSOR
T-4	= MODULATING CAPILLARY DUCT THERMOSTAT
T-5	= MODULATING CAPILLARY DUCT THERMOSTAT
T-6	= MODULATING CAPILLARY SENSOR
V-1	= N.O. MODULATING 2-WAY HEATING COIL VALVE
D-1	= OUTSIDE AIR DAMPER MOTOR
D-2	= RETURN AIR DAMPER MOTOR
D-3	= RELIEF AIR DAMPER MOTOR
D-4	= ZONE DAMPER MOTOR
EP-1	= SOLENOID AIR VALVE
PE-1	= PRESSURE/ELECTRIC SWITCH
RC-1	= RECEIVER CONTROLLER
S-1	= MINIMUM POSITION SWITCH

Heating and Cooling Multizone Unit, Steam Heating and DX Cooling, O.A., R.A., and REL Dampers Using Economizer Control, up to 14 Zone Dampers

Whenever the fan runs, the EP-1 is energized and O.A., R.A., and REL dampers are placed under automatic controls. When the fan stops, all dampers return to their normal positions.

Room thermostats T-1 control the multizone dampers to maintain space conditions. Submaster receiver controller RC-1 with sensors T-6 and T-3 maintain a variable hot deck temperature in accordance with O.A. temperature. Thermostat T-5 with sensing bulb in the cold deck controls PE switch PE-1 on the DX cooling coil. Duct thermostats T-2 and T-4 with sensing bulbs in the mix chamber control the O.A., R.A., and REL dampers to operate as economizer thermostats. T-2 and T-4 allow for up to 100% O.A. when conditions are favorable. Minimum position switch S-1 allows for a minimum position of the O.A. damper as long as the fan is running.

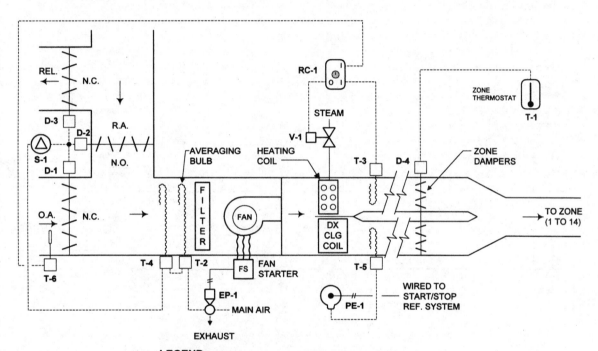

LEGEND:

T-1	=	MODULATING ROOM ZONE THERMOSTAT
T-2	=	MODULATING CAPILLARY DUCT THERMOSTAT
T-3	=	MODULATING CAPILLARY DUCT SENSOR
T-4	=	MODULATING CAPILLARY DUCT THERMOSTAT
T-5	=	MODULATING CAPILLARY DUCT THERMOSTAT
T-6	=	MODULATING CAPILLARY SENSOR
V-1	=	N.O. MODULATING STEAM COIL VALVE
D-1	=	OUTSIDE AIR DAMPER MOTOR
D-2	=	RETURN AIR DAMPER MOTOR
D-3	=	RELIEF AIR DAMPER MOTOR
D-4	=	ZONE DAMPER MOTOR
EP-1	=	SOLENOID AIR VALVE
PE-1	=	PRESSURE/ELECTRIC SWITCH
RC-1	=	RECEIVER CONTROLLER
S-1	=	MINIMUM POSITION SWITCH

Heating and Cooling Multizone Unit, Hot Water Heating and Chilled Water Cooling, O.A., R.A., and REL Dampers Using Economizer Control Cycle, up to 14 Zone Dampers

Whenever the fan runs, the EP-1 is energized and O.A., R.A., and REL dampers are placed under automatic controls. When the fan stops, all dampers return to their normal positions.

Room thermostats T-1 control the multizone dampers to maintain space conditions. Submaster receiver controller RC-1 with sensors T-6 and T-3 maintain a variable hot deck temperature in accordance with O.A. temperature.

Receiver controller RC-2 with sensor T-5 in the cold deck controls valve V-2 on the cooling coil. Duct thermostats T-2 and T-4 with sensing bulbs in the mix chamber control the O.A., R.A., and REL dampers to operate as economizer thermostats. T-2 and T-4 allow for up to 100% O.A. when conditions are favorable. Minimum position switch S-1 allows for a minimum position of the O.A. damper as long as the fan is running.

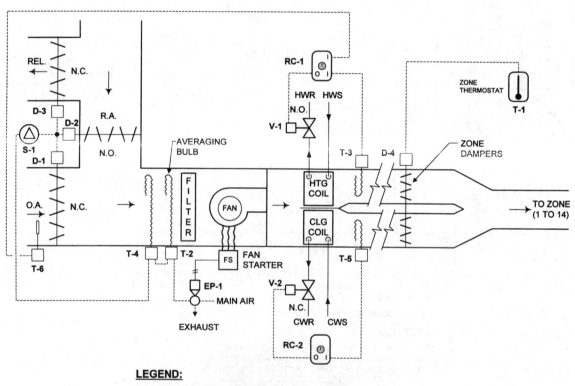

LEGEND:

T-1	= MODULATING ROOM ZONE THERMOSTAT
T-2	= MODULATING CAPILLARY DUCT THERMOSTAT
T-3	= MODULATING CAPILLARY DUCT SENSOR
T-4	= MODULATING CAPILLARY DUCT THERMOSTAT
T-5	= MODULATING CAPILLARY DUCT SENSOR
T-6	= MODULATING CAPILLARY SENSOR
V-1	= N.O. MODULATING 2-WAY HEATING COIL VALVE
V-2	= N.C. MODULATING 2-WAY COOLING COIL VALVE
D-1	= OUTSIDE AIR DAMPER MOTOR
D-2	= RETURN AIR DAMPER MOTOR
D-3	= RELIEF AIR DAMPER MOTOR
D-4	= ZONE DAMPER MOTOR
EP-1	= SOLENOID AIR VALVE
RC-1	= RECEIVER CONTROLLER
RC-2	= RECEIVER CONTROLLER
S-1	= MINIMUM POSITION SWITCH

Heating and Cooling Multizone Unit, Steam Heating and Chilled Water Cooling, O.A., R.A., and REL Dampers Using Economizer Control Cycle, up to 14 Zone Dampers

Whenever the fan runs, the EP-1 is energized and O.A., R.A., and REL dampers are placed under automatic controls. When the fan stops, all dampers return to their normal positions.

Room thermostats T-1 control the multizone dampers to maintain space conditions. Submaster receiver controller RC-1 with sensors T-6 and T-3 maintain a variable hot deck temperature in accordance with O.A. temperature. Receiver controller RC-2 with sensor T-5 in the cold deck controls valve V-2 on the cooling coil. Duct thermostats T-2 and T-4 with sensing bulbs in the mix chamber control the O.A., R.A., and REL dampers to operate as economizer thermostats. T-2 and T-4 allow for up to 100% O.A. when conditions are favorable. Minimum position switch S-1 allows for a minimum position of the O.A. damper as long as the fan is running.

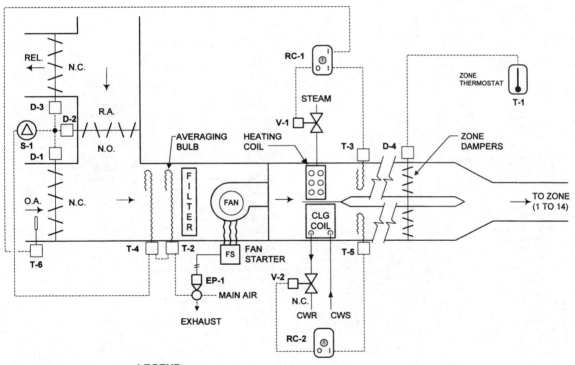

LEGEND:

T-1	=	MODULATING ROOM ZONE THERMOSTAT
T-2	=	MODULATING CAPILLARY DUCT THERMOSTAT
T-3	=	MODULATING CAPILLARY DUCT SENSOR
T-4	=	MODULATING CAPILLARY DUCT THERMOSTAT
T-5	=	MODULATING CAPILLARY DUCT SENSOR
T-6	=	MODULATING CAPILLARY SENSOR
V-1	=	N.O. MODULATING STEAM COIL VALVE
V-2	=	N.C. MODULATING 2-WAY COOLING COIL VALVE
D-1	=	OUTSIDE AIR DAMPER MOTOR
D-2	=	RETURN AIR DAMPER MOTOR
D-3	=	RELIEF AIR DAMPER MOTOR
D-4	=	ZONE DAMPER MOTOR
EP-1	=	SOLENOID AIR VALVE
PE-1	=	PRESSURE/ELECTRIC SWITCH
RC-1	=	RECEIVER CONTROLLER
RC-2	=	RECEIVER CONTROLLER
S-1	=	MINIMUM POSITION SWITCH

Heating and Cooling Multizone Unit, Steam Heating and Chilled Water Cooling, O.A., R.A., and REL Dampers Using Economizer Control Cycle, up to 14 Zone Dampers

Whenever the fan runs, the EP-1 is energized and O.A., R.A., and REL dampers are placed under automatic controls. When the fan stops, all dampers return to their normal positions.

Room thermostats T-1 control the multizone dampers to maintain space conditions. Submaster receiver controller RC-1 with sensors T-5 and T-3 maintain a variable hot deck temperature in accordance with O.A. temperature.

Discriminator relay R-1 with signals from all the zone thermostats controls the DX coil through PE switch PE-1. If any one zone calls for cooling, the R-1 operates the PE switch to turn on the cooling. Duct thermostats T-2 and T-4 with sensing bulbs in the mix chamber control the O.A., R.A., and REL dampers to operate as economizer thermostats. T-2 and T-4 allow for up to 100% O.A. when conditions are favorable. Minimum position switch S-1 allows for a minimum position of the O.A. damper as long as the fan is running.

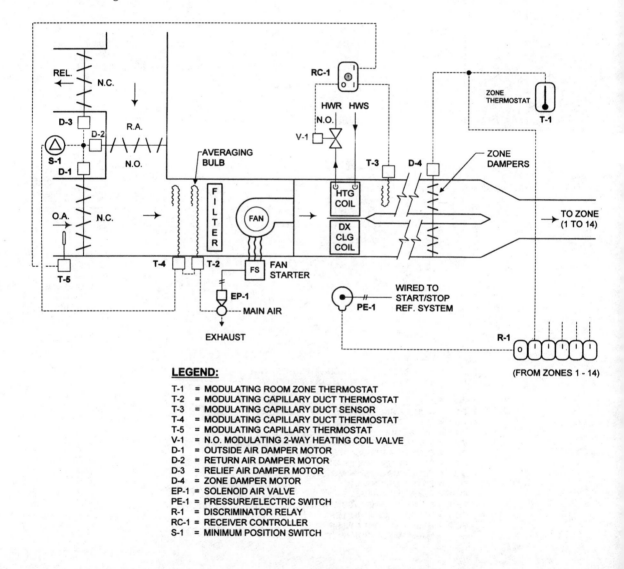

LEGEND:

T-1	=	MODULATING ROOM ZONE THERMOSTAT
T-2	=	MODULATING CAPILLARY DUCT THERMOSTAT
T-3	=	MODULATING CAPILLARY DUCT SENSOR
T-4	=	MODULATING CAPILLARY DUCT THERMOSTAT
T-5	=	MODULATING CAPILLARY THERMOSTAT
V-1	=	N.O. MODULATING 2-WAY HEATING COIL VALVE
D-1	=	OUTSIDE AIR DAMPER MOTOR
D-2	=	RETURN AIR DAMPER MOTOR
D-3	=	RELIEF AIR DAMPER MOTOR
D-4	=	ZONE DAMPER MOTOR
EP-1	=	SOLENOID AIR VALVE
PE-1	=	PRESSURE/ELECTRIC SWITCH
R-1	=	DISCRIMINATOR RELAY
RC-1	=	RECEIVER CONTROLLER
S-1	=	MINIMUM POSITION SWITCH

Heating and Cooling Multizone Unit, Steam Heating and DX Cooling, O.A., R.A., and REL Dampers Using Economizer Control Cycle, up to 14 Zone Dampers

Whenever the fan runs, the EP-1 is energized and O.A., R.A., and REL dampers are placed under automatic controls. When the fan stops, all dampers return to their normal positions.

Room thermostats T-1 control the multizone dampers to maintain space conditions. Submaster receiver controller RC-1 with sensors T-5 and T-3 maintain a variable hot deck temperature in accordance with O.A. temperature. Discriminator relay R-1 with signals from all the zone thermostats controls the DX coil through PE switch PE-1. If any one zone calls for cooling, the R-1 operates the PE switch to turn on the cooling. Duct thermostats T-2 and T-4 with sensing bulbs in the mix chamber control the O.A., R.A., and REL dampers to operate as economizer thermostats. T-2 and T-4 allow for up to 100% O.A. when conditions are favorable. Minimum position switch S-1 allows for a minimum position of the O.A. damper as long as the fan is running.

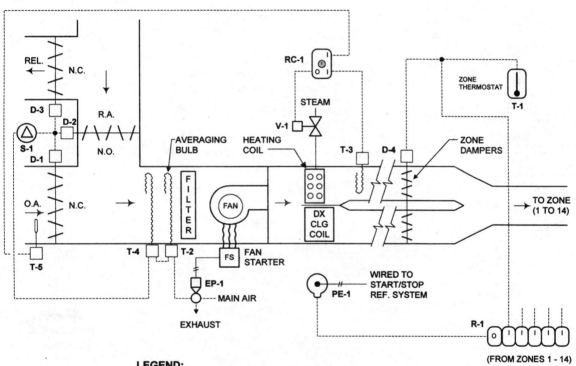

LEGEND:

T-1 = MODULATING ROOM ZONE THERMOSTAT
T-2 = MODULATING CAPILLARY DUCT THERMOSTAT
T-3 = MODULATING CAPILLARY DUCT SENSOR
T-4 = MODULATING CAPILLARY DUCT THERMOSTAT
T-5 = MODULATING CAPILLARY SENSOR
V-1 = N.O. MODULATING STEAM COIL VALVE
D-1 = OUTSIDE AIR DAMPER MOTOR
D-2 = RETURN AIR DAMPER MOTOR
D-3 = RELIEF AIR DAMPER MOTOR
D-4 = ZONE DAMPER MOTOR
EP-1 = SOLENOID AIR VALVE
PE-1 = PRESSURE/ELECTRIC SWITCH
R-1 = DISCRIMINATOR RELAY
RC-1 = RECEIVER CONTROLLER
S-1 = MINIMUM POSITION SWITCH

Heating and Cooling Multizone Unit, Steam Heating and DX Cooling, O.A., R.A., and REL Dampers Using Mixed Air Control Cycle, up to 14 Zone Dampers

Whenever the fan runs, the EP-1 is energized and O.A., R.A., and REL dampers are placed under automatic controls. When the fan stops, all dampers return to their normal positions.

   Room thermostats T-1 control the multizone dampers to maintain space conditions. Submaster receiver controller RC-1, with sensors T-2 and T-3, maintain a variable hot deck temperature in accordance with O.A. temperature. Discriminator relay R-1 with signals from all the zone thermostats controls the DX coil through PE switch PE-1. If any one zone calls for cooling, the R-1 operates the PE switch to turn on the cooling. Duct thermostat T-4 with sensing bulb in the mix chamber controls the O.A., R.A., and REL dampers to maintain a fixed temperature at that point. Minimum position switch S-1 allows for a minimum position of the O.A. damper as long as the fan is running.

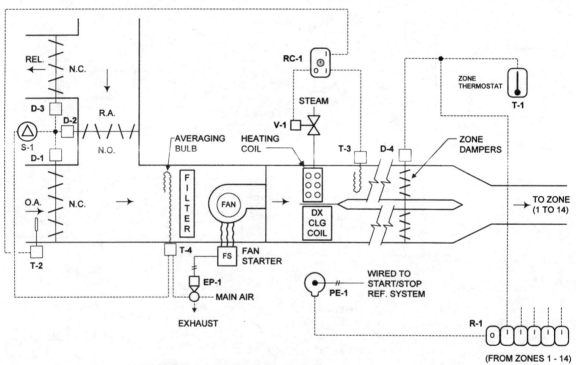

(FROM ZONES 1 - 14)

LEGEND:

T-1	= MODULATING ROOM ZONE THERMOSTAT
T-2	= OUTSIDE AIR MASTER TEMPERATURE SENSOR
T-3	= HOT DECK TEMPERATURE SENSOR
T-4	= MIXED AIR CONTROLLER
V-1	= N.O. MODULATING STEAM COIL VALVE
D-1	= OUTSIDE AIR DAMPER MOTOR
D-2	= RETURN AIR DAMPER MOTOR
D-3	= RELIEF AIR DAMPER MOTOR
D-4	= ZONE DAMPER MOTOR
EP-1	= SOLENOID AIR VALVE
PE-1	= PRESSURE/ELECTRIC SWITCH
R-1	= DISCRIMINATOR RELAY
RC-1	= RECEIVER CONTROLLER
S-1	= MINIMUM POSITION SWITCH

Heating and Cooling Multizone Unit, Hot Water Heating and DX Cooling, O.A., R.A., and REL Dampers Using Mixed Air Control Cycle, up to 14 Zone Dampers

Whenever the fan runs, the EP-1 is energized and O.A., R.A., and REL dampers are placed under automatic controls. When the fan stops, all dampers return to their normal positions.

Room thermostats T-1 control the multizone dampers to maintain space conditions. Submaster receiver controller RC-1, with sensors T-2 and T-3, maintain a variable hot deck temperature in accordance with O.A. temperature. Discriminator relay R-1 with signals from all the zone thermostats controls the DX coil through PE switch PE-1. If any one zone calls for cooling, the R-1 operates the PE switch to turn on the cooling. Duct thermostat T-4 with sensing bulb in the mix chamber controls the O.A., R.A., and REL dampers to maintain a fixed temperature at that point. Minimum position switch S-1 allows for a minimum position of the O.A. damper as long as the fan is running.

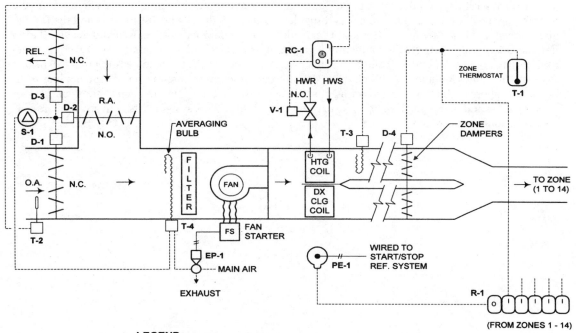

LEGEND:

T-1 = MODULATING ROOM ZONE THERMOSTAT
T-2 = OUTSIDE AIR MASTER TEMPERATURE SENSOR
T-3 = HOT DECK TEMPERATURE SENSOR
T-4 = MIXED AIR CONTROLLER
V-1 = N.O. MODULATING 2-WAY HEATING COIL VALVE
D-1 = OUTSIDE AIR DAMPER MOTOR
D-2 = RETURN AIR DAMPER MOTOR
D-3 = RELIEF AIR DAMPER MOTOR
D-4 = ZONE DAMPER MOTOR
EP-1 = SOLENOID AIR VALVE
PE-1 = PRESSURE/ELECTRIC SWITCH
R-1 = DISCRIMINATOR RELAY
RC-1 = RECEIVER CONTROLLER
S-1 = MINIMUM POSITION SWITCH

Three-Deck Heating and Cooling Multizone Unit, Hot Water Heating and Chilled Water Cooling, O.A., R.A., and REL Dampers Using Mixed Air Controls

Whenever the fan runs, the EP-1 is energized and O.A., R.A., and REL dampers are placed under automatic controls. When the fan stops, the dampers return to their normal positions.

Duct thermostat T-2 controls the O.A., R.A., and REL dampers through minimum position switch S-1, which can be set to maintain a minimum amount of O.A. as long as the fan is running, regardless of the actions of T-2.

Receiver controller RC-2, with sensors T-5 in the O.A. and sensor T-3 in the hot deck, controls valve V-1. Receiver controller RC-3, with sensor T-4 in the cold deck, controls the chilled water coil valve V-2 to maintain a fixed temperature. The zone dampers D-4, D-5, and D-6 are sequenced through relays R and R-2 to operate as follows:

On a call for heating by the room thermostat T-1, room thermostat T-1 has the hot deck damper D-4 wide open with the cold deck damper D-6 closed and the ventilation damper D-5 also closed. As the temperature reaches the set point, the first thing that happens is the D-4 begins to close and the D-5 begins to open (the D-6 stays as is—closed). After a while the hot damper D-4 is closed and the ventilation damper D-5 is wide open. As the temperature begins to rise, the ventilation damper D-5 reverses its action (through relays R-1 and R-2) and begins to close. At the same time the cold deck damper D-6 begins to open. The relays R-1 and R-2 switch the action at a certain point so that the ventilation damper is direct acting at one point and reverse acting at another point. That is so that the ventilation damper can open on a rise and then also close on a rise in room temperature. This system is legal in areas where normal multizone units are not.

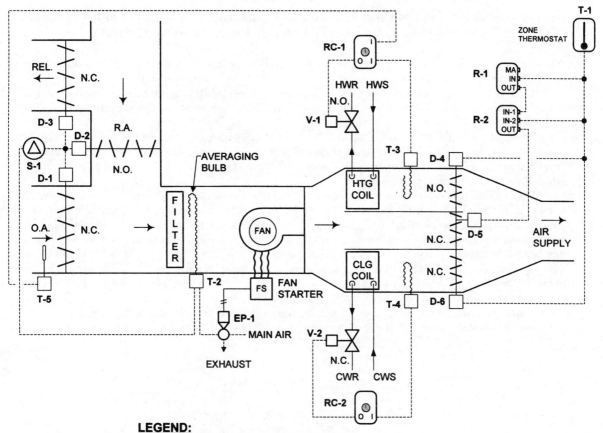

LEGEND:

T-1 = MODULATING ROOM ZONE THERMOSTAT
T-2 = MODULATING CAPILLARY DUCT THERMOSTAT
T-3 = HOT DECK TEMPERATURE SENSOR
T-4 = COLD DECK TEMPERATURE SENSOR
T-5 = OUTSIDE AIR MASTER TEMPERATURE SENSOR
V-1 = N.O. MODULATING 2-WAY HEATING COIL VALVE
V-2 = N.C. MODULATING 2-WAY COOLING COIL VALVE
D-1 = OUTSIDE AIR DAMPER MOTOR
D-2 = RETURN AIR DAMPER MOTOR
D-3 = RELIEF AIR DAMPER MOTOR
D-4 = HOT DECK ZONE DAMPER MOTOR
D-5 = VENTILATION DECK ZONE DAMPER MOTOR
D-6 = COLD DECK ZONE DAMPER MOTOR
EP-1 = SOLENOID AIR VALVE
R-1 = REVERSING RELAY
R-2 = SWITCHING RELAY
RC-1 = RECEIVER CONTROLLER
RC-2 = RECEIVER CONTROLLER
S-1 = MINIMUM POSITION SWITCH

Three-Deck Heating and Cooling Multizone Unit, Steam Heating and Chilled Water Cooling, O.A., R.A., and REL Dampers Using Mixed Air Control

Whenever the fan runs, the EP-1 is energized and O.A., R.A., and REL dampers are placed under automatic controls. When the fan stops, the dampers return to their normal positions.

Duct thermostat T-2 controls the O.A., R.A., and REL dampers through minimum position switch S-1, which can be set to maintain a minimum amount of O.A. as long as the fan is running, regardless of the actions of T-2.

Receiver controller RC-2, with sensors T-5 in the O.A. and sensor T-3 in the hot deck, controls valve V-1. Receiver controller RC-3, with sensor T-4 in the cold deck, controls the chilled water coil valve V-2 to maintain a fixed temperature. The zone dampers D-4, D-5, and D-6 are sequenced through relays R and R-2 to operate as follows:

On a call for heating by the room thermostat T-1, room thermostat T-1 has the hot deck damper D-4 wide open with the cold deck damper D-6 closed and the ventilation damper D-5 also closed. As the temperature reaches the set point, the first thing that happens is the D-4 begins to close and the D-5 begins to open (the D-6 stays as is—closed). After a while the hot damper D-4 is closed and the ventilation damper D-5 is wide open. As the temperature begins to rise, the ventilation damper D-5 reverses its action (through relays R-1 and R-2) and begins to close. At the same time the cold deck damper D-6 begins to open. The relays R-1 and R-2 switch the action at a certain point so that the ventilation damper is direct acting at one point and reverse acting at another point.

That is so that the ventilation damper can open on a rise and then also close on a rise in room temperature. This system is legal in areas where normal multizone units are not.

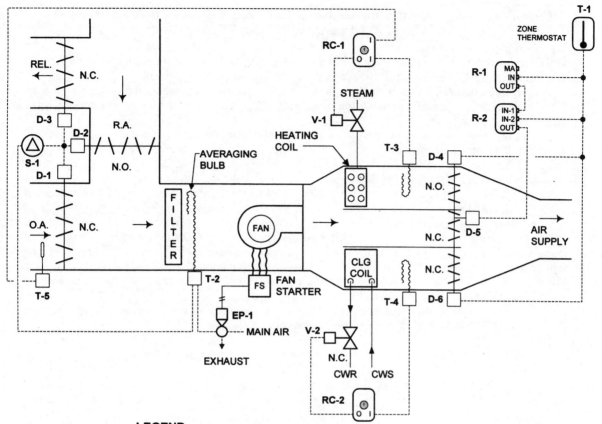

LEGEND:

T-1 = MODULATING ROOM ZONE THERMOSTAT
T-2 = MODULATING CAPILLARY DUCT THERMOSTAT
T-3 = HOT DECK TEMPERATURE SENSOR
T-4 = COLD DECK TEMPERATURE SENSOR
T-5 = OUTSIDE AIR MASTER TEMPERATURE SENSOR
V-1 = N.O. MODULATING STEAM COIL VALVE
V-2 = N.C. MODULATING 2-WAY COOLING COIL VALVE
D-1 = OUTSIDE AIR DAMPER MOTOR
D-2 = RETURN AIR DAMPER MOTOR
D-3 = RELIEF AIR DAMPER MOTOR
D-4 = HOT DECK ZONE DAMPER MOTOR
D-5 = VENTILATION DECK ZONE DAMPER MOTOR
D-6 = COLD DECK ZONE DAMPER MOTOR
EP-1 = SOLENOID AIR VALVE
R-1 = REVERSING RELAY
R-2 = SWITCHING RELAY
RC-1 = RECEIVER CONTROLLER
RC-2 = RECEIVER CONTROLLER
S-1 = MINIMUM POSITION SWITCH

Three-Deck Heating and Cooling Multizone Unit, Steam Heating and DX Cooling, O.A., R.A., and REL Dampers Using Mixed Air Control

Whenever the fan runs, the EP-1 is energized and O.A., R.A., and REL dampers are placed under automatic controls. When the fan stops, the dampers return to their normal positions.

Duct thermostat T-2 controls the O.A., R.A., and REL dampers through minimum position switch S-1, which can be set to maintain a minimum amount of O.A. as long as the fan is running, regardless of the actions of T-2.

Receiver controller RC-2, with sensors T-5 in the O.A. and sensor T-3 in the hot deck, controls valve V-1. Receiver controller RC-3, with sensor T-4 in the cold deck, controls the DX coil through PE switch PE-1 to maintain cold deck temperature. The zone dampers D-4, D-5, and D-6 are sequenced through relays R and R-2 to operate as follows:

On a call for heating by the room thermostat T-1, room thermostat T-1 has the hot deck damper D-4 wide open with the cold deck damper D-6 closed and the ventilation damper D-5 also closed. As the temperature reaches the set point, the first thing that happens is the D-4 begins to close and the D-5 begins to open (the D-6 stays as is—closed). After a while the hot damper D-4 is closed and the ventilation damper D-5 is wide open. As the temperature begins to rise, the ventilation damper D-5 reverses its action (through relays R-1 and R-2) and begins to close. At the same time the cold deck damper D-6 begins to open. The relays R-1 and R-2 switch the action at a certain point so that the ventilation damper is direct acting at one point and reverse acting at another point.

That is so that the ventilation damper can open on a rise and then also close on a rise in room temperature. This system is legal in areas where normal multizone units are not.

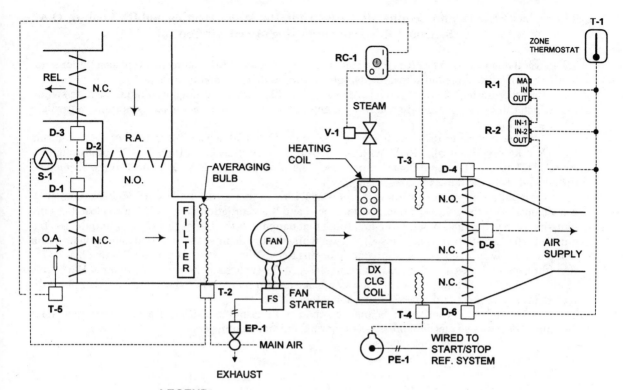

LEGEND:

T-1 = MODULATING ROOM ZONE THERMOSTAT
T-2 = MODULATING CAPILLARY DUCT THERMOSTAT
T-3 = HOT DECK TEMPERATURE SENSOR
T-4 = COLD DECK TEMPERATURE THERMOSTAT
T-5 = OUTSIDE AIR MASTER TEMPERATURE SENSOR
V-1 = N.O. MODULATING STEAM COIL VALVE
D-1 = OUTSIDE AIR DAMPER MOTOR
D-2 = RETURN AIR DAMPER MOTOR
D-3 = RELIEF AIR DAMPER MOTOR
D-4 = HOT DECK ZONE DAMPER MOTOR
D-5 = VENTILATION DECK ZONE DAMPER MOTOR
D-6 = COLD DECK ZONE DAMPER MOTOR
EP-1 = SOLENOID AIR VALVE
PE-1 = PRESSURE/ELECTRIC SWITCH
R-1 = REVERSING RELAY
R-2 = SWITCHING RELAY
RC-1 = RECEIVER CONTROLLER
S-1 = MINIMUM POSITION SWITCH

Three-Deck Heating and Cooling Multizone Unit, Hot Water Heating and DX Cooling, O.A., R.A., and REL Dampers Using Mixed Air Control

Whenever the fan runs, the EP-1 is energized and O.A., R.A., and REL dampers are placed under automatic controls. When the fan stops, the dampers return to their normal positions.

Duct thermostat T-2 controls the O.A., R.A., and REL dampers through minimum position switch S-1, which can be set to maintain a minimum amount of O.A. as long as the fan is running, regardless of the actions of T-2.

Receiver controller RC-2, with sensors T-5 in the O.A. and sensor T-3 in the hot deck, controls valve V-1. Receiver controller RC-3, with sensor T-4 in the cold deck, controls the DX coil through PE switch PE-1 to maintain cold deck temperature. The zone dampers D-4, D-5, and D-6 are sequenced through relays R-1 and R-2 to operate as follows:

On a call for heating by the room thermostat T-1, room thermostat T-1 has the hot deck damper D-4 wide open with the cold deck damper D-6 closed and the ventilation damper D-5 also closed. As the temperature reaches the set point, the first thing that happens is the D-4 begins to close and the D-5 begins to open (the D-6 stays as is—closed). After a while the hot damper D-4 is closed and the ventilation damper D-5 is wide open. As the temperature begins to rise, the ventilation damper D-5 reverses its action (through relays R-1 and R-2) and begins to close. At the same time the cold deck damper D-6 begins to open. The relays R-1 and R-2 switch the action at a certain point so that the ventilation damper is direct acting at one point and reverse acting at another point.

That is so that the ventilation damper can open on a rise and then also close on a rise in room temperature. This system is legal in areas where normal multizone units are not.

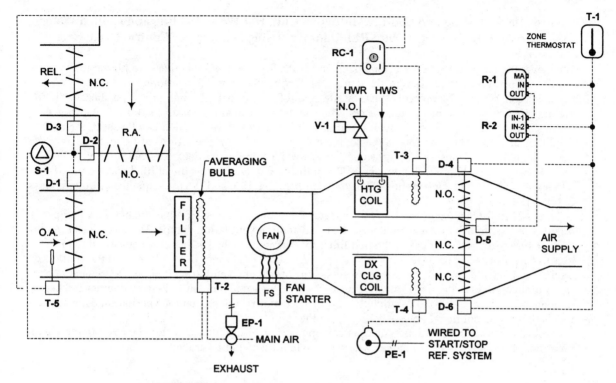

LEGEND:

T-1	=	MODULATING ROOM ZONE THERMOSTAT
T-2	=	MODULATING CAPILLARY DUCT THERMOSTAT
T-3	=	HOT DECK TEMPERATURE SENSOR
T-4	=	COLD DECK TEMPERATURE THERMOSTAT
T-5	=	OUTSIDE AIR MASTER TEMPERATURE SENSOR
V-1	=	N.O. MODULATING 2-WAY HEATING COIL VALVE
D-1	=	OUTSIDE AIR DAMPER MOTOR
D-2	=	RETURN AIR DAMPER MOTOR
D-3	=	RELIEF AIR DAMPER MOTOR
D-4	=	HOT DECK ZONE DAMPER MOTOR
D-5	=	VENTILATION DECK ZONE DAMPER MOTOR
D-6	=	COLD DECK ZONE DAMPER MOTOR
EP-1	=	SOLENOID AIR VALVE
PE-1	=	PRESSURE/ELECTRIC SWITCH
R-1	=	REVERSING RELAY
R-2	=	SWITCHING RELAY
RC-1	=	RECEIVER CONTROLLER
S-1	=	MINIMUM POSITION SWITCH

Three-Deck Heating and Cooling Multizone Unit, Hot Water Heating and Chilled Water Cooling, O.A., R.A., and REL Dampers Using Economizer Control Cycle

Whenever the fan runs, the EP-1 is energized and O.A., R.A., and REL dampers are placed under automatic controls. When the fan stops, the dampers return to their normal positions.

Duct thermostat T-2 controls the O.A., R.A., and REL dampers through mixed air thermostat T-6. T-2 and T-6 act as economizer thermostats. They act together to allow for up to 100% O.A. when conditions allow. T-2 and T-6 act through minimum position switch S-1, which can be set to maintain a minimum amount of O.A. as long as the fan is running, regardless of the actions of T-2 and T-6. Receiver controller RC-2 with sensors T-5 in the O.A. and sensor T-3 in the hot deck controls valve V-1.

Receiver controller RC-3 with sensor T-4 in the cold deck controls the chilled water coil valve V-2 to maintain cold deck temperature. The zone dampers D-4, D-5, and D-6 are sequenced through relays R-1 and R-2 to operate as follows:

On a call for heating by the room thermostat T-1, room thermostat T-1 has the hot deck damper D-4 wide open with the cold deck damper D-6 closed and the ventilation damper D-5 also closed. As the temperature reaches the set point, the first thing that happens is the D-4 begins to close and the D-5 begins to open (the D-6 stays as is—closed). After a while the hot damper D-4 is closed and the ventilation damper D-5 is wide open. As the temperature begins to rise, the ventilation damper D-5 reverses its action (through relays R-1 and R-2) and begins to close. At the same time the cold deck damper D-6 begins to open. The relays R-1 and R-2 switch the action at a certain point so that the ventilation damper is direct acting at one point and reverse acting at another point.

That is so that the ventilation damper can open on a rise and then also close on a rise in room temperature. This system is legal in areas where normal multizone units are not.

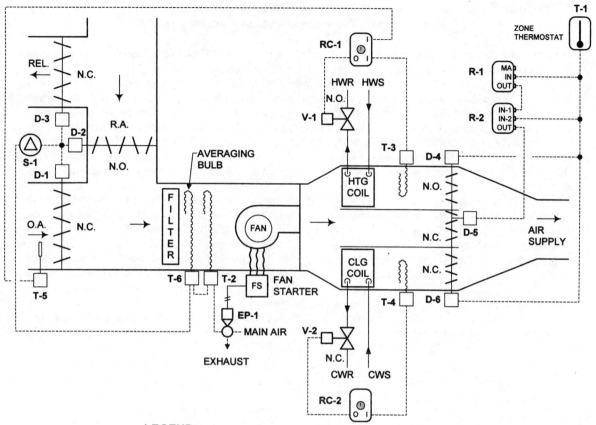

LEGEND:

T-1 = MODULATING ROOM ZONE THERMOSTAT
T-2 = MODULATING CAPILLARY DUCT THERMOSTAT
T-3 = HOT DECK TEMPERATURE SENSOR
T-4 = COLD DECK TEMPERATURE SENSOR
T-5 = OUTSIDE AIR MASTER TEMPERATURE SENSOR
T-6 = MODULATING CAPILLARY DUCT THERMOSTAT
V-1 = N.O. MODULATING 2-WAY HEATING COIL VALVE
V-2 = N.C. MODULATING 2-WAY COOLING COIL VALVE
D-1 = OUTSIDE AIR DAMPER MOTOR
D-2 = RETURN AIR DAMPER MOTOR
D-3 = RELIEF AIR DAMPER MOTOR
D-4 = HOT DECK ZONE DAMPER MOTOR
D-5 = VENTILATION DECK ZONE DAMPER MOTOR
D-6 = COLD DECK ZONE DAMPER MOTOR
EP-1 = SOLENOID AIR VALVE
R-1 = REVERSING RELAY
R-2 = SWITCHING RELAY
RC-1 = RECEIVER CONTROLLER
RC-2 = RECEIVER CONTROLLER
S-1 = MINIMUM POSITION SWITCH

Three-Deck Heating and Cooling Multizone Unit, Hot Water Heating and DX Cooling, O.A., R.A., and REL Dampers Using Economizer Cycle

Whenever the fan runs, the EP-1 is energized and O.A., R.A., and REL dampers are placed under automatic controls. When the fan stops, the dampers return to their normal positions.

Duct thermostat T-2 controls the O.A., R.A., and REL dampers through mixed air thermostat T-6. T-2 and T-6 act as economizer thermostats. They act together to allow for up to 100% O.A. when conditions allow. T-2 and T-6 act through minimum position switch S-1, which can be set to maintain a minimum amount of O.A. as long as the fan is running, regardless of the actions of T-2 and T-6.

Receiver controller RC-2, with sensors T-5 in the O.A. and sensor T-3 in the hot deck, controls valve V-1. Thermostat T-4 in the cold deck controls the DX coil through PE switch PE-1 to maintain cold deck temperature. The zone dampers D-4, D-5, and D-6 are sequenced through relays R-1 and R-2 to operate as follows:

On a call for heating by the room thermostat T-1, room thermostat T-1 has the hot deck damper D-4 wide open with the cold deck damper D-6 closed and the ventilation damper D-5 also closed. As the temperature reaches the set point, the first thing that happens is the D-4 begins to close and the D-5 begins to open (the D-6 stays as is—closed). After a while the hot damper D-4 is closed and the ventilation damper D-5 is wide open. As the temperature begins to rise, the ventilation damper D-5 reverses its action (through relays R-1 and R-2) and begins to close. At the same time the cold deck damper D-6 begins to open. The relays R-1 and R-2 switch the action at a certain point so that the ventilation damper is direct acting at one point and reverse acting at another point.

That is so that the ventilation damper can open on a rise and then also close on a rise in room temperature. This system is legal in areas where normal multizone units are not.

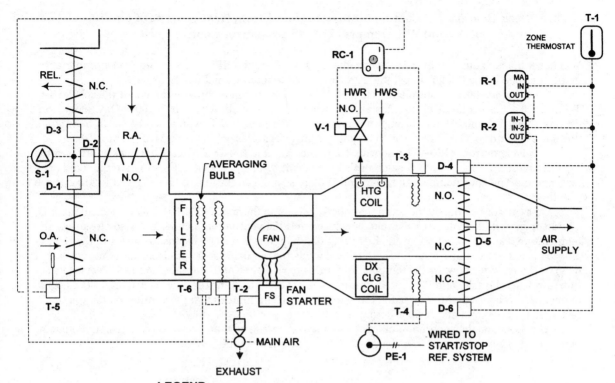

LEGEND:

T-1 = MODULATING ROOM ZONE THERMOSTAT
T-2 = MODULATING CAPILLARY DUCT THERMOSTAT
T-3 = HOT DECK TEMPERATURE SENSOR
T-4 = COLD DECK TEMPERATURE THERMOSTAT
T-5 = OUTSIDE AIR MASTER TEMPERATURE SENSOR
T-6 = MODULATING CAPILLARY DUCT THERMOSTAT
V-1 = N.O. MODULATING 2-WAY HEATING COIL VALVE
D-1 = OUTSIDE AIR DAMPER MOTOR
D-2 = RETURN AIR DAMPER MOTOR
D-3 = RELIEF AIR DAMPER MOTOR
D-4 = HOT DECK ZONE DAMPER MOTOR
D-5 = VENTILATION DECK ZONE DAMPER MOTOR
D-6 = COLD DECK ZONE DAMPER MOTOR
EP-1 = SOLENOID AIR VALVE
PE-1 = PRESSURE/ELECTRIC SWITCH
R-1 = REVERSING RELAY
R-2 = SWITCHING RELAY
RC-1 = RECEIVER CONTROLLER
S-1 = MINIMUM POSITION SWITCH

Three-Deck Heating and Cooling Multizone Unit, Steam Heating and DX Cooling, O.A., R.A., and REL Dampers Using Economizer Control Cycle

Whenever the fan runs, the EP-1 is energized and O.A., R.A., and REL dampers are placed under automatic controls. When the fan stops, the dampers return to their normal positions.

Duct thermostat T-2 controls the O.A., R.A., and REL dampers through mixed air thermostat T-6. T-2 and T-6 act as economizer thermostats. They act together to allow for up to 100% O.A. when conditions allow. T-2 and T-6 act through minimum position switch S-1, which can be set to maintain a minimum amount of O.A. as long as the fan is running, regardless of the actions of T-2 and T-6.

Receiver controller RC-2 with sensors T-5 in the O.A. and sensor T-3 in the hot deck controls valve V-1. Thermostat T-4 in the cold deck controls the DX coil through PE switch PE-1 to maintain cold deck temperature. The zone dampers D-4, D-5, and D-6 are sequenced through relays R-1 and R-2 to operate as follows:

On a call for heating by the room thermostat T-1, room thermostat T-1 has the hot deck damper D-4 wide open with the cold deck damper D-6 closed and the ventilation damper D-5 also closed. As the temperature reaches the set point, the first thing that happens is the D-4 begins to close and the D-5 begins to open (the D-6 stays as is—closed). After a while the hot damper D-4 is closed and the ventilation damper D-5 is wide open. As the temperature begins to rise, the ventilation damper D-5 reverses its action (through relays R-1 and R-2) and begins to close. At the same time the cold deck damper D-6 begins to open. The relays R-1 and R-2 switch the action at a certain point so that the ventilation damper is direct acting at one point and reverse acting at another point.

That is so that the ventilation damper can open on a rise and then also close on a rise in room temperature. This system is legal in areas where normal multizone units are not.

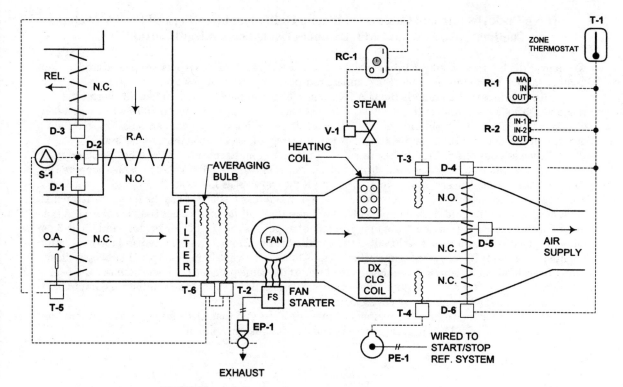

LEGEND:

T-1 = MODULATING ROOM ZONE THERMOSTAT
T-2 = MODULATING CAPILLARY DUCT THERMOSTAT
T-3 = HOT DECK TEMPERATURE SENSOR
T-4 = COLD DECK TEMPERATURE THERMOSTAT
T-5 = OUTSIDE AIR MASTER TEMPERATURE SENSOR
T-6 = MODULATING CAPILLARY DUCT THERMOSTAT
V-1 = N.O. MODULATING STEAM COIL VALVE
D-1 = OUTSIDE AIR DAMPER MOTOR
D-2 = RETURN AIR DAMPER MOTOR
D-3 = RELIEF AIR DAMPER MOTOR
D-4 = HOT DECK ZONE DAMPER MOTOR
D-5 = VENTILATION DECK ZONE DAMPER MOTOR
D-6 = COLD DECK ZONE DAMPER MOTOR
EP-1 = SOLENOID AIR VALVE
PE-1 = PRESSURE/ELECTRIC SWITCH
R-1 = REVERSING RELAY
R-2 = SWITCHING RELAY
RC-1 = RECEIVER CONTROLLER
S-1 = MINIMUM POSITION SWITCH

Three-Deck Heating and Cooling Multizone Unit, Steam Heating and Chilled Water Cooling, O.A., R.A., and REL Dampers Using Economizer Control Cycle

Whenever the fan runs, the EP-1 is energized and O.A., R.A., and REL dampers are placed under automatic controls. When the fan stops, the dampers return to their normal positions.

Duct thermostat T-2 controls the O.A., R.A., and REL dampers through mixed air thermostat T-6. T-2 and T-6 act as economizer thermostats. They act together to allow for up to 100% O.A. when conditions allow. T-2 and T-6 act through minimum position switch S-1, which can be set to maintain a minimum amount of O.A. as long as the fan is running, regardless of the actions of T-2 and T-6.

Receiver controller RC-2, with sensors T-5 in the O.A. and sensor T-3 in the hot deck, controls valve V-1. Thermostat T-4 in the cold deck controls the cooling coil valve V-2. The zone dampers D-4, D-5, and D-6 are sequenced through relays R-1 and R-2 to operate as follows:

On a call for heating by the room thermostat T-1, room thermostat T-1 has the hot deck damper D-4 wide open with the cold deck damper D-6 closed and the ventilation damper D-5 also closed. As the temperature reaches the set point, the first thing that happens is the D-4 begins to close and the D-5 begins to open (the D-6 stays as is—closed). After a while the hot damper D-4 is closed and the ventilation damper D-5 is wide open. As the temperature begins to rise, the ventilation damper D-5 reverses its action (through relays R-1 and R-2) and begins to close. At the same time the cold deck damper D-6 begins to open. The relays R-1 and R-2 switch the action at a certain point so that the ventilation damper is direct acting at one point and reverse acting at another point.

That is so that the ventilation damper can open on a rise and then also close on a rise in room temperature. This system is legal in areas where normal multizone units are not.

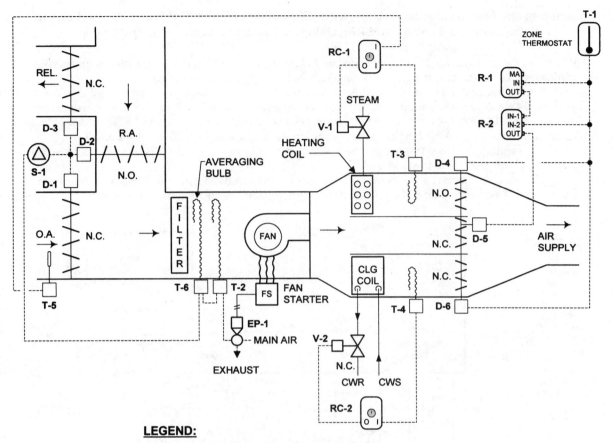

LEGEND:

T-1 = MODULATING ROOM ZONE THERMOSTAT
T-2 = MODULATING CAPILLARY DUCT THERMOSTAT
T-3 = HOT DECK TEMPERATURE SENSOR
T-4 = COLD DECK TEMPERATURE SENSOR
T-5 = OUTSIDE AIR MASTER TEMPERATURE SENSOR
T-6 = MODULATING CAPILLARY DUCT THERMOSTAT
V-1 = N.O. MODULATING 2-WAY HEATING COIL VALVE
V-2 = N.C. MODULATING 2-WAY COOLING COIL VALVE
D-1 = OUTSIDE AIR DAMPER MOTOR
D-2 = RETURN AIR DAMPER MOTOR
D-3 = RELIEF AIR DAMPER MOTOR
D-4 = HOT DECK ZONE DAMPER MOTOR
D-5 = VENTILATION DECK ZONE DAMPER MOTOR
D-6 = COLD DECK ZONE DAMPER MOTOR
EP-1 = SOLENOID AIR VALVE
R-1 = REVERSING RELAY
R-2 = SWITCHING RELAY
RC-1 = RECEIVER CONTROLLER
RC-2 = RECEIVER CONTROLLER
S-1 = MINIMUM POSITION SWITCH

Heating and Cooling Multizone Unit, with DX Cooling Coil and Hot Water Reheat Coils in the Zones, O.A., R.A., and REL Dampers Using Economizer Control Cycle

Whenever the fan runs, the EP-1 is energized and O.A., R.A., and REL dampers are placed under automatic controls. When the fan stops, the dampers return to their normal positions.

Duct thermostat T-2 controls the O.A., R.A., and REL dampers through mixed air thermostat T-4. T-2 and T-4 act as economizer thermostats. They act together to allow for up to 100% O.A. when conditions allow. T-2 and T-4 act through minimum position switch S-1, which can be set to maintain a minimum amount of O.A. as long as the fan is running, regardless of the actions of T-2 and T-4.

The ventilation comes through the deck where normally the warm air is passed. Thermostat T-3 controls PE switch PE-1 on the cold deck.

The zone dampers D-4 are controlled by room thermostat(s) T-1 in sequence with reheat coil valve(s) V-1. There can be up to 14 damper zones with 14 reheat or more coils.

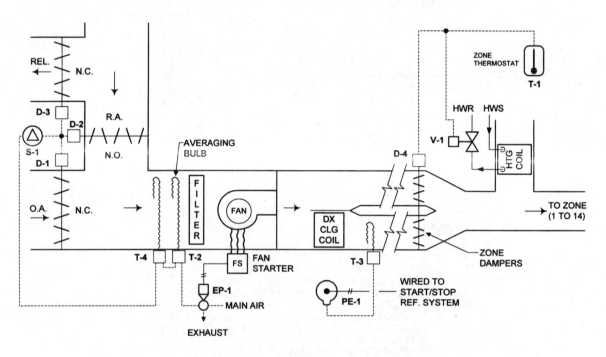

LEGEND:

T-1	=	MODULATING ROOM ZONE THERMOSTAT
T-2	=	MODULATING CAPILLARY DUCT THERMOSTAT
T-3	=	MODULATING CAPILLARY DUCT THERMOSTAT
T-4	=	MODULATING CAPILLARY DUCT THERMOSTAT
V-1	=	MODULATING ZONE HW REHEAT COIL VALVE
D-1	=	OUTSIDE AIR DAMPER MOTOR
D-2	=	RETURN AIR DAMPER MOTOR
D-3	=	RELIEF AIR DAMPER MOTOR
D-4	=	ZONE DAMPER MOTOR
EP-1	=	SOLENOID AIR VALVE
PE-1	=	PRESSURE/ELECTRIC SWITCH
S-1	=	MINIMUM POSITION SWITCH

Heating and Cooling Multizone Unit, with DX Cooling Coil and Steam Reheat Coils in the Zones, O.A., R.A., and REL Dampers Using Economizer Control Cycle

Whenever the fan runs, the EP-1 is energized and O.A., R.A., and REL dampers are placed under automatic controls. When the fan stops, the dampers return to their normal positions.

Duct thermostat T-2 controls the O.A., R.A., and REL dampers through mixed air thermostat T-4. T-2 and T-4 act as economizer thermostats. They act together to allow for up to 100% O.A. when conditions allow. T-2 and T-4 act through minimum position switch S-1, which can be set to maintain a minimum amount of O.A. as long as the fan is running, regardless of the actions of T-2 and T-4.

The ventilation comes through the deck where normally the warm air is passed. Thermostat T-3 controls PE switch PE-1 on the cold deck.

The zone dampers D-4 are controlled by room thermostat(s) T-1 in sequence with reheat coil valve(s) V-1. There can be up to 14 damper zones with 14 reheat or more coils.

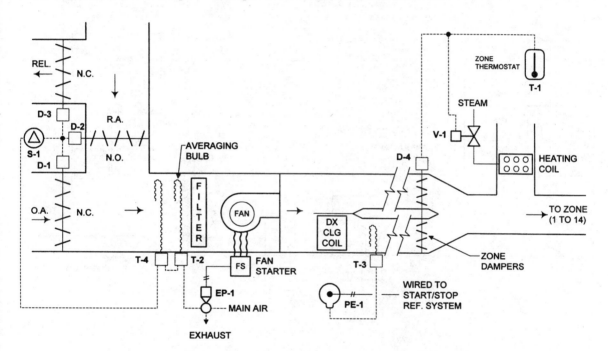

LEGEND:

T-1	=	MODULATING ROOM ZONE THERMOSTAT
T-2	=	MODULATING CAPILLARY DUCT THERMOSTAT
T-3	=	MODULATING CAPILLARY DUCT THERMOSTAT
T-4	=	MODULATING CAPILLARY DUCT THERMOSTAT
V-1	=	MODULATING ZONE STEAM REHEAT COIL VALVE
D-1	=	OUTSIDE AIR DAMPER MOTOR
D-2	=	RETURN AIR DAMPER MOTOR
D-3	=	RELIEF AIR DAMPER MOTOR
D-4	=	ZONE DAMPER MOTOR
EP-1	=	SOLENOID AIR VALVE
PE-1	=	PRESSURE/ELECTRIC SWITCH
S-1	=	MINIMUM POSITION SWITCH

Heating and Cooling Multizone Unit, with Chilled Water Cooling Coil and Hot Water Reheat Coils in the Zones, O.A., R.A., and REL Dampers Using Economizer Control Cycle

Whenever the fan runs, the EP-1 is energized and O.A., R.A., and REL dampers are placed under automatic controls. When the fan stops, the dampers return to their normal positions.

Duct thermostat T-2 controls the O.A., R.A., and REL dampers through mixed air thermostat T-4. T-2 and T-4 act as economizer thermostats. They act together to allow for up to 100% O.A. when conditions allow. T-2 and T-4 act through minimum position switch S-1, which can be set to maintain a minimum amount of O.A. as long as the fan is running, regardless of the actions of T-2 and T-4.

The ventilation comes through the deck where normally the warm air is passed. Thermostat T-3 controls chilled water coil, through valve V-2.

The zone dampers D-4 are controlled by room thermostat(s) T-1 in sequence with reheat coil valve(s) V-1. There can be up to 14 damper zones with 14 reheat or more coils.

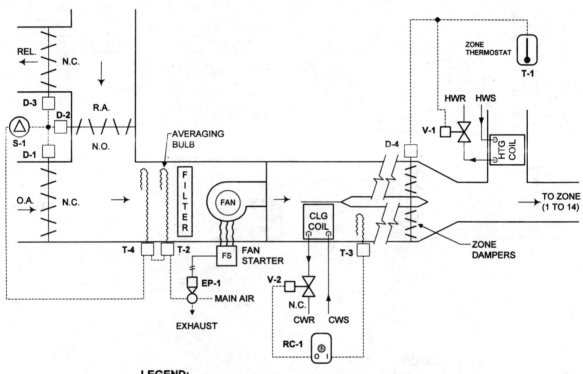

LEGEND:

T-1 = MODULATING ROOM ZONE THERMOSTAT
T-2 = MODULATING CAPILLARY DUCT THERMOSTAT
T-3 = MODULATING CAPILLARY DUCT SENSOR
T-4 = MODULATING CAPILLARY DUCT THERMOSTAT
V-1 = MODULATING ZONE HW REHEAT COIL VALVE
V-2 = N.C. MODULATING 2-WAY COOLING COIL VALVE
D-1 = OUTSIDE AIR DAMPER MOTOR
D-2 = RETURN AIR DAMPER MOTOR
D-3 = RELIEF AIR DAMPER MOTOR
D-4 = ZONE DAMPER MOTOR
EP-1 = SOLENOID AIR VALVE
RC-1 = RECEIVER CONTROLLER
S-1 = MINIMUM POSITION SWITCH

Heating and Cooling Multizone Unit, with Chilled Water Coil and Steam Reheat Coils in the Zones, O.A., R.A., and REL Dampers Using Economizer Control Cycle

Whenever the fan runs, the EP-1 is energized and O.A., R.A., and REL dampers are placed under automatic controls. When the fan stops, the dampers return to their normal positions.

Duct thermostat T-2 controls the O.A., R.A., and REL dampers through mixed air thermostat T-4. T-2 and T-4 act as economizer thermostats. They act together to allow for up to 100% O.A. when conditions allow. T-2 and T-4 act through minimum position switch S-1, which can be set to maintain a minimum amount of O.A. as long as the fan is running, regardless of the actions of T-2 and T-4.

The ventilation comes through the deck where normally the warm air is passed. Thermostat T-3 controls chilled water coil, through valve V-2.

The zone dampers D-4 are controlled by room thermostat(s) T-1 in sequence with reheat coil valve(s) V-1. There can be up to 14 damper zones with 14 reheat or more coils.

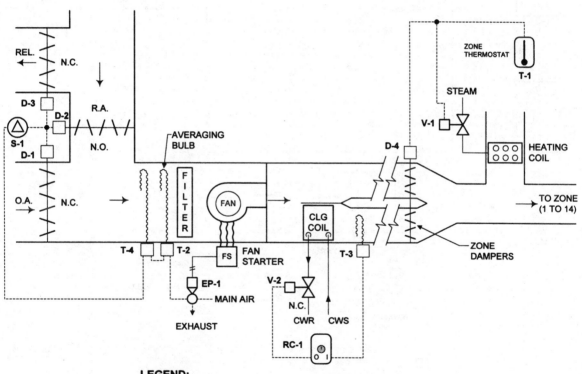

LEGEND:

T-1	= MODULATING ROOM ZONE THERMOSTAT
T-2	= MODULATING CAPILLARY DUCT THERMOSTAT
T-3	= MODULATING CAPILLARY DUCT SENSOR
T-4	= MODULATING CAPILLARY DUCT THERMOSTAT
V-1	= MODULATING ZONE STEAM REHEAT COIL VALVE
V-2	= N.C. MODULATING 2-WAY COOLING COIL VALVE
D-1	= OUTSIDE AIR DAMPER MOTOR
D-2	= RETURN AIR DAMPER MOTOR
D-3	= RELIEF AIR DAMPER MOTOR
D-4	= ZONE DAMPER MOTOR
EP-1	= SOLENOID AIR VALVE
RC-1	= RECEIVER CONTROLLER
S-1	= MINIMUM POSITION SWITCH

Heating and Cooling Multizone Unit, with DX Cooling and Hot Water Reheat Coils in the Zones, O.A., R.A., and REL Dampers Using a Mixed Air Cycle

Whenever the fan runs, the EP-1 is energized and O.A., R.A., and REL dampers are placed under automatic controls. When the fan stops, the dampers return to their normal positions.

Duct thermostat T-2 controls the O.A., R.A., and REL dampers. T-2 acts through minimum position switch S-1, which can be set to maintain a minimum amount of O.A. as long as the fan is running, regardless of the actions of T-2.

The ventilation comes through the deck where normally the warm air is passed. Thermostat T-3 controls PE switch PE-1 on the DX coil.

The zone dampers D-4 are controlled by room thermostat(s) T-1 in sequence with reheat coil valve(s) V-1. There can be up to 14 damper zones with 14 reheat or more coils.

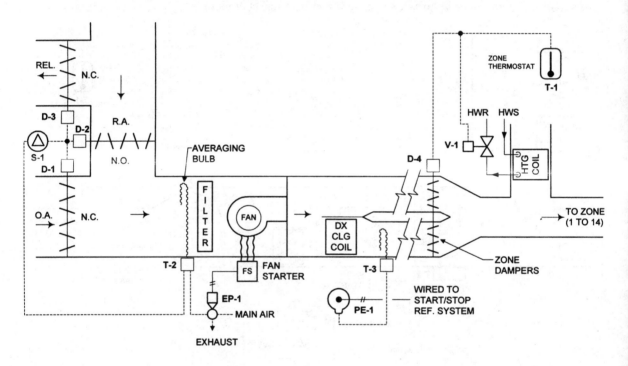

LEGEND:

T-1	=	MODULATING ROOM ZONE THERMOSTAT
T-2	=	MODULATING CAPILLARY DUCT THERMOSTAT
T-3	=	MODULATING CAPILLARY DUCT THERMOSTAT
V-1	=	MODULATING ZONE HW REHEAT COIL VALVE
D-1	=	OUTSIDE AIR DAMPER MOTOR
D-2	=	RETURN AIR DAMPER MOTOR
D-3	=	RELIEF AIR DAMPER MOTOR
D-4	=	ZONE DAMPER MOTOR
EP-1	=	SOLENOID AIR VALVE
PE-1	=	PRESSURE/ELECTRIC SWITCH
S-1	=	MINIMUM POSITION SWITCH

Heating and Cooling Multizone Unit, with DX Coil and Steam Heating Reheat Coils in the Zones, O.A., R.A., and REL Dampers Using Mixed Air Control Cycle

Whenever the fan runs, the EP-1 is energized and O.A., R.A., and REL dampers are placed under automatic controls. When the fan stops, the dampers return to their normal positions.

Duct thermostat T-2 controls the O.A., R.A., and REL dampers. T-2 acts through minimum position switch S-1, which can be set to maintain a minimum amount of O.A. as long as the fan is running, regardless of the actions of T-2.

The ventilation comes through the deck where normally the warm air is passed. Thermostat T-3 controls PE switch PE-1 on the DX coil.

The zone dampers D-4 are controlled by room thermostat(s) T-1 in sequence with reheat coil valve(s) V-1. There can be up to 14 damper zones with 14 reheat or more coils.

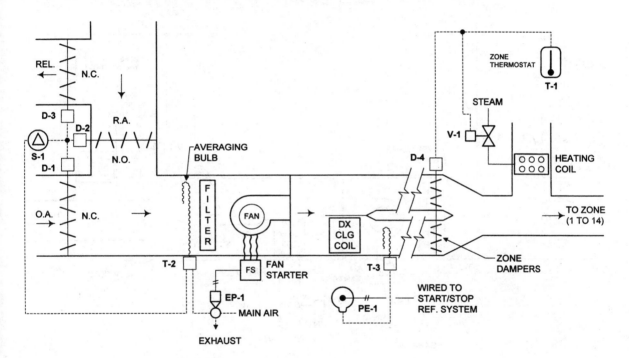

LEGEND:

T-1 = MODULATING ROOM ZONE THERMOSTAT
T-2 = MODULATING CAPILLARY DUCT THERMOSTAT
T-3 = MODULATING CAPILLARY DUCT THERMOSTAT
V-1 = MODULATING ZONE STEAM REHEAT COIL VALVE
D-1 = OUTSIDE AIR DAMPER MOTOR
D-2 = RETURN AIR DAMPER MOTOR
D-3 = RELIEF AIR DAMPER MOTOR
D-4 = ZONE DAMPER MOTOR
EP-1 = SOLENOID AIR VALVE
PE-1 = PRESSURE/ELECTRIC SWITCH
S-1 = MINIMUM POSITION SWITCH

Heating and Cooling Multizone Unit, with Chilled Water coil and Hot Water Reheat Coils in the Zones,, O.A., R.A., and REL Dampers Using Mixed Air Cycle

Whenever the fan runs, the EP-1 is energized and O.A., R.A., and REL dampers are placed under automatic controls. When the fan stops, the dampers return to their normal positions.

Duct thermostat T-2 controls the O.A., R.A., and REL dampers. T-2 acts through minimum position switch S-1, which can be set to maintain a minimum amount of O.A. as long as the fan is running, regardless of the actions of T-2.

The ventilation comes through the deck where normally the warm air is passed. Receiver controller RC-1 with sensor T-3 in the cold deck controls valve V-2 on the chilled water coil.

The zone dampers D-4 are controlled by room thermostat(s) T-1 in sequence with reheat coil valve(s) V-1. There can be up to 14 damper zones with 14 reheat or more coils.

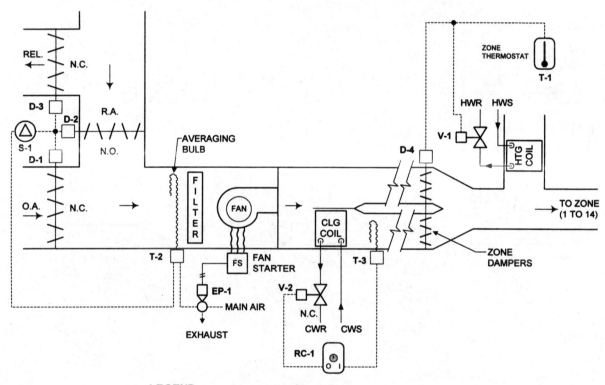

LEGEND:

T-1 = MODULATING ROOM ZONE THERMOSTAT
T-2 = MODULATING CAPILLARY DUCT THERMOSTAT
T-3 = MODULATING CAPILLARY DUCT SENSOR
V-1 = MODULATING ZONE HW REHEAT COIL VALVE
V-2 = N.C. MODULATING 2-WAY COOLING COIL VALVE
D-1 = OUTSIDE AIR DAMPER MOTOR
D-2 = RETURN AIR DAMPER MOTOR
D-3 = RELIEF AIR DAMPER MOTOR
D-4 = ZONE DAMPER MOTOR
EP-1 = SOLENOID AIR VALVE
RC-1 = RECEIVER CONTROLLER
S-1 = MINIMUM POSITION SWITCH

Heating and Cooling Multizone Unit, with Chilled Water Coil and Steam Reheat Coils in the Zones,, O.A., R.A., and REL Dampers Using Mixed Air Cycle

Whenever the fan runs, the EP-1 is energized and O.A., R.A., and REL dampers are placed under automatic controls. When the fan stops, the dampers return to their normal positions.

Duct thermostat T-2 controls the O.A., R.A., and REL dampers. T-2 acts through minimum position switch S-1, which can be set to maintain a minimum amount of O.A. as long as the fan is running, regardless of the actions of T-2.

The ventilation comes through the deck where normally the warm air is passed. Receiver controller RC-1 with sensor T-3 in the cold deck controls valve V-2 on the chilled water coil.

The zone dampers D-4 are controlled by room thermostat(s) T-1 in sequence with reheat coil valve(s) V-1. There can be up to 14 damper zones with 14 reheat or more coils.

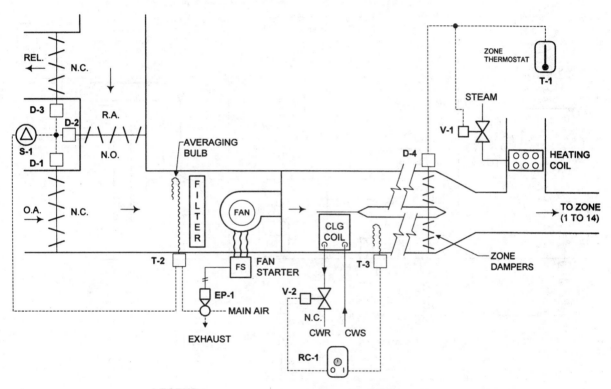

LEGEND:

T-1	=	MODULATING ROOM ZONE THERMOSTAT
T-2	=	MODULATING CAPILLARY DUCT THERMOSTAT
T-3	=	MODULATING CAPILLARY DUCT SENSOR
V-1	=	MODULATING ZONE STEAM REHEAT COIL VALVE
V-2	=	N.C. MODULATING 2-WAY COOLING COIL VALVE
D-1	=	OUTSIDE AIR DAMPER MOTOR
D-2	=	RETURN AIR DAMPER MOTOR
D-3	=	RELIEF AIR DAMPER MOTOR
D-4	=	ZONE DAMPER MOTOR
EP-1	=	SOLENOID AIR VALVE
RC-1	=	RECEIVER CONTROLLER
S-1	=	MINIMUM POSITION SWITCH

Dual-Duct Heating and Cooling Air-Handling Unit, Hot Water Heating Coil, Chilled Water Cooling Coil, O.A., R.A., and REL Dampers Using Mixed Air Control Cycle

Whenever the fan runs, the EP-1 is energized and O.A., R.A., and REL dampers are placed under automatic controls. When the fan stops, the dampers return to their normal positions.

Receiver controller RC-1 with O.A. sensor T-5 and hot deck sensor T-3 controls the hot deck valve V-1. Receiver controller RC-2 with sensor T-4 controls valve V-2 on the cooling coil. Duct thermostat T-2 controls the O.A., R.A., and REL dampers. T-2 acts through minimum position switch S-1, which can be set to maintain a minimum amount of O.A. as long as the fan is running, regardless of the actions of T-2.

The zone mixing box dampers D-4 are controlled by room thermostats T-1. The mixing boxes are factory-fabricated boxes with dampers that modulate the cold air and hot air supplies to the box to maintain space conditions. There are an unlimited number of boxes that can be used with dual-duct systems.

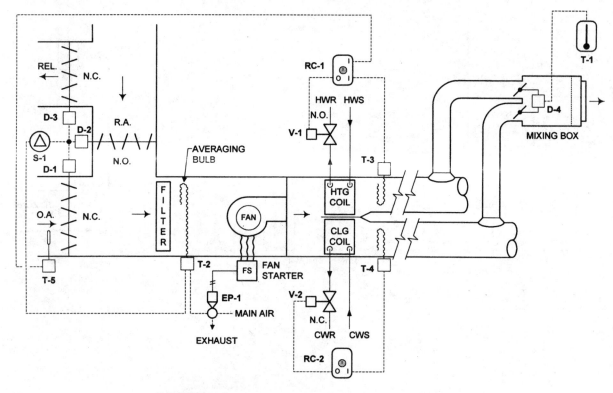

LEGEND:

T-1	=	MODULATING ROOM ZONE THERMOSTAT
T-2	=	MODULATING CAPILLARY DUCT THERMOSTAT
T-3	=	HOT DECK TEMPERATURE SENSOR
T-4	=	COLD DECK TEMPERATURE SENSOR
T-5	=	OUTSIDE AIR MASTER TEMPERATURE SENSOR
V-1	=	N.O. MODULATING 2-WAY HEATING COIL VALVE
V-2	=	N.C. MODULATING 2-WAY COOLING COIL VALVE
D-1	=	OUTSIDE AIR DAMPER MOTOR
D-2	=	RETURN AIR DAMPER MOTOR
D-3	=	RELIEF AIR DAMPER MOTOR
D-4	=	MIXING BOX DAMPER MOTOR
EP-1	=	SOLENOID AIR VALVE
RC-1	=	RECEIVER CONTROLLER
RC-2	=	RECEIVER CONTROLLER
S-1	=	MINIMUM POSITION SWITCH

Dual-Duct Heating and Cooling Air-Handling Unit, Steam Heating Coil, Chilled Water Cooling Coil, O.A., R.A., and REL Dampers Using Mixed Air Control Cycle

Whenever the fan runs, the EP-1 is energized and O.A., R.A., and REL dampers are placed under automatic controls. When the fan stops, the dampers return to their normal positions.

Receiver controller RC-1 with O.A. sensor T-5 and hot deck sensor T-3 controls the hot deck valve V-1. Receiver controller RC-2 with sensor T-4 controls valve V-2 on the cooling coil. Duct thermostat T-2 controls the O.A., R.A., and REL dampers. T-2 acts through minimum position switch S-1, which can be set to maintain a minimum amount of O.A. as long as the fan is running, regardless of the actions of T-2.

The zone mixing box dampers D-4 are controlled by room thermostats T-1. The mixing boxes are factory-fabricated boxes with dampers that modulate the cold air and hot air supplies to the box to maintain space conditions. There are an unlimited number of boxes that can be used with dual-duct systems.

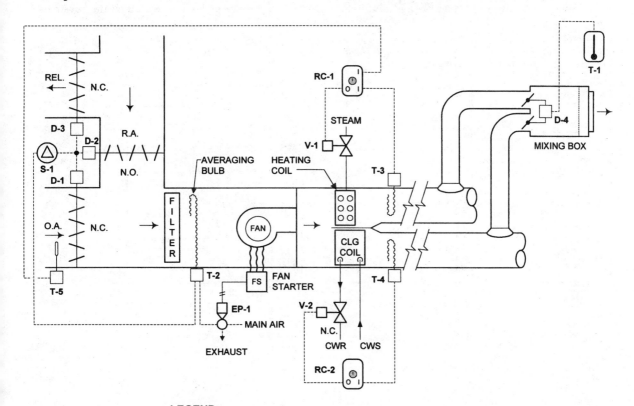

LEGEND:

T-1	=	MODULATING ROOM ZONE THERMOSTAT
T-2	=	MODULATING CAPILLARY DUCT THERMOSTAT
T-3	=	HOT DECK TEMPERATURE SENSOR
T-4	=	COLD DECK TEMPERATURE SENSOR
T-5	=	OUTSIDE AIR MASTER TEMPERATURE SENSOR
V-1	=	N.O. MODULATING STEAM COIL VALVE
V-2	=	N.C. MODULATING 2-WAY COOLING COIL VALVE
D-1	=	OUTSIDE AIR DAMPER MOTOR
D-2	=	RETURN AIR DAMPER MOTOR
D-3	=	RELIEF AIR DAMPER MOTOR
D-4	=	MIXING BOX DAMPER MOTOR
EP-1	=	SOLENOID AIR VALVE
RC-1	=	RECEIVER CONTROLLER
RC-2	=	RECEIVER CONTROLLER
S-1	=	MINIMUM POSITION SWITCH

Dual-Duct Heating and Cooling Air-Handling Unit, Steam Heating Coil, DX Cooling Coil. O.A., R.A., and REL Dampers Using Mixed Air Control Cycle

Whenever the fan runs, the EP-1 is energized and O.A., R.A., and REL dampers are placed under automatic controls. When the fan stops, the dampers return to their normal positions.

Receiver controller RC-1 with O.A. sensor T-5 and hot deck sensor T-3 controls the hot deck valve V-1.

Thermostat T-4 controls PE switch PE-1 on the DX cooling coil. Duct thermostat T-2 controls the O.A., R.A., and REL dampers. T-2 acts through minimum position switch S-1, which can be set to maintain a minimum amount of O.A. as long as the fan is running, regardless of the actions of T-2.

The zone mixing box dampers D-4 are controlled by room thermostats T-1. The mixing boxes are factory-fabricated boxes with dampers that modulate the cold air and hot air supplies to the box to maintain space conditions. There are an unlimited number of boxes that can be used with dual-duct systems.

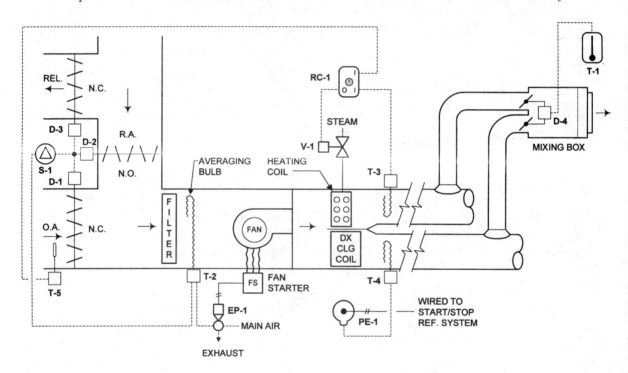

LEGEND:

T-1 = MODULATING ROOM ZONE THERMOSTAT
T-2 = MODULATING CAPILLARY DUCT THERMOSTAT
T-3 = HOT DECK TEMPERATURE SENSOR
T-4 = COLD DECK TEMPERATURE SENSOR
T-5 = OUTSIDE AIR MASTER TEMPERATURE SENSOR
V-1 = N.O. MODULATING STEAM COIL VALVE
D-1 = OUTSIDE AIR DAMPER MOTOR
D-2 = RETURN AIR DAMPER MOTOR
D-3 = RELIEF AIR DAMPER MOTOR
D-4 = MIXING BOX DAMPER MOTOR
EP-1 = SOLENOID AIR VALVE
PE-1 = PRESSURE/ELECTRIC SWITCH
RC-1 = RECEIVER CONTROLLER
S-1 = MINIMUM POSITION SWITCH

Dual-Duct Heating and Cooling Air-Handling Unit, Hot Water Heating Coil, DX Cooling Coil, O.A., R.A., and REL Dampers Using Mixed Air Cycle

Whenever the fan runs, the EP-1 is energized and O.A., R.A., and REL dampers are placed under automatic controls. When the fan stops, the dampers return to their normal positions.

Receiver controller RC-1 with O.A. sensor T-5 and hot deck sensor T-3 controls the hot deck valve V-1. Thermostat T-4 controls PE switch PE-1 on the DX cooling coil. Duct thermostat T-2 controls the O.A., R.A., and REL dampers. T-2 acts through minimum position switch S-1, which can be set to maintain a minimum amount of O.A. as long as the fan is running, regardless of the actions of T-2.

The zone mixing box dampers D-4 are controlled by room thermostats T-1. The mixing boxes are factory-fabricated boxes with dampers that modulate the cold air and hot air supplies to the box to maintain space conditions. There are an unlimited number of boxes that can be used with dual-duct systems.

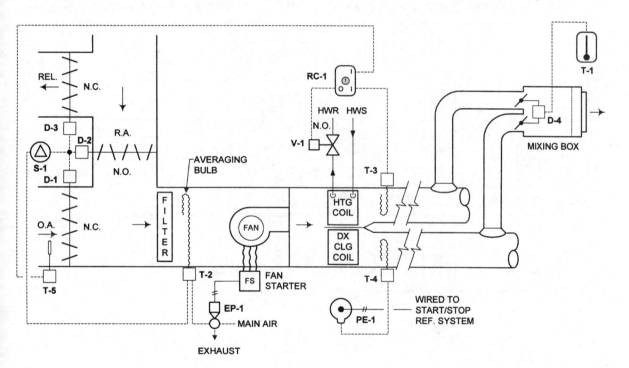

LEGEND:

T-1	= MODULATING ROOM ZONE THERMOSTAT
T-2	= MODULATING CAPILLARY DUCT THERMOSTAT
T-3	= HOT DECK TEMPERATURE SENSOR
T-4	= COLD DECK TEMPERATURE THERMOSTAT
T-5	= OUTSIDE AIR MASTER TEMPERATURE SENSOR
V-1	= N.O. MODULATING 2-WAY HEATING COIL VALVE
D-1	= OUTSIDE AIR DAMPER MOTOR
D-2	= RETURN AIR DAMPER MOTOR
D-3	= RELIEF AIR DAMPER MOTOR
D-4	= MIXING BOX DAMPER MOTOR
EP-1	= SOLENOID AIR VALVE
PE-1	= PRESSURE/ELECTRIC SWITCH
RC-1	= RECEIVER CONTROLLER
S-1	= MINIMUM POSITION SWITCH

Dual-Duct Heating and Cooling Air-Handling Unit, Steam Heating Coil, Chilled Water Cooling Coil, O.A., R.A., and REL Dampers Using Economizer Control Cycle

Whenever the fan runs, the EP-1 is energized and O.A., R.A., and REL dampers are placed under automatic controls. When the fan stops, the dampers return to their normal positions.

Receiver controller RC-1 with O.A. sensor T-6 and hot deck sensor T-3 controls the hot deck valve V-1. Receiver controller RC-2 with sensor T-4 in the cold deck controls valve V-2 on the chilled water coil. Duct thermostat T-2 controls through thermostat T-5 the O.A., R.A., and REL dampers. T-2 and T-5 are economizer thermostats and can allow up to 100% O.A. when conditions are favorable. T-2 and T-5 act through minimum position switch S-1, which can be set to maintain a minimum amount of O.A. as long as the fan is running, regardless of the actions of T-2 and T-5.

The zone mixing box dampers D-4 are controlled by room thermostats T-1. The mixing boxes are factory-fabricated boxes with dampers that modulate the cold air and hot air supplies to the box to maintain space conditions. There are an unlimited number of boxes that can be used with dual-duct systems.

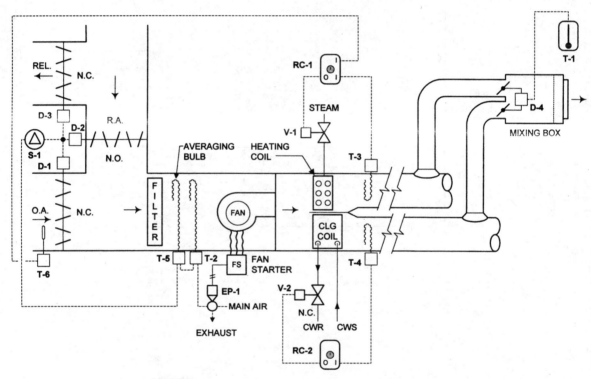

LEGEND:

T-1	= MODULATING ROOM ZONE THERMOSTAT
T-2	= MODULATING CAPILLARY DUCT THERMOSTAT
T-3	= HOT DECK TEMPERATURE SENSOR
T-4	= COLD DECK TEMPERATURE SENSOR
T-5	= MODULATING CAPILLARY DUCT THERMOSTAT
T-6	= OUTSIDE AIR MASTER TEMPERATURE SENSOR
V-1	= N.O. MODULATING STEAM COIL VALVE
V-2	= N.C. MODULATING 2-WAY COOLING COIL VALVE
D-1	= OUTSIDE AIR DAMPER MOTOR
D-2	= RETURN AIR DAMPER MOTOR
D-3	= RELIEF AIR DAMPER MOTOR
D-4	= MIXING BOX DAMPER MOTOR
EP-1	= SOLENOID AIR VALVE
RC-1	= RECEIVER CONTROLLER
RC-2	= RECEIVER CONTROLLER
S-1	= MINIMUM POSITION SWITCH

Dual-Duct Heating and Cooling Air-Handling Unit, Hot Water Heating Coil, DX Cooling Coil, O.A., R.A., and REL Dampers Using Economizer Cycle

Whenever the fan runs, the EP-1 is energized and O.A., R.A., and REL dampers are placed under automatic controls. When the fan stops, the dampers return to their normal positions.

Receiver controller RC-1 with O.A. sensor T-6 and hot deck sensor T-3 controls the hot deck valve V-1. Thermostat T-4 in the cold deck controls PE switch PE-1 on the DX cooling coil. Duct thermostat T-2 controls through thermostat T-5 the O.A., R.A., and REL dampers. T-2 and T-5 are economizer thermostats and can allow up to 100% O.A. when conditions are favorable. T-2 and T-5 act through minimum position switch S-1, which can be set to maintain a minimum amount of O.A. as long as the fan is running, regardless of the actions of T-2 and T-5.

The zone mixing box dampers D-4 are controlled by room thermostats T-1. The mixing boxes are factory-fabricated boxes with dampers that modulate the cold air and hot air supplies to the box to maintain space conditions. There are an unlimited number of boxes that can be used with dual-duct systems.

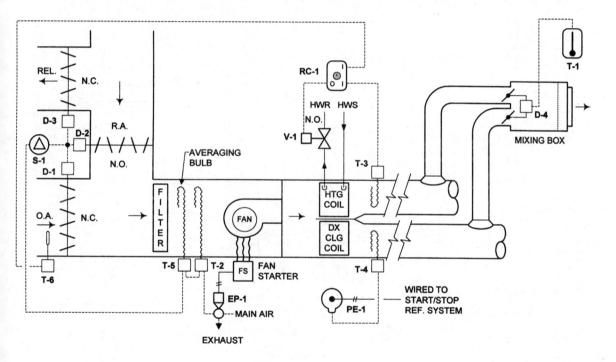

LEGEND:

T-1 = MODULATING ROOM ZONE THERMOSTAT
T-2 = MODULATING CAPILLARY DUCT THERMOSTAT
T-3 = HOT DECK TEMPERATURE SENSOR
T-4 = COLD DECK TEMPERATURE THERMOSTAT
T-5 = MODULATING CAPILLARY DUCT THERMOSTAT
T-6 = OUTSIDE AIR MASTER TEMPERATURE SENSOR
V-1 = N.O. MODULATING 2-WAY HEATING COIL VALVE
D-1 = OUTSIDE AIR DAMPER MOTOR
D-2 = RETURN AIR DAMPER MOTOR
D-3 = RELIEF AIR DAMPER MOTOR
D-4 = MIXING BOX DAMPER MOTOR
EP-1 = SOLENOID AIR VALVE
PE-1 = PRESSURE/ELECTRIC SWITCH
RC-1 = RECEIVER CONTROLLER
S-1 = MINIMUM POSITION SWITCH

Dual-Duct Heating and Cooling Air-Handling Unit, Hot Water Heating Coil, Chilled Water Cooling Coil, O.A., R.A., and REL Dampers Using Economizer Cycle

Whenever the fan runs, the EP-1 is energized and O.A., R.A., and REL dampers are placed under automatic controls. When the fan stops, the dampers return to their normal positions.

Receiver controller RC-1 with O.A. sensor T-6 and hot deck sensor T-3 controls the hot deck valve V-1. Receiver controller RC-2 with sensor T-4 in the cold deck controls valve V-2 on the chilled water coil. Duct thermostat T-2 controls through thermostat T-5 the O.A., R.A., and REL dampers. T-2 and T-5 are economizer thermostats and can allow up to 100% O.A. when conditions are favorable. T-2 and T-5 act through minimum position switch S-1, which can be set to maintain a minimum amount of O.A. as long as the fan is running, regardless of the actions of T-2 and T-5.

The zone mixing box dampers D-4 are controlled by room thermostats T-1. The mixing boxes are factory-fabricated boxes with dampers that modulate the cold air and hot air supplies to the box to maintain space conditions. There are an unlimited number of boxes that can be used with dual-duct systems.

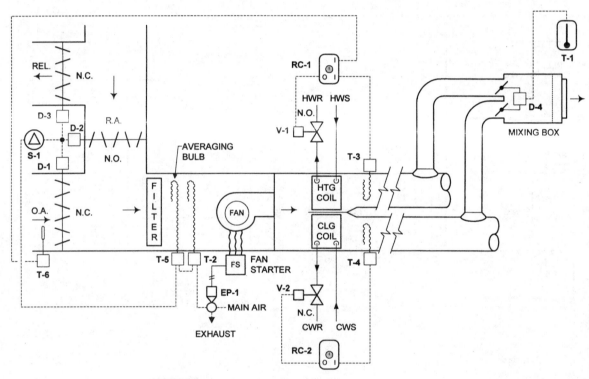

LEGEND:

T-1	=	MODULATING ROOM ZONE THERMOSTAT
T-2	=	MODULATING CAPILLARY DUCT THERMOSTAT
T-3	=	HOT DECK TEMPERATURE SENSOR
T-4	=	COLD DECK TEMPERATURE SENSOR
T-5	=	MODULATING CAPILLARY DUCT THERMOSTAT
T-6	=	OUTSIDE AIR MASTER TEMPERATURE SENSOR
V-1	=	N.O. MODULATING 2-WAY HEATING COIL VALVE
V-2	=	N.C. MODULATING 2-WAY COOLING COIL VALVE
D-1	=	OUTSIDE AIR DAMPER MOTOR
D-2	=	RETURN AIR DAMPER MOTOR
D-3	=	RELIEF AIR DAMPER MOTOR
D-4	=	MIXING BOX DAMPER MOTOR
EP-1	=	SOLENOID AIR VALVE
RC-1	=	RECEIVER CONTROLLER
RC-2	=	RECEIVER CONTROLLER
S-1	=	MINIMUM POSITION SWITCH

Dual-Duct Heating and Cooling Air-Handling Unit, Steam Heating Coil, DX Cooling Coil, O.A., R.A., and REL Dampers Using Economizer Control Cycle

Whenever the fan runs, the EP-1 is energized and O.A., R.A., and REL dampers are placed under automatic controls. When the fan stops, the dampers return to their normal positions.

Receiver controller RC-1 with O.A. sensor T-6 and hot deck sensor T-3 controls the hot deck valve V-1. Thermostat T-4 in the cold deck controls PE switch PE-1 on the DX cooling coil. Duct thermostat T-2 controls through thermostat T-5 the O.A., R.A., and REL dampers. T-2 and T-5 are economizer thermostats and can allow up to 100% O.A. when conditions are favorable. T-2 and T-5 act through minimum position switch S-1, which can be set to maintain a minimum amount of O.A. as long as the fan is running, regardless of the actions of T-2 and T-5.

The zone mixing box dampers D-4 are controlled by room thermostats T-1. The mixing boxes are factory-fabricated boxes with dampers that modulate the cold air and hot air supplies to the box to maintain space conditions. There are an unlimited number of boxes that can be used with dual-duct systems.

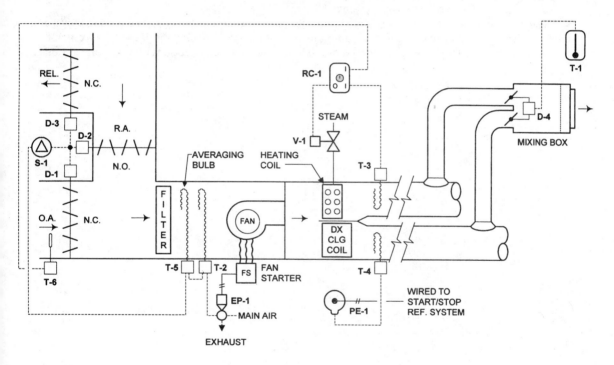

LEGEND:

T-1 = MODULATING ROOM ZONE THERMOSTAT
T-2 = MODULATING CAPILLARY DUCT THERMOSTAT
T-3 = HOT DECK TEMPERATURE SENSOR
T-4 = COLD DECK TEMPERATURE THERMOSTAT
T-5 = MODULATING CAPILLARY DUCT THERMOSTAT
T-6 = OUTSIDE AIR MASTER TEMPERATURE SENSOR
V-1 = N.O. MODULATING STEAM COIL VALVE
D-1 = OUTSIDE AIR DAMPER MOTOR
D-2 = RETURN AIR DAMPER MOTOR
D-3 = RELIEF AIR DAMPER MOTOR
D-4 = MIXING BOX DAMPER MOTOR
EP-1 = SOLENOID AIR VALVE
PE-1 = PRESSURE/ELECTRIC SWITCH
RC-1 = RECEIVER CONTROLLER
S-1 = MINIMUM POSITION SWITCH

Dual-Duct Heating and Cooling Air-Handling Unit, Steam Heating Coil, DX Cooling Coil, O.A., R.A., and REL Dampers Using Economizer Control Cycle Reset of Heating Coil Based upon Flow during Warm-up

Whenever the fan runs, the EP-1 is energized and O.A., R.A., and REL dampers are placed under automatic controls. When the fan stops, the dampers return to their normal positions.

Receiver controller RC-1 with O.A. sensor T-6 and hot deck sensor T-3 control the hot deck valve V-1. A flow controller F-1 sensing the flow in the hot duct can, after being reset by switch S-2 (normal warm-up), control the hot duct valve V-1 during the warm-up period. This action during warm-up tends to balance the flow through the two ducts so that not too much air goes down the hot duct during that part of the cycle. Thermostat T-4 in the cold deck controls PE switch PE-1 on the DX cooling coil. Duct thermostat T-2 controls through thermostat T-5 the O.A., R.A., and REL dampers. T-2 and T-5 are economizer thermostats and can allow up to 100% O.A. when conditions are favorable. T-2 and T-5 act through minimum position switch S-1, which can be set to maintain a minimum amount of O.A. as long as the fan is running, regardless of the actions of T-2 and T-5.

The zone mixing box dampers D-4 are controlled by room thermostats T-1. The mixing boxes are factory-fabricated boxes with dampers that modulate the cold air and hot air supplies to the box to maintain space conditions. There are an unlimited number of boxes that can be used with dual-duct systems.

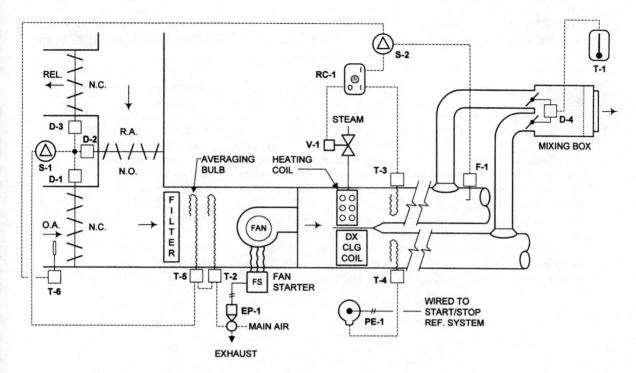

LEGEND:

T-1 = MODULATING ROOM ZONE THERMOSTAT
T-2 = MODULATING CAPILLARY DUCT THERMOSTAT
T-3 = HOT DECK TEMPERATURE SENSOR
T-4 = COLD DECK TEMPERATURE THERMOSTAT
T-5 = MODULATING CAPILLARY DUCT THERMOSTAT
T-6 = OUTSIDE AIR MASTER TEMPERATURE SENSOR
V-1 = N.O. MODULATING STEAM COIL VALVE
D-1 = OUTSIDE AIR DAMPER MOTOR
D-2 = RETURN AIR DAMPER MOTOR
D-3 = RELIEF AIR DAMPER MOTOR
D-4 = MIXING BOX DAMPER MOTOR
EP-1 = SOLENOID AIR VALVE
F-1 = HOT DECK FLOW SENSOR
PE-1 = PRESSURE/ELECTRIC SWITCH
RC-1 = RECEIVER CONTROLLER
S-1 = MINIMUM POSITION SWITCH
S-2 = "NORMAL WARM-UP" SWITCH

Dual-Duct Heating and Cooling Unit. Hot Water Heating Coil, DX Cooling Coil, O.A., R.A., and REL Dampers Using Economizer Control Cycle Reset of Heating Coil Based upon Flow during Warm-up

Whenever the fan runs, the EP-1 is energized and O.A., R.A., and REL dampers are placed under automatic controls. When the fan stops, the dampers return to their normal positions.

Receiver controller RC-1 with O.A. sensor T-6 and hot deck sensor T-3 controls the hot deck valve V-1. A flow controller F-1 sensing the flow in the hot duct can, after being reset by switch S-2 (normal warm-up), control the hot duct valve V-1 during the warm-up period. This action during warm-up tends to balance the flow through the two ducts so that not too much air goes down the hot duct during that part of the cycle. Thermostat T-4 in the cold deck controls PE switch PE-1 on the DX cooling coil. Duct thermostat T-2 controls through thermostat T-5 the O.A., R.A., and REL dampers. T-2 and T-5 are economizer thermostats and can allow up to 100% O.A. when conditions are favorable. T-2 and T-5 act through minimum position switch S-1, which can be set to maintain a minimum amount of O.A. as long as the fan is running, regardless of the actions of T-2 and T-5.

The zone mixing box dampers D-4 are controlled by room thermostats T-1. The mixing boxes are factory-fabricated boxes with dampers that modulate the cold air and hot air supplies to the box to maintain space conditions. There are an unlimited number of boxes that can be used with dual-duct systems.

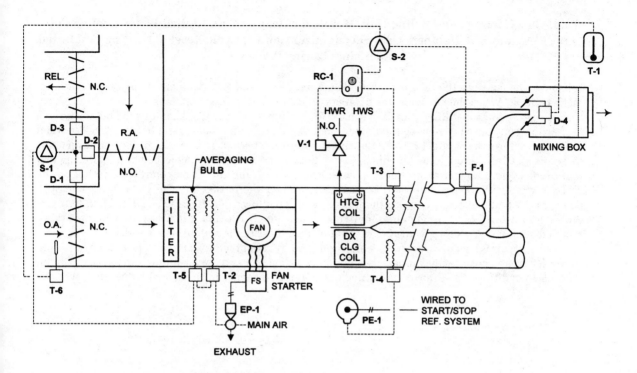

LEGEND:

T-1 = MODULATING ROOM ZONE THERMOSTAT
T-2 = MODULATING CAPILLARY DUCT THERMOSTAT
T-3 = HOT DECK TEMPERATURE SENSOR
T-4 = COLD DECK TEMPERATURE THERMOSTAT
T-5 = MODULATING CAPILLARY DUCT THERMOSTAT
T-6 = OUTSIDE AIR MASTER TEMPERATURE SENSOR
V-1 = N.O. MODULATING 2-WAY HEATING COIL VALVE
D-1 = OUTSIDE AIR DAMPER MOTOR
D-2 = RETURN AIR DAMPER MOTOR
D-3 = RELIEF AIR DAMPER MOTOR
D-4 = MIXING BOX DAMPER MOTOR
EP-1 = SOLENOID AIR VALVE
F-1 = HOT DECK FLOW SENSOR
PE-1 = PRESSURE/ELECTRIC SWITCH
RC-1 = RECEIVER CONTROLLER
S-1 = MINIMUM POSITION SWITCH
S-2 = "NORMAL WARM-UP" SWITCH

Dual-Duct Heating and Cooling Unit. Hot Water Heating Coil, Chilled Water Cooling Coil, O.A., R.A., and REL Dampers Using Economizer Control Cycle Reset of Heating Coil Based upon Flow During Warm-up

Whenever the fan runs, the EP-1 is energized and O.A., R.A., and REL dampers are placed under automatic controls. When the fan stops, the dampers return to their normal positions.

Receiver controller RC-1 with O.A. sensor T-6 and hot deck sensor T-3 controls the hot deck valve V-1. A flow controller F-1 sensing the flow in the hot duct can, after being reset by switch S-2 (normal warm-up), control the hot duct valve V-1 during the warm-up period. This action during warm-up tends to balance the flow through the two ducts so that not too much air goes down the hot duct during that part of the cycle. Receiver controller RC-2 with sensor T-4 in the cold deck controls valve V-2 on the chilled water coil. Duct thermostat T-2 controls through thermostat T-5 the O.A., R.A., and REL dampers. T-2 and T-5 are economizer thermostats and can allow up to 100% O.A. when conditions are favorable. T-2 and T-5 act through minimum position switch S-1, which can be set to maintain a minimum amount of O.A. as long as the fan is running, regardless of the actions of T-2 and T-5.

The zone mixing box dampers D-4 are controlled by room thermostats T-1. The mixing boxes are factory-fabricated boxes with dampers that modulate the cold air and hot air supplies to the box to maintain space conditions. There are an unlimited number of boxes that can be used with dual-duct systems.

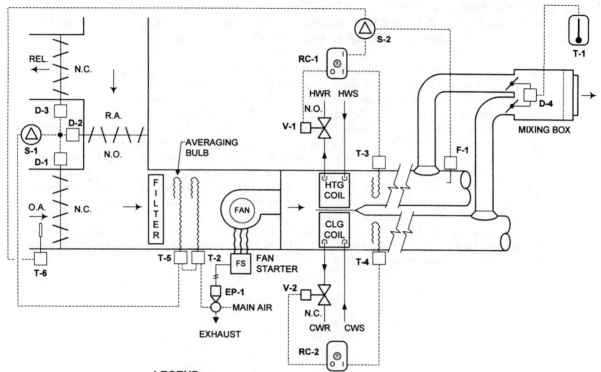

LEGEND:

T-1 = MODULATING ROOM ZONE THERMOSTAT
T-2 = MODULATING CAPILLARY DUCT THERMOSTAT
T-3 = HOT DECK TEMPERATURE SENSOR
T-4 = COLD DECK TEMPERATURE SENSOR
T-5 = MODULATING CAPILLARY DUCT THERMOSTAT
T-6 = OUTSIDE AIR MASTER TEMPERATURE SENSOR
V-1 = N.O. MODULATING 2-WAY HEATING COIL VALVE
V-2 = N.C. MODULATING 2-WAY COOLING COIL VALVE
D-1 = OUTSIDE AIR DAMPER MOTOR
D-2 = RETURN AIR DAMPER MOTOR
D-3 = RELIEF AIR DAMPER MOTOR
D-4 = MIXING BOX DAMPER MOTOR
EP-1 = SOLENOID AIR VALVE
F-1 = HOT DECK FLOW SENSOR
RC-1 = RECEIVER CONTROLLER
RC-2 = RECEIVER CONTROLLER
S-1 = MINIMUM POSITION SWITCH
S-2 = "NORMAL WARM-UP" SWITCH

Dual-Duct Heating and Cooling Air-Handling Unit, Steam Heating Coil, Chilled Water Cooling Coil, O.A., R.A., and REL Dampers Using Economizer Control Cycle Reset of Heating Coil Based upon Flow During Warm-up

Whenever the fan runs, the EP-1 is energized and O.A., R.A., and REL dampers are placed under automatic controls. When the fan stops, the dampers return to their normal positions.

Receiver controller RC-1 with O.A. sensor T-6 and hot deck sensor T-3 controls the hot deck valve V-1. A flow controller F-1 sensing the flow in the hot duct can, after being reset by switch S-2 (normal warm-up), control the hot duct valve V-1 during the warm-up period. This action during warm-up tends to balance the flow through the two ducts so that not too much air goes down the hot duct during that part of the cycle. Receiver controller RC-2 with sensor T-4 in the cold deck controls valve V-2 on the chilled water coil. Duct thermostat T-2 controls through thermostat T-5 the O.A., R.A., and REL dampers. T-2 and T-5 are economizer thermostats and can allow up to 100% O.A. when conditions are favorable. T-2 and T-5 act through minimum position switch S-1, which can be set to maintain a minimum amount of O.A. as long as the fan is running, regardless of the actions of T-2 and T-5.

The zone mixing box dampers D-4 are controlled by room thermostats T-1. The mixing boxes are factory-fabricated boxes with dampers that modulate the cold air and hot air supplies to the box to maintain space conditions. There are an unlimited number of boxes that can be used with dual-duct systems.

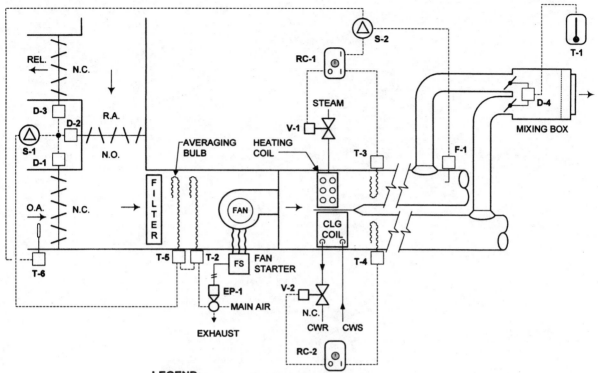

LEGEND:

T-1 = MODULATING ROOM ZONE THERMOSTAT
T-2 = MODULATING CAPILLARY DUCT THERMOSTAT
T-3 = HOT DECK TEMPERATURE SENSOR
T-4 = COLD DECK TEMPERATURE SENSOR
T-5 = MODULATING CAPILLARY DUCT THERMOSTAT
T-6 = OUTSIDE AIR MASTER TEMPERATURE SENSOR
V-1 = N.O. MODULATING STEAM COIL VALVE
V-2 = N.C. MODULATING 2-WAY COOLING COIL VALVE
D-1 = OUTSIDE AIR DAMPER MOTOR
D-2 = RETURN AIR DAMPER MOTOR
D-3 = RELIEF AIR DAMPER MOTOR
D-4 = MIXING BOX DAMPER MOTOR
EP-1 = SOLENOID AIR VALVE
F-1 = HOT DECK FLOW SENSOR
RC-1 = RECEIVER CONTROLLER
RC-2 = RECEIVER CONTROLLER
S-1 = MINIMUM POSITION SWITCH
S-2 = "NORMAL WARM-UP" SWITCH

CHAPTER 18

VAV Systems

VAV Heating and Cooling Air-Handling Unit, Steam Heating, Chilled Water Cooling, with R.A. Fan and Inlet Vane Controls on Supply and Return Fan, O.A., R.A., and REL Dampers Using Economizer Control Cycle, Variable Volume Fan Capacity Control through Volumetric Matching of Fans

Whenever the fan runs, the EP-1 is energized and O.A., R.A., and REL dampers are placed under automatic controls. When the fan stops, the dampers return to their normal positions.

Receiver controller RC-1 with sensors T-1 in the O.A. and sensor T-2 in the mix chamber controls the O.A., R.A., and REL dampers. The system acts as an economizer control cycle. R-1 relay allows the O.A. sensor T-6 to override the actions of T-2. Minimum pressure selector switch S-1 through relay R-2 controls the dampers to allow a minimum amount of ventilation air, regardless of the actions of the economizer thermostats. Receiver controller RC-3 with static pressure sensor SP-1 controls the supply fan inlet vanes. Receiver controller RC-2 with sensor T-3 in the fan discharge controls valves V-1 and V-2 on the heating and cooling coils in sequence.

Flow controller F-1 resets the control point of receiver controller RC-4, which through flow controller F-2 controls the inlet vanes on the return fan.

This above system is designed to "match" the supply and return fans so that they act as a team and maintain space static pressure while at the same time modulating down from full capacity to reduced capacity as the VAV boxes in the system become satisfied.

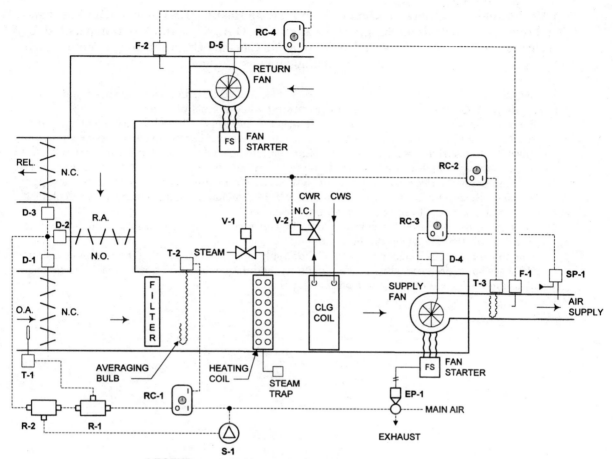

LEGEND:

T-1   = MODULATING CAPILLARY THERMOSTAT
T-2   = MODULATING CAPILLARY DUCT SENSOR
T-3   = MODULATING CAPILLARY DUCT SENSOR
V-1   = N.O. MODULATING STEAM COIL VALVE
V-2   = N.C. MODULATING 2-WAY COOLING COIL VALVE
D-1   = OUTSIDE AIR DAMPER MOTOR
D-2   = RETURN AIR DAMPER MOTOR
D-3   = RELIEF AIR DAMPER MOTOR
D-4   = SUPPLY FAN INLET VANE DAMPER MOTOR
D-5   = RETURN FAN INLET VANE DAMPER MOTOR
EP-1 = SOLENOID AIR VALVE
F-1   = SUPPLY AIR FLOW SENSOR
F-2   = RETURN AIR FLOW SENSOR
R-1   = HIGH SIGNAL SELECTOR
R-2   = HIGH SIGNAL SELECTOR
RC-1 = RECEIVER CONTROLLER
RC-2 = RECEIVER CONTROLLER
RC-3 = RECEIVER CONTROLLER
RC-4 = RECEIVER CONTROLLER
S-1   = MINIMUM POSITION SWITCH
SP-1 = STATIC PRESSURE SENSOR

VAV Heating and Cooling Air-Handling Unit, Steam Heating, DX Cooling, with R.A. Fan and Inlet Vane Controls on Supply and Return Fan, O.A., R.A., and REL Dampers Using Economizer Control Cycle, Variable Volume Fan Capacity Control through Volumetric Matching of Fans

Whenever the fan runs, the EP-1 is energized and O.A., R.A., and REL dampers are placed under automatic controls. When the fan stops, the dampers return to their normal positions.

Receiver controller RC-1 with sensors T-1 in the O.A. and sensor T-2 in the mix chamber controls the O.A., R.A., and REL dampers. The system acts as an economizer control cycle. R-1 relay allows the O.A. sensor T-6 to override the actions of T-2. Minimum pressure selector switch S-1 through relay R-2 controls the dampers to allow a minimum amount of ventilation air, regardless of the actions of the economizer thermostats. Receiver controller RC-3 with static pressure sensor SP-1 controls the supply fan inlet vanes. Receiver controller RC-2 with sensor T-3 in the fan discharge controls PE switch PE-1 on the DX cooling coil and heating coil valve V-1 in sequence.

Flow controller F-1 resets the control point of receiver controller RC-4, which through flow controller F-2 controls the inlet vanes on the return fan.

This above system is designed to "match" the supply and return fans so that they act as a team and maintain space static pressure while at the same time modulating down from full capacity to reduced capacity as the VAV boxes in the system become satisfied.

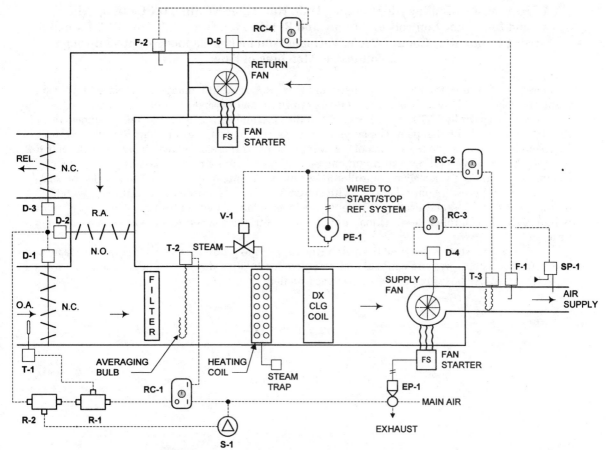

LEGEND:

T-1	=	MODULATING CAPILLARY THERMOSTAT
T-2	=	MODULATING CAPILLARY DUCT SENSOR
T-3	=	MODULATING CAPILLARY DUCT SENSOR
V-1	=	N.O. MODULATING STEAM COIL VALVE
D-1	=	OUTSIDE AIR DAMPER MOTOR
D-2	=	RETURN AIR DAMPER MOTOR
D-3	=	RELIEF AIR DAMPER MOTOR
D-4	=	SUPPLY FAN INLET VANE DAMPER MOTOR
D-5	=	RETURN FAN INLET VANE DAMPER MOTOR
EP-1	=	SOLENOID AIR VALVE
F-1	=	SUPPLY AIR FLOW SENSOR
F-2	=	RETURN AIR FLOW SENSOR
PE-1	=	PRESSURE/ELECTRIC SWITCH
R-1	=	HIGH SIGNAL SELECTOR
R-2	=	HIGH SIGNAL SELECTOR
RC-1	=	RECEIVER CONTROLLER
RC-2	=	RECEIVER CONTROLLER
RC-3	=	RECEIVER CONTROLLER
RC-4	=	RECEIVER CONTROLLER
S-1	=	MINIMUM POSITION SWITCH
SP-1	=	STATIC PRESSURE SENSOR

VAV Heating and Cooling Air-Handling Unit, Hot Water Heating, DX Cooling, with R.A. Fan and Inlet Vane Controls on Supply and Return Fan, O.A., R.A., and REL Dampers Using Economizer Control Cycle, Variable Volume Fan Capacity Control through Volumetric Matching of Fans

Whenever the fan runs, the EP-1 is energized and O.A., R.A., and REL dampers are placed under automatic controls. When the fan stops, the dampers return to their normal positions.

Receiver controller RC-1 with sensors T-1 in the O.A. and sensor T-2 in the mix chamber controls the O.A., R.A., and REL dampers. The system acts as an economizer control cycle. R-1 relay allows the O.A. sensor T-1 to override the actions of T-2. Minimum pressure selector switch S-1 through relay R-2 controls the dampers to allow a minimum amount of ventilation air, regardless of the actions of the economizer thermostats. Receiver controller RC-3 with static pressure sensor SP-1 controls the supply fan inlet vanes. Receiver controller RC-2 with sensor T-3 in the fan discharge controls PE switch PE-1 on the DX cooling coil and heating coil valve V-1 in sequence. Flow controller F-1 resets the control point of receiver controller RC-4, which through flow controller F-2 controls the inlet vanes on the return fan.

This above system is designed to "match" the supply and return fans so that they act as a team and maintain space static pressure while at the same time modulating down from full capacity to reduced capacity as the VAV boxes in the system become satisfied.

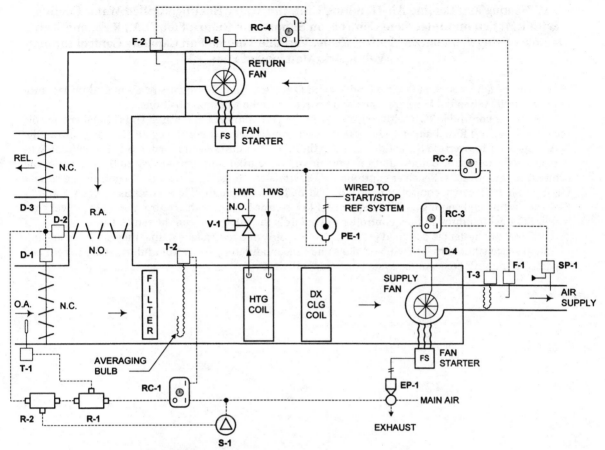

LEGEND:

T-1 = MODULATING CAPILLARY THERMOSTAT
T-2 = MODULATING CAPILLARY DUCT SENSOR
T-3 = MODULATING CAPILLARY DUCT SENSOR
V-1 = N.O. MODULATING 2-WAY HEATING COIL VALVE
D-1 = OUTSIDE AIR DAMPER MOTOR
D-2 = RETURN AIR DAMPER MOTOR
D-3 = RELIEF AIR DAMPER MOTOR
D-4 = SUPPLY FAN INLET VANE DAMPER MOTOR
D-5 = RETURN FAN INLET VANE DAMPER MOTOR
EP-1 = SOLENOID AIR VALVE
F-1 = SUPPLY AIR FLOW SENSOR
F-2 = RETURN AIR FLOW SENSOR
PE-1 = PRESSURE/ELECTRIC SWITCH
R-1 = HIGH SIGNAL SELECTOR
R-2 = HIGH SIGNAL SELECTOR
RC-1 = RECEIVER CONTROLLER
RC-2 = RECEIVER CONTROLLER
RC-3 = RECEIVER CONTROLLER
RC-4 = RECEIVER CONTROLLER
S-1 = MINIMUM POSITION SWITCH
SP-1 = STATIC PRESSURE SENSOR

VAV Heating and Cooling Air-Handling Unit, Hot Water Heating, Chilled Water Cooling, with R.A. Fan and Inlet Vane Controls on Supply and Return Fan, O.A., R.A., and REL Dampers Using Economizer Control Cycle, Variable Volume Fan Capacity Control through Volumetric Matching of Fans

Whenever the fan runs, the EP-1 is energized and O.A., R.A., and REL dampers are placed under automatic controls. When the fan stops, the dampers return to their normal positions.

Receiver controller RC-1 with sensors T-1 in the O.A. and sensor T-2 in the mix chamber controls the O.A., R.A., and REL dampers. The system acts as an economizer control cycle. R-1 relay allows the O.A. sensor T-1 to override the actions of T-2. Minimum pressure selector switch S-1, through relay R-2, controls the dampers to allow a minimum amount of ventilation air, regardless of the actions of the economizer thermostats. Receiver controller RC-3 with static pressure sensor SP-1 controls the supply fan inlet vanes. Receiver controller RC-2 with sensor T-3 in the fan discharge controls valves V-1 and V-2 on the heating and cooling coils in sequence. Flow sensor F-1 resets the control point of receiver controller RC-4, which through flow controller F-2 controls the inlet vanes on the return fan.

This above system is designed to "match" the supply and return fans so that they act as a team and maintain space static pressure while at the same time modulating down from full capacity to reduced capacity as the VAV boxes in the system become satisfied.

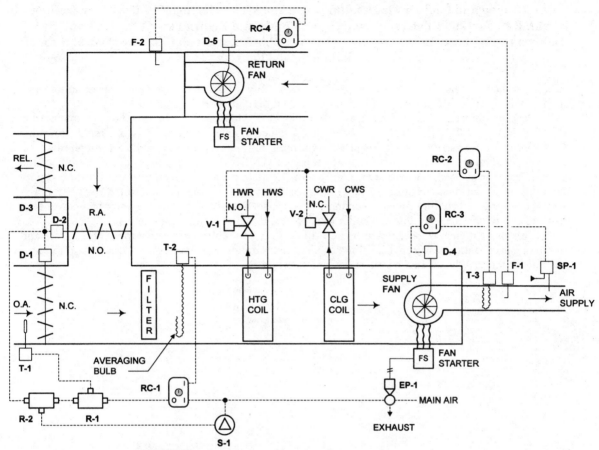

LEGEND:

T-1 = MODULATING CAPILLARY THERMOSTAT
T-2 = MODULATING CAPILLARY DUCT SENSOR
T-3 = MODULATING CAPILLARY DUCT SENSOR
V-1 = N.O. MODULATING 2-WAY HEATING COIL VALVE
V-2 = N.C. MODULATING 2-WAY COOLING COIL VALVE
D-1 = OUTSIDE AIR DAMPER MOTOR
D-2 = RETURN AIR DAMPER MOTOR
D-3 = RELIEF AIR DAMPER MOTOR
D-4 = SUPPLY FAN INLET VANE DAMPER MOTOR
D-5 = RETURN FAN INLET VANE DAMPER MOTOR
EP-1 = SOLENOID AIR VALVE
F-1 = SUPPLY AIR FLOW SENSOR
F-2 = RETURN AIR FLOW SENSOR
R-1 = HIGH SIGNAL SELECTOR
R-2 = HIGH SIGNAL SELECTOR
RC-1 = RECEIVER CONTROLLER
RC-2 = RECEIVER CONTROLLER
RC-3 = RECEIVER CONTROLLER
RC-4 = RECEIVER CONTROLLER
S-1 = MINIMUM POSITION SWITCH
SP-1 = STATIC PRESSURE SENSOR

VAV Heating and Cooling Air-Handling Unit, Hot Water Heating, Chilled Water Cooling, with R.A. Fan and Inlet Vane Controls on Supply and Return Fan, O.A., R.A., and REL Dampers Using Economizer Control Cycle, Variable Volume Fan Capacity Control through Balancing the Fans

Whenever the fan runs, the EP-1 is energized and O.A., R.A., and REL dampers are placed under automatic controls. When the fan stops, the dampers return to their normal positions.

Receiver controller RC-1 with sensors T-1 in the O.A. and sensor T-2 in the mix chamber controls the O.A., R.A., and REL dampers. The system acts as an economizer control cycle. R-1 relay allows the O.A. sensor T-1 to override the actions of T-2. Minimum pressure selector switch S-1 through relay R-2 controls the dampers to allow a minimum amount of ventilation air, regardless of the actions of the economizer thermostats. Receiver controller RC-3 with static pressure sensor SP-1 controls the supply fan inlet vanes. Receiver controller RC-2 with sensor T-3 in the fan discharge controls valves V-1 and V-2 on the heating and cooling coils in sequence. Flow sensor F-1 resets the control point of receiver controller RC-4, which through flow controller F-2 controls the inlet vanes on the return fan.

This above system is designed to "match" the supply and return fans so that they act as a team and maintain space static pressure while at the same time modulating down from full capacity to reduced capacity as the VAV boxes in the system become satisfied.

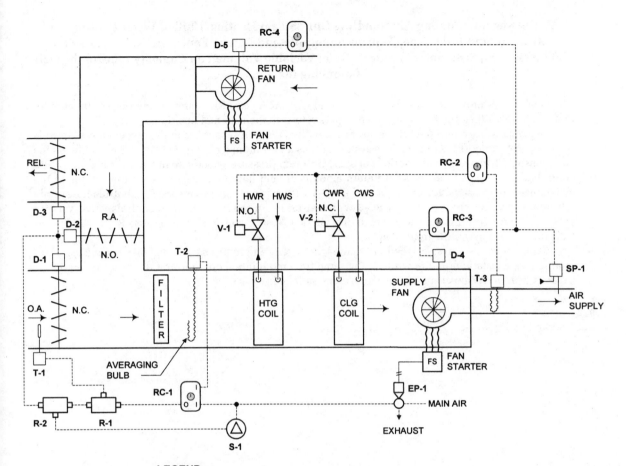

LEGEND:

T-1	=	MODULATING CAPILLARY THERMOSTAT
T-2	=	MODULATING CAPILLARY DUCT SENSOR
T-3	=	MODULATING CAPILLARY DUCT SENSOR
V-1	=	N.O. MODULATING 2-WAY HEATING COIL VALVE
V-2	=	N.C. MODULATING 2-WAY COOLING COIL VALVE
D-1	=	OUTSIDE AIR DAMPER MOTOR
D-2	=	RETURN AIR DAMPER MOTOR
D-3	=	RELIEF AIR DAMPER MOTOR
D-4	=	SUPPLY FAN INLET VANE DAMPER MOTOR
D-5	=	RETURN FAN INLET VANE DAMPER MOTOR
EP-1	=	SOLENOID AIR VALVE
R-1	=	HIGH SIGNAL SELECTOR
R-2	=	HIGH SIGNAL SELECTOR
RC-1	=	RECEIVER CONTROLLER
RC-2	=	RECEIVER CONTROLLER
RC-3	=	RECEIVER CONTROLLER
RC-4	=	RECEIVER CONTROLLER
S-1	=	MINIMUM POSITION SWITCH
SP-1	=	STATIC PRESSURE SENSOR

VAV Heating and Cooling Air-Handling Unit, Steam Heating, Chilled Water Cooling, with R.A. Fan and Inlet Vane Controls on Supply and Return Fan, O.A., R.A., and REL Dampers Using Economizer Control Cycle, Variable Volume Fan Capacity Control through Balancing the Fans

Whenever the fan runs, the EP-1 is energized and O.A., R.A., and REL dampers are placed under automatic controls. When the fan stops, the dampers return to their normal positions.

Receiver controller RC-1 with sensors T-1 in the O.A. and sensor T-2 in the mix chamber controls the O.A., R.A., and REL dampers. The system acts as an economizer control cycle. R-1 relay allows the O.A. sensor T-1 to override the actions of T-2. Minimum pressure selector switch S-1 through relay R-2 controls the dampers to allow a minimum amount of ventilation air, regardless of the actions of the economizer thermostats. Receiver controller RC-3 with static pressure sensor SP-1 controls the supply fan inlet vanes. Receiver controller RC-2 with sensor T-3 in the fan discharge controls valve V-1 and valve V-2 on the heating and cooling coils. The static pressure sensor SP-1 also resets RC-4, which controls the inlet vanes on the return fan.

This above system is designed to "match" the supply and return fans so that they act as a team and maintain space static pressure while at the same time modulating down from full capacity to reduced capacity as the VAV boxes in the system become satisfied.

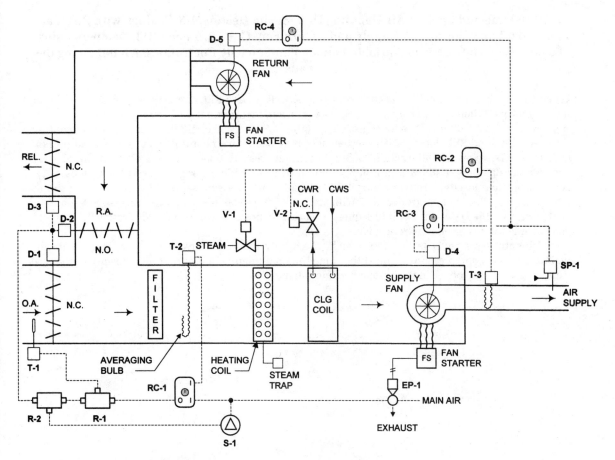

LEGEND:

T-1	= MODULATING CAPILLARY THERMOSTAT
T-2	= MODULATING CAPILLARY DUCT SENSOR
T-3	= MODULATING CAPILLARY DUCT SENSOR
V-1	= N.O. MODULATING STEAM COIL VALVE
V-2	= N.C. MODULATING 2-WAY COOLING COIL VALVE
D-1	= OUTSIDE AIR DAMPER MOTOR
D-2	= RETURN AIR DAMPER MOTOR
D-3	= RELIEF AIR DAMPER MOTOR
D-4	= SUPPLY FAN INLET VANE DAMPER MOTOR
D-5	= RETURN FAN INLET VANE DAMPER MOTOR
EP-1	= SOLENOID AIR VALVE
R-1	= HIGH SIGNAL SELECTOR
R-2	= HIGH SIGNAL SELECTOR
RC-1	= RECEIVER CONTROLLER
RC-2	= RECEIVER CONTROLLER
RC-3	= RECEIVER CONTROLLER
RC-4	= RECEIVER CONTROLLER
S-1	= MINIMUM POSITION SWITCH
SP-1	= STATIC PRESSURE SENSOR

VAV Heating and Cooling Air-Handling Unit, Steam Heating, DX Cooling, with R.A. Fan and Inlet Vane Controls on Supply and Return Fan, O.A., R.A., and REL Dampers Using Economizer Control Cycle, Variable Volume Fan Capacity Control through Balancing the Fans

Whenever the fan runs, the EP-1 is energized and O.A., R.A., and REL dampers are placed under automatic controls. When the fan stops, the dampers return to their normal positions.

Receiver controller RC-1 with sensors T-1 in the O.A. and sensor T-2 in the mix chamber controls the O.A., R.A., and REL dampers. The system acts as an economizer control cycle. R-1 relay allows the O.A. sensor T-1 to override the actions of T-2. Minimum pressure selector switch S-1 through relay R-2 controls the dampers to allow a minimum amount of ventilation air, regardless of the actions of the economizer thermostats. Receiver controller RC-3 with static pressure sensor SP-1 controls the supply fan inlet vanes. Receiver controller RC-2 with sensor T-3 in the fan discharge controls valve V-1 and PE switch PE-1 on the DX cooling coil in sequence. The static pressure sensor SP-1 also resets RC-4, which controls the inlet vanes on the return fan.

This above system is designed to "match" the supply and return fans so that they act as a team and maintain space static pressure while at the same time modulating down from full capacity to reduced capacity as the VAV boxes in the system become satisfied.

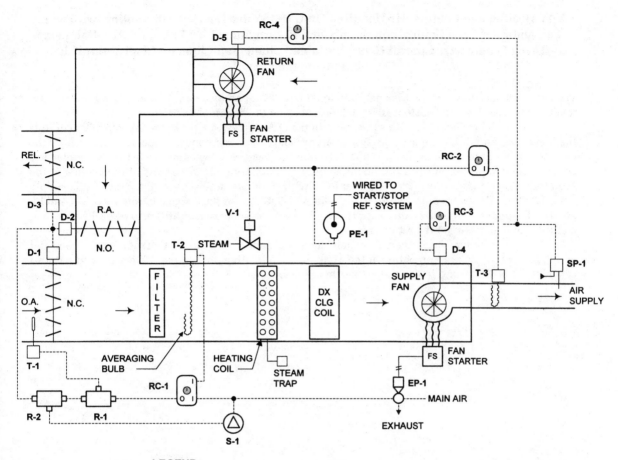

LEGEND:

T-1   = MODULATING CAPILLARY THERMOSTAT
T-2   = MODULATING CAPILLARY DUCT SENSOR
T-3   = MODULATING CAPILLARY DUCT SENSOR
V-1   = N.O. MODULATING STEAM COIL VALVE
D-1   = OUTSIDE AIR DAMPER MOTOR
D-2   = RETURN AIR DAMPER MOTOR
D-3   = RELIEF AIR DAMPER MOTOR
D-4   = SUPPLY FAN INLET VANE DAMPER MOTOR
D-5   = RETURN FAN INLET VANE DAMPER MOTOR
EP-1 = SOLENOID AIR VALVE
PE-1 = PRESSURE/ELECTRIC SWITCH
R-1   = HIGH SIGNAL SELECTOR
R-2   = HIGH SIGNAL SELECTOR
RC-1 = RECEIVER CONTROLLER
RC-2 = RECEIVER CONTROLLER
RC-3 = RECEIVER CONTROLLER
RC-4 = RECEIVER CONTROLLER
S-1   = MINIMUM POSITION SWITCH
SP-1 = STATIC PRESSURE SENSOR

VAV Heating and Cooling Air-Handling Unit, Hot Water Heating, DX Cooling, with R.A. Fan and Inlet Vane Controls on Supply and Return Fan, O.A., R.A., and REL Dampers Using Economizer Control Cycle, Variable Volume Fan Capacity Control through Balancing the Fans

Whenever the fan runs, the EP-1 is energized and O.A., R.A., and REL dampers are placed under automatic controls. When the fan stops, the dampers return to their normal positions.

Receiver controller RC-1 with sensors T-1 in the O.A. and sensor T-2 in the mix chamber controls the O.A., R.A., and REL dampers. The system acts as an economizer control cycle. R-1 relay allows the O.A. sensor T-1 to override the actions of T-2. Minimum pressure selector switch S-1 through relay R-2 controls the dampers to allow a minimum amount of ventilation air, regardless of the actions of the economizer thermostats. Receiver controller RC-3 with static pressure sensor SP-1 controls the supply fan inlet vanes. Receiver controller RC-2 with sensor T-3 in the fan discharge controls valve V-1 and PE switch PE-1 on the DX cooling coil in sequence. The static pressure sensor SP-1 also resets RC-4, which controls the inlet vanes on the return fan.

This above system is designed to "match" the supply and return fans so that they act as a team and maintain space static pressure while at the same time modulating down from full capacity to reduced capacity as the VAV boxes in the system become satisfied.

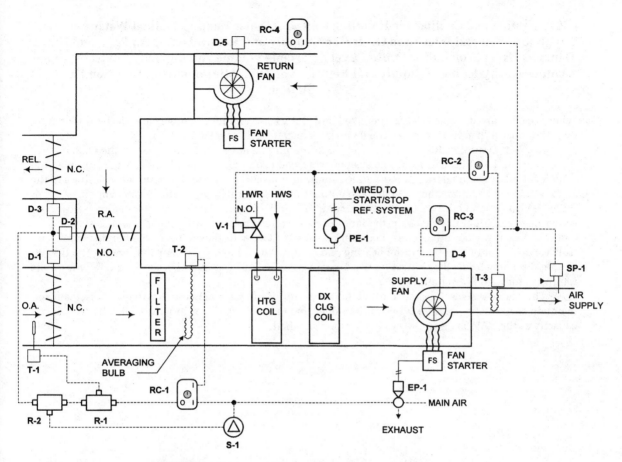

LEGEND:

T-1 = MODULATING CAPILLARY THERMOSTAT
T-2 = MODULATING CAPILLARY DUCT SENSOR
T-3 = MODULATING CAPILLARY DUCT SENSOR
V-1 = N.O. MODULATING 2-WAY HEATING COIL VALVE
D-1 = OUTSIDE AIR DAMPER MOTOR
D-2 = RETURN AIR DAMPER MOTOR
D-3 = RELIEF AIR DAMPER MOTOR
D-4 = SUPPLY FAN INLET VANE DAMPER MOTOR
D-5 = RETURN FAN INLET VANE DAMPER MOTOR
EP-1 = SOLENOID AIR VALVE
PE-1 = PRESSURE/ELECTRIC SWITCH
R-1 = HIGH SIGNAL SELECTOR
R-2 = HIGH SIGNAL SELECTOR
RC-1 = RECEIVER CONTROLLER
RC-2 = RECEIVER CONTROLLER
RC-3 = RECEIVER CONTROLLER
RC-4 = RECEIVER CONTROLLER
S-1 = MINIMUM POSITION SWITCH
SP-1 = STATIC PRESSURE SENSOR

VAV Heating and Cooling Air-Handling Unit, Hot Water Heating, Chilled Water Cooling, with R.A. Fan and Inlet Vane Controls on Supply and Return Fan, O.A., R.A., and REL Dampers Using Economizer Control Cycle, Variable Volume Fan Capacity Control through Volumetric Matching of Supply and Return Fans, with Minimum O.A. Injection Fan for Ventilation

Whenever the fan runs, the EP-1 is energized and O.A., R.A., and REL dampers are placed under automatic controls. When the fan stops, the dampers return to their normal positions.

Receiver controller RC-1 with sensor T-2 in the mix chamber and sensor T-1 in the O.A. control the O.A., R.A., and REL dampers. The system acts as an economizer control cycle. Minimum position switch S-1 allows through relays R-1 and R-2 a minimum setting for the O.A. damper regardless of the actions of the economizer thermostats. The outdoor air is split into two sections. One contains a minimum O.A. injection fan to ensure a fixed amount of ventilation air. The fan is operated by a flow controller operating the inlet vanes on the fan. Receiver controller RC-3 with static pressure sensor SP-1 controls the supply fan inlet vanes. Receiver controller RC-2 with sensor T-3 in the fan discharge controls valve V-1 and valve V-2 on the heating and cooling coils. Flow controller F-3 sensing the flow in the fan discharge resets the control point of receiver controller RC-4 that also senses the flow from the return fan through flow controller F-2.

This above system is designed to "match" the supply and return fans so that they act as a team and maintain space static pressure while at the same time modulating down from full capacity to reduced capacity as the VAV boxes in the system become satisfied.

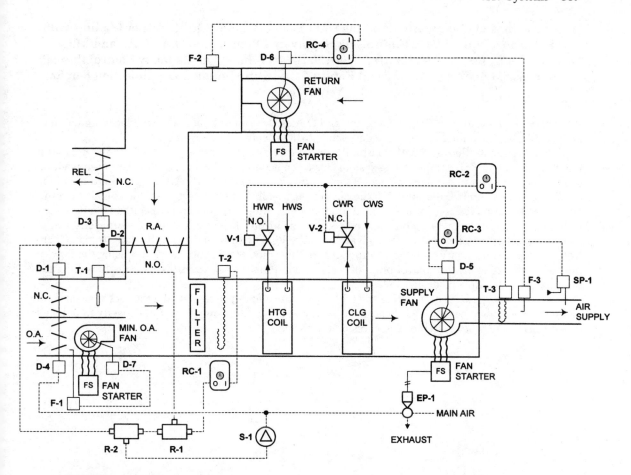

LEGEND:

T-1	= MODULATING CAPILLARY THERMOSTAT		EP-1	=	SOLENOID AIR VALVE
T-2	= MODULATING CAPILLARY DUCT SENSOR		F-1	=	FLOW CONTROLLER
T-3	= MODULATING CAPILLARY DUCT SENSOR		F-2	=	RETURN AIR FLOW SENSOR
V-1	= N.O. MODULATING 2-WAY HEATING COIL VALVE		F-3	=	SUPPLY AIR FLOW SENSOR
V-2	= N.C. MODULATING 2-WAY COOLING COIL VALVE		R-1	=	HIGH SIGNAL SELECTOR
D-1	= MAXIMUM OUTSIDE AIR DAMPER MOTOR		R-2	=	HIGH SIGNAL SELECTOR
D-2	= RETURN AIR DAMPER MOTOR		RC-1	=	RECEIVER CONTROLLER
D-3	= RELIEF AIR DAMPER MOTOR		RC-2	=	RECEIVER CONTROLLER
D-4	= MINIMUM OUTSIDE AIR DAMPER MOTOR		RC-3	=	RECEIVER CONTROLLER
D-5	= SUPPLY FAN INLET VANE DAMPER MOTOR		RC-4	=	RECEIVER CONTROLLER
D-6	= RETURN FAN INLET VANE DAMPER MOTOR		S-1	=	MINIMUM POSITION SWITCH
D-7	= MIN. O.A. INJECTION FAN INLET VANE DAMPER MOTOR		SP-1	=	STATIC PRESSURE SENSOR

VAV Heating and Cooling Air-Handling Unit, Steam Heating, Chilled Water Cooling. with R.A. Fan and Inlet Vane Controls on Supply and Return Fan, O.A., R.A., and REL Dampers Using Economizer Control Cycle, Variable Volume Fan Capacity Control through Volumetric Matching of Supply and Return Fans, with Minimum O.A. Injection Fan for Ventilation

Whenever the fan runs, the EP-1 is energized and O.A., R.A., and REL dampers are placed under automatic controls. When the fan stops, the dampers return to their normal positions.

Receiver controller RC-1 with sensor T-2 in the mix chamber and sensor T-1 in the O.A. control the O.A., R.A., and REL dampers. The system acts as an economizer control cycle. Minimum position switch S-1 allows through relays R-1 and R-2 a minimum setting for the O.A. damper regardless of the actions of the economizer thermostats. The outdoor air is split into two sections. One contains a minimum O.A. injection fan to insure a fixed amount of ventilation air. The fan is operated by a flow controller operating the inlet vanes on the fan. Receiver controller RC-3 with static pressure sensor SP-1 controls the supply fan inlet vanes. Receiver controller RC-2 with sensor T-3 in the fan discharge controls valve V-1 and valve V-2 on the heating and cooling coils. Flow controller F-3 sensing the flow in the fan discharge resets the control point of receiver controller RC-4 that also senses the flow from the return fan through flow controller F-2.

This above system is designed to "match" the supply and return fans so that they act as a team and maintain space static pressure while at the same time modulating down from full capacity to reduced capacity as the VAV boxes in the system become satisfied.

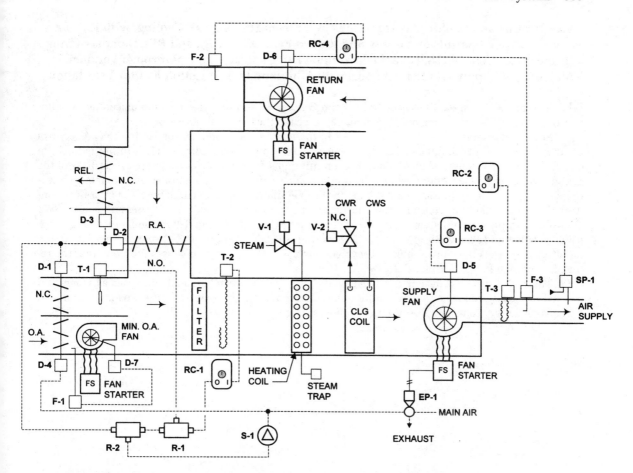

LEGEND:

T-1	=	MODULATING CAPILLARY THERMOSTAT
T-2	=	MODULATING CAPILLARY DUCT SENSOR
T-3	=	MODULATING CAPILLARY DUCT SENSOR
V-1	=	N.O. MODULATING STEAM COIL VALVE
V-2	=	N.C. MODULATING 2-WAY COOLING COIL VALVE
D-1	=	MAXIMUM OUTSIDE AIR DAMPER MOTOR
D-2	=	RETURN AIR DAMPER MOTOR
D-3	=	RELIEF AIR DAMPER MOTOR
D-4	=	MINIMUM OUTSIDE AIR DAMPER MOTOR
D-5	=	SUPPLY FAN INLET VANE DAMPER MOTOR
D-6	=	RETURN FAN INLET VANE DAMPER MOTOR
D-7	=	MIN. O.A. INJECTION FAN INLET VANE DAMPER MOTOR

EP-1	=	SOLENOID AIR VALVE
F-1	=	FLOW CONTROLLER
F-2	=	RETURN AIR FLOW SENSOR
F-3	=	SUPPLY AIR FLOW SENSOR
R-1	=	HIGH SIGNAL SELECTOR
R-2	=	HIGH SIGNAL SELECTOR
RC-1	=	RECEIVER CONTROLLER
RC-2	=	RECEIVER CONTROLLER
RC-3	=	RECEIVER CONTROLLER
RC-4	=	RECEIVER CONTROLLER
S-1	=	MINIMUM POSITION SWITCH
SP-1	=	STATIC PRESSURE SENSOR

VAV Heating and Cooling Air-Handling Unit, Steam Heating, DX Cooling, with R.A. Fan and Inlet Vane Controls on Supply and Return Fan, O.A., R.A., and REL Dampers Using Economizer Control Cycle, Variable Volume Fan Capacity Control through Volumetric Matching of Supply and Return Fans, with Minimum O.A. Injection Fan for Ventilation

Whenever the fan runs, the EP-1 is energized and O.A., R.A., and REL dampers are placed under automatic controls. When the fan stops, the dampers return to their normal positions.

Receiver controller RC-1 with sensor T-2 in the mix chamber and sensor T-1 in the O.A. control the O.A., R.A., and REL dampers. The system acts as an economizer control cycle. Minimum position switch S-1 allows through relays R-1 and R-2 a minimum setting for the O.A. damper regardless of the actions of the economizer thermostats. The outdoor air is split into two sections. One contains a minimum O.A. injection fan to insure a fixed amount of ventilation air. The fan is operated by a flow controller operating the inlet vanes on the fan. Receiver controller RC-3 with static pressure sensor SP-1 controls the supply fan inlet vanes. Receiver controller RC-2 with sensor T-3 in the fan discharge controls valve V-1 and PE switch PE-1 on the DX cooling coil. Flow controller F-3 sensing the flow in the fan discharge resets the control point of receiver controller RC-4 that also senses the flow from the return fan through flow controller F-2.

This above system is designed to "match" the supply and return fans so that they act as a team and maintain space static pressure while at the same time modulating down from full capacity to reduced capacity as the VAV boxes in the system become satisfied.

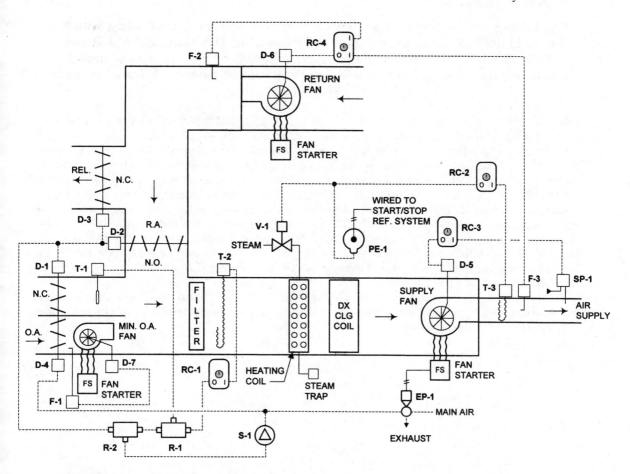

LEGEND:

T-1 = MODULATING CAPILLARY THERMOSTAT
T-2 = MODULATING CAPILLARY DUCT SENSOR
T-3 = MODULATING CAPILLARY DUCT SENSOR
V-1 = N.O. MODULATING STEAM COIL VALVE
D-1 = MAXIMUM OUTSIDE AIR DAMPER MOTOR
D-2 = RETURN AIR DAMPER MOTOR
D-3 = RELIEF AIR DAMPER MOTOR
D-4 = MINIMUM OUTSIDE AIR DAMPER MOTOR
D-5 = SUPPLY FAN INLET VANE DAMPER MOTOR
D-6 = RETURN FAN INLET VANE DAMPER MOTOR
D-7 = MIN. O.A. INJECTION FAN INLET VANE DAMPER MOTOR
EP-1 = SOLENOID AIR VALVE

PE-1 = PRESSURE/ELECTRIC SWITCH
F-1 = FLOW CONTROLLER
F-2 = RETURN AIR FLOW SENSOR
F-3 = SUPPLY AIR FLOW SENSOR
R-1 = HIGH SIGNAL SELECTOR
R-2 = HIGH SIGNAL SELECTOR
RC-1 = RECEIVER CONTROLLER
RC-2 = RECEIVER CONTROLLER
RC-3 = RECEIVER CONTROLLER
RC-4 = RECEIVER CONTROLLER
S-1 = MINIMUM POSITION SWITCH
SP-1 = STATIC PRESSURE SENSOR

VAV Heating and Cooling Air-Handling Unit, Hot Water Heating, DX Cooling, with R.A. Fan and Inlet Vane Controls on Supply and Return Fan, O.A., R.A., and REL Dampers Using Economizer Control Cycle, Variable Volume Fan Capacity Control through Volumetric Matching of Supply and Return Fans, with Minimum O.A. Injection Fan for Ventilation

Whenever the fan runs, the EP-1 is energized and O.A., R.A., and REL dampers are placed under automatic controls. When the fan stops, the dampers return to their normal positions.

Receiver controller RC-1 with sensor T-2 in the mix chamber and sensor T-1 in the O.A. control the O.A., R.A., and REL dampers. The system acts as an economizer control cycle. Minimum position switch S-1 allows, through relays R-1 and R-2, a minimum setting for the O.A. damper, regardless of the actions of the economizer thermostats. The outdoor air is split into two sections. One contains a minimum O.A. injection fan to insure a fixed amount of ventilation air. The fan is operated by a flow controller operating the inlet vanes on the fan. Receiver controller RC-3 with static pressure sensor SP-1 controls the supply fan inlet vanes. Receiver controller RC-2 with sensor T-3 in the fan discharge controls valve V-1 and PE switch PE-1 on the DX cooling coil. Flow controller F-3 sensing the flow in the fan discharge resets the control point of receiver controller RC-4 that also senses the flow from the return fan through flow controller F-2.

This above system is designed to "match" the supply and return fans so that they act as a team and maintain space static pressure while at the same time modulating down from full capacity to reduced capacity as the VAV boxes in the system become satisfied.

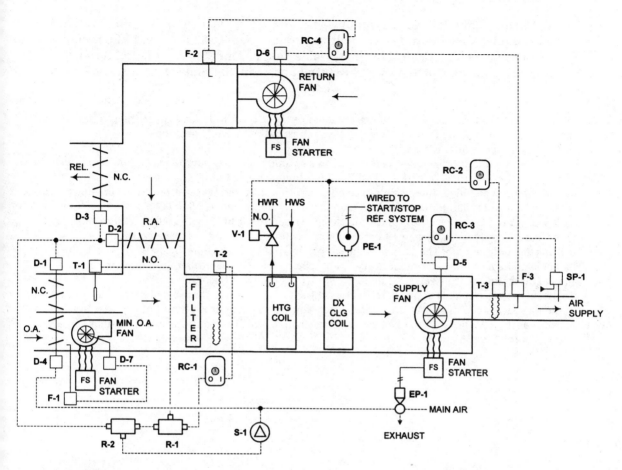

LEGEND:

T-1	=	MODULATING CAPILLARY THERMOSTAT
T-2	=	MODULATING CAPILLARY DUCT SENSOR
T-3	=	MODULATING CAPILLARY DUCT SENSOR
V-1	=	N.O. MODULATING 2-WAY HEATING COIL VALVE
D-1	=	MAXIMUM OUTSIDE AIR DAMPER MOTOR
D-2	=	RETURN AIR DAMPER MOTOR
D-3	=	RELIEF AIR DAMPER MOTOR
D-4	=	MINIMUM OUTSIDE AIR DAMPER MOTOR
D-5	=	SUPPLY FAN INLET VANE DAMPER MOTOR
D-6	=	RETURN FAN INLET VANE DAMPER MOTOR
D-7	=	MIN. O.A. INJECTION FAN INLET VANE DAMPER MOTOR
EP-1	=	SOLENOID AIR VALVE

PE-1	=	PRESSURE/ELECTRIC SWITCH
F-1	=	FLOW CONTROLLER
F-2	=	RETURN AIR FLOW SENSOR
F-3	=	SUPPLY AIR FLOW SENSOR
R-1	=	HIGH SIGNAL SELECTOR
R-2	=	HIGH SIGNAL SELECTOR
RC-1	=	RECEIVER CONTROLLER
RC-2	=	RECEIVER CONTROLLER
RC-3	=	RECEIVER CONTROLLER
RC-4	=	RECEIVER CONTROLLER
S-1	=	MINIMUM POSITION SWITCH
SP-1	=	STATIC PRESSURE SENSOR

VAV Heating and Cooling Air-Handling Unit, Hot Water Heating, DX Cooling, with R.A. Fan and Inlet Vane Controls on Supply and Return Fan, O.A., R.A., and REL Dampers Using Economizer Control Cycle, Variable Volume Fan Capacity Control through Volumetric Matching of Supply and Return Fans, with Minimum O.A. Injection Fan for Ventilation

Whenever the fan runs, the EP-1 is energized and O.A., R.A., and REL dampers are placed under automatic controls. When the fan stops, the dampers return to their normal positions.

Receiver controller RC-1 with sensor T-2 in the mix chamber and sensor T-1 in the O.A. control the O.A., R.A., and REL dampers. The system acts as an economizer control cycle. Minimum position switch S-1 allows through relays R-1 and R-2 a minimum setting for the O.A. damper regardless of the actions of the economizer thermostats. The outdoor air is split into two sections. One contains a minimum O.A. injection fan to insure a fixed amount of ventilation air. The fan is operated by a flow controller operating the inlet vanes on the fan. Receiver controller RC-3 with static pressure sensor SP-1 controls the supply fan inlet vanes. Sensor SP-1 resets the control point of receiver controller RC-4 controlling the return air fan. The control of the supply and return fans is done by matching the fan through balancing. The fans must be similar in size and capacity. Receiver controller RC-2 through sensor T-3 in the fan discharge controls valve V-1 and PE switch PE-1 on the DX cooling coil.

This above system is designed to "match" the supply and return fans so that they act as a team and maintain space static pressure while at the same time modulating down from full capacity to reduced capacity as the VAV boxes in the system become satisfied. The fans require accurate balancing by a balancing contractor.

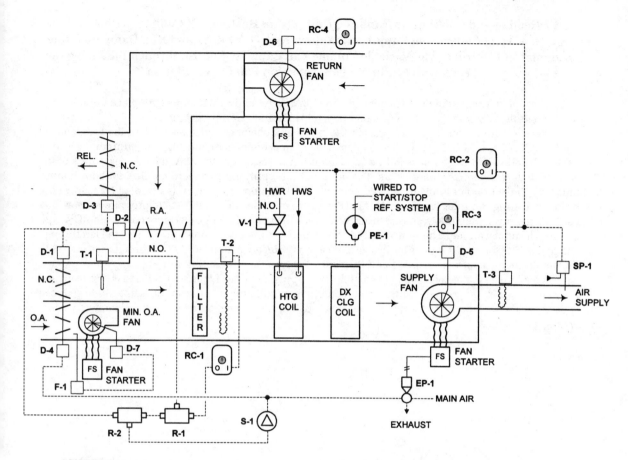

LEGEND:

T-1 = MODULATING CAPILLARY THERMOSTAT
T-2 = MODULATING CAPILLARY DUCT SENSOR
T-3 = MODULATING CAPILLARY DUCT SENSOR
V-1 = N.O. MODULATING 2-WAY HEATING COIL VALVE
D-1 = MAXIMUM OUTSIDE AIR DAMPER MOTOR
D-2 = RETURN AIR DAMPER MOTOR
D-3 = RELIEF AIR DAMPER MOTOR
D-4 = MINIMUM OUTSIDE AIR DAMPER MOTOR
D-5 = SUPPLY FAN INLET VANE DAMPER MOTOR
D-6 = RETURN FAN INLET VANE DAMPER MOTOR
D-7 = MIN. O.A. INJECTION FAN INLET VANE DAMPER MOTOR
EP-1 = SOLENOID AIR VALVE

PE-1 = PRESSURE/ELECTRIC SWITCH
F-1 = FLOW CONTROLLER
R-1 = HIGH SIGNAL SELECTOR
R-2 = HIGH SIGNAL SELECTOR
RC-1 = RECEIVER CONTROLLER
RC-2 = RECEIVER CONTROLLER
RC-3 = RECEIVER CONTROLLER
RC-4 = RECEIVER CONTROLLER
S-1 = MINIMUM POSITION SWITCH
SP-1 = STATIC PRESSURE SENSOR

VAV Heating and Cooling Air-Handling Unit, Steam Heating, DX Cooling, with R.A. Fan and Inlet Vane Controls on Supply and Return Fan, O.A., R.A., and REL Dampers Using Economizer Control Cycle, Variable Volume Fan Capacity Control through Balancing the Fans with Minimum O.A. Injection Fan for Ventilation

Whenever the fan runs, the EP-1 is energized and O.A., R.A., and REL dampers are placed under automatic controls. When the fan stops, the dampers return to their normal positions.

Receiver controller RC-1 with sensor T-2 in the mix chamber and sensor T-1 in the O.A. control the O.A., R.A., and REL dampers. The system acts as an economizer control cycle. Minimum position switch S-1 allows through relays R-1 and R-2 a minimum setting for the O.A. damper regardless of the actions of the economizer thermostats. The outdoor air is split into two sections. One contains a minimum O.A. injection fan to insure a fixed amount of ventilation air. The fan is operated by a flow controller operating the inlet vanes on the fan. Receiver controller RC-3 with static pressure sensor SP-1 controls the supply fan inlet vanes. Sensor SP-1 resets the control point of receiver controller RC-4 controlling the return air fan. The control of the supply and return fans is done by matching the fan through balancing. The fans must be similar in size and capacity. Receiver controller RC-2 through sensor T-3 in the fan discharge controls valve V-1 and PE switch PE-1 on the DX cooling coil.

This above system is designed to "match" the supply and return fans so that they act as a team and maintain space static pressure while at the same time modulating down from full capacity to reduced capacity as the VAV boxes in the system become satisfied. The fans require accurate balancing by a balancing contractor.

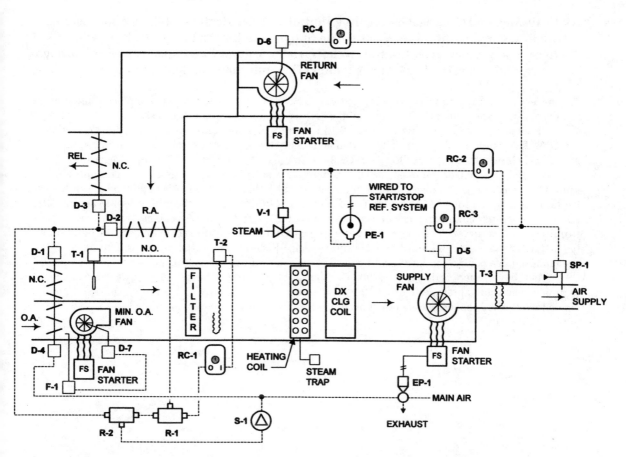

LEGEND:

T-1	=	MODULATING CAPILLARY THERMOSTAT
T-2	=	MODULATING CAPILLARY DUCT SENSOR
T-3	=	MODULATING CAPILLARY DUCT SENSOR
V-1	=	N.O. MODULATING STEAM COIL VALVE
D-1	=	MAXIMUM OUTSIDE AIR DAMPER MOTOR
D-2	=	RETURN AIR DAMPER MOTOR
D-3	=	RELIEF AIR DAMPER MOTOR
D-4	=	MINIMUM OUTSIDE AIR DAMPER MOTOR
D-5	=	SUPPLY FAN INLET VANE DAMPER MOTOR
D-6	=	RETURN FAN INLET VANE DAMPER MOTOR
D-7	=	MIN. O.A. INJECTION FAN INLET VANE DAMPER MOTOR
EP-1	=	SOLENOID AIR VALVE

PE-1	=	PRESSURE/ELECTRIC SWITCH
F-1	=	FLOW CONTROLLER
R-1	=	HIGH SIGNAL SELECTOR
R-2	=	HIGH SIGNAL SELECTOR
RC-1	=	RECEIVER CONTROLLER
RC-2	=	RECEIVER CONTROLLER
RC-3	=	RECEIVER CONTROLLER
RC-4	=	RECEIVER CONTROLLER
S-1	=	MINIMUM POSITION SWITCH
SP-1	=	STATIC PRESSURE SENSOR

VAV Heating and Cooling Air-Handling Unit, Hot Water Heating, Chilled Water Cooling, with R.A. Fan and Inlet Vane Controls on Supply and Return Fan, O.A., R.A., and REL Dampers Using Economizer Control Cycle, Variable Volume Fan Capacity Control through Balancing the Fans with Minimum O.A. Injection Fan for Ventilation

Whenever the fan runs, the EP-1 is energized and O.A., R.A., and REL dampers are placed under automatic controls. When the fan stops, the dampers return to their normal positions.

Receiver controller RC-1 with sensor T-2 in the mix chamber and sensor T-1 in the O.A. control the O.A., R.A., and REL dampers. The system acts as an economizer control cycle. Minimum position switch S-1 allows through relays R-1 and R-2 a minimum setting for the O.A. damper regardless of the actions of the economizer thermostats. The outdoor air is split into two sections. One contains a minimum O.A. injection fan to insure a fixed amount of ventilation air. The fan is operated by a flow controller operating the inlet vanes on the fan. Receiver controller RC-3 with static pressure sensor SP-1 controls the supply fan inlet vanes. Sensor SP-1 resets the control point of receiver controller RC-4 controlling the return air fan. The control of the supply and return fans is done by matching the fan through balancing. The fans must be similar in size and capacity. Receiver controller RC-2 through sensor T-3 in the fan discharge controls valve V-1 and valve V-2 on the heating and cooling coils.

This above system is designed to "match" the supply and return fans so that they act as a team and maintain space static pressure while at the same time modulating down from full capacity to reduced capacity as the VAV boxes in the system become satisfied. The fans require accurate balancing by a balancing contractor.

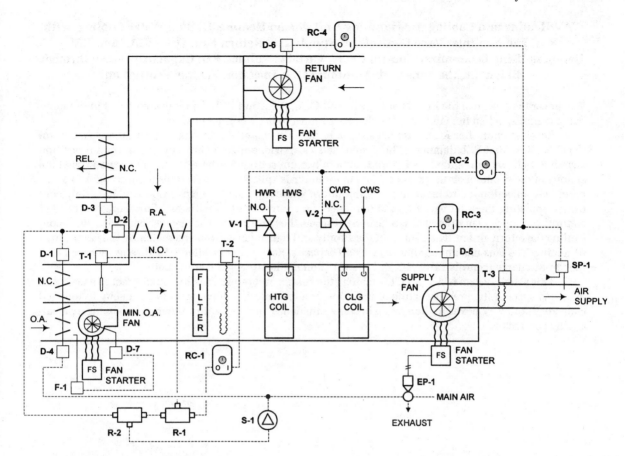

LEGEND:

T-1	=	MODULATING CAPILLARY THERMOSTAT
T-2	=	MODULATING CAPILLARY DUCT SENSOR
T-3	=	MODULATING CAPILLARY DUCT SENSOR
V-1	=	N.O. MODULATING 2-WAY HEATING COIL VALVE
V-2	=	N.C. MODULATING 2-WAY COOLING COIL VALVE
D-1	=	MAXIMUM OUTSIDE AIR DAMPER MOTOR
D-2	=	RETURN AIR DAMPER MOTOR
D-3	=	RELIEF AIR DAMPER MOTOR
D-4	=	MINIMUM OUTSIDE AIR DAMPER MOTOR
D-5	=	SUPPLY FAN INLET VANE DAMPER MOTOR
D-6	=	RETURN FAN INLET VANE DAMPER MOTOR
D-7	=	MIN. O.A. INJECTION FAN INLET VANE DAMPER MOTOR

EP-1	=	SOLENOID AIR VALVE
F-1	=	FLOW CONTROLLER
R-1	=	HIGH SIGNAL SELECTOR
R-2	=	HIGH SIGNAL SELECTOR
RC-1	=	RECEIVER CONTROLLER
RC-2	=	RECEIVER CONTROLLER
RC-3	=	RECEIVER CONTROLLER
RC-4	=	RECEIVER CONTROLLER
S-1	=	MINIMUM POSITION SWITCH
SP-1	=	STATIC PRESSURE SENSOR

VAV Heating and Cooling Air-Handling Unit, Steam Heating, Chilled Water Cooling, with R.A. Fan and Inlet Vane Controls on Supply and Return Fan, O.A., R.A., and REL Dampers Using Economizer Control Cycle, Variable Volume Fan Capacity Control through Balancing the Fans with Minimum O.A. Injection Fan for Ventilation

Whenever the fan runs, the EP-1 is energized and O.A., R.A., and REL dampers are placed under automatic controls. When the fan stops, the dampers return to their normal positions.

Receiver controller RC-1 with sensor T-2 in the mix chamber and sensor T-1 in the O.A. control the O.A., R.A., and REL dampers. The system acts as an economizer control cycle. Minimum position switch S-1 allows through relays R-1 and R-2 a minimum setting for the O.A. damper regardless of the actions of the economizer thermostats. The outdoor air is split into two sections. One contains a minimum O.A. injection fan to insure a fixed amount of ventilation air. The fan is operated by a flow controller operating the inlet vanes on the fan. Receiver controller RC-3 with static pressure sensor SP-1 controls the supply fan inlet vanes. Sensor SP-1 resets the control point of receiver controller RC-4 controlling the return air fan. The control of the supply and return fans is done by matching the fan through balancing. The fans must be similar in size and capacity. Receiver controller RC-2 through sensor T-3 in the fan discharge controls valve V-1 and valve V-2 on the heating and cooling coils.

This above system is designed to "match" the supply and return fans so that they act as a team and maintain space static pressure while at the same time modulating down from full capacity to reduced capacity as the VAV boxes in the system become satisfied. The fans require accurate balancing by a balancing contractor.

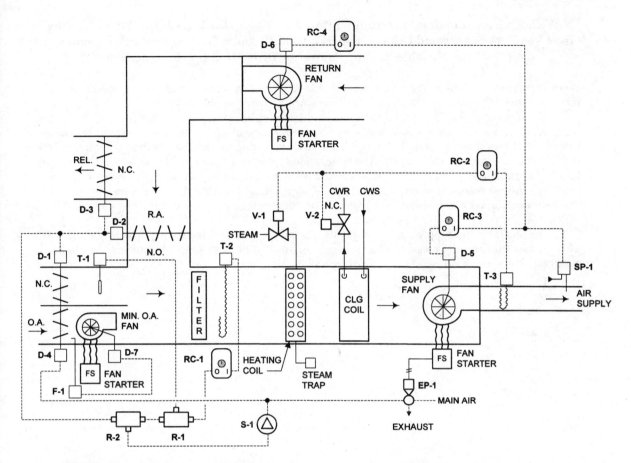

LEGEND:

T-1	= MODULATING CAPILLARY THERMOSTAT		EP-1	= SOLENOID AIR VALVE	
T-2	= MODULATING CAPILLARY DUCT SENSOR		F-1	= FLOW CONTROLLER	
T-3	= MODULATING CAPILLARY DUCT SENSOR		R-1	= HIGH SIGNAL SELECTOR	
V-1	= N.O. MODULATING STEAM COIL VALVE		R-2	= HIGH SIGNAL SELECTOR	
V-2	= N.C. MODULATING 2-WAY COOLING COIL VALVE		RC-1	= RECEIVER CONTROLLER	
D-1	= MAXIMUM OUTSIDE AIR DAMPER MOTOR		RC-2	= RECEIVER CONTROLLER	
D-2	= RETURN AIR DAMPER MOTOR		RC-3	= RECEIVER CONTROLLER	
D-3	= RELIEF AIR DAMPER MOTOR		RC-4	= RECEIVER CONTROLLER	
D-4	= MINIMUM OUTSIDE AIR DAMPER MOTOR		S-1	= MINIMUM POSITION SWITCH	
D-5	= SUPPLY FAN INLET VANE DAMPER MOTOR		SP-1	= STATIC PRESSURE SENSOR	
D-6	= RETURN FAN INLET VANE DAMPER MOTOR				
D-7	= MIN. O.A. INJECTION FAN INLET VANE DAMPER MOTOR				

VAV Heating and Cooling Air-Handling Unit, Hot Water Heating, Chilled Water Cooling, Inlet Vanes on Supply and Exhaust Fans, O.A., R.A., and REL Dampers with Economizer Control Cycle, Variable Volume Control by Direct Building Pressure Control

Whenever the fan runs, the EP-1 is energized and O.A., R.A., and REL dampers are placed under automatic controls. When the fan stops, the dampers return to their normal positions.

Receiver controller RC-1 with sensor T-2 in the mix chamber and sensor T-1 in the O.A. control the O.A., R.A., and REL dampers. The system acts as an economizer control cycle. Minimum position switch S-1 allows through relays R-1 and R-2 a minimum setting for the O.A. damper regardless of the actions of the economizer thermostats. The system does not have a return air fan but has an exhaust fan. Receiver controller RC-3 with static pressure sensor SP-1 controls the supply fan inlet vanes. Sensor SP-1 resets the control point of receiver controller RC-4 controlling the exhaust air fan. Receiver controller RC-2 with sensor T-3 controls valves V-1 and V-2 on the heating and cooling coils. The damper for the exhaust fan is operated in conjunction with the other dampers.

The control of the supply and exhaust fans is done through building static pressure controls with SP-1 sensing the building pressure. Receiver controller RC-2 through sensor T-3 in the fan discharge controls valve V-1 and valve V-2 on the heating and cooling coils.

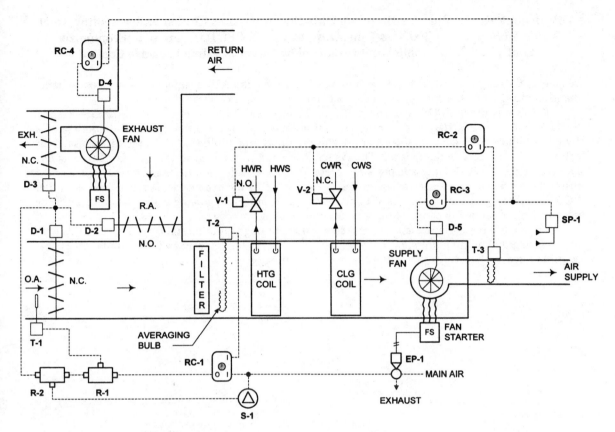

LEGEND:

T-1 = MODULATING CAPILLARY THERMOSTAT
T-2 = MODULATING CAPILLARY DUCT SESNOR
T-3 = MODULATING CAPILLARY DUCT SENSOR
V-1 = N.O. MODULATING 2-WAY HEATING COIL VALVE
V-2 = N.C. MODULATING 2-WAY COOLING COIL VALVE
D-1 = OUTSIDE AIR DAMPER MOTOR
D-2 = RETURN AIR DAMPER MOTOR
D-3 = EXHAUST AIR DAMPER MOTOR
D-4 = EXHAUST FAN INLET VANE DAMPER MOTOR
D-5 = SUPPLY FAN INLET VANE DAMPER MOTOR
EP-1 = SOLENOID AIR VALVE
R-1 = HIGH SIGNAL SELECTOR
R-2 = HIGH SIGNAL SELECTOR
RC-1 = RECEIVER CONTROLLER
RC-2 = RECEIVER CONTROLLER
RC-3 = RECEIVER CONTROLLER
RC-4 = RECEIVER CONTROLLER
S-1 = MINIMUM POSITION SWITCH
SP-1 = STATIC PRESSURE SENSOR

VAV Heating and Cooling Air-Handling Unit, Steam Heating, Chilled Water Cooling, Inlet Vanes on Supply and Exhaust Fans, O.A., R.A., and REL Dampers with Economizer Control Cycle, Variable Volume Control by Direct Building Pressure Control

Whenever the fan runs, the EP-1 is energized and O.A., R.A., and REL dampers are placed under automatic controls. When the fan stops, the dampers return to their normal positions.

Receiver controller RC-1 with sensor T-2 in the mix chamber and sensor T-1 in the O.A. control the O.A., R.A., and REL dampers. The system acts as an economizer control cycle. Minimum position switch S-1 allows through relays R-1 and R-2 a minimum setting for the O.A. damper regardless of the actions of the economizer thermostats. The system does not have a return air fan but has an exhaust fan. Receiver controller RC-3 with static pressure sensor SP-1 controls the supply fan inlet vanes. Sensor SP-1 resets the control point of receiver controller RC-4 controlling the exhaust air fan. Receiver controller RC-2 with sensor T-3 controls valves V-1 and V-2 on the heating and cooling coils. The damper for the exhaust fan is operated in conjunction with the other dampers.

The control of the supply and exhaust fans is done through building static pressure controls with SP-1 sensing the building pressure. Receiver controller RC-2 through sensor T-3 in the fan discharge controls valve V-1 and valve V-2 on the heating and cooling coils.

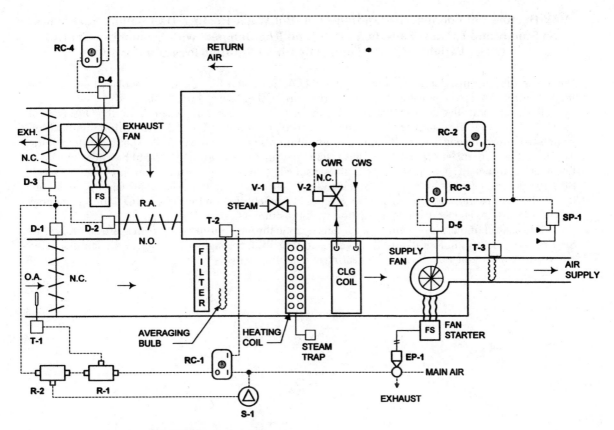

LEGEND:

T-1	=	MODULATING CAPILLARY THERMOSTAT
T-2	=	MODULATING CAPILLARY DUCT SENSOR
T-3	=	MODULATING CAPILLARY DUCT SENSOR
V-1	=	N.O. MODULATING STEAM COIL VALVE
V-2	=	N.C. MODULATING 2-WAY COOLING COIL VALVE
D-1	=	OUTSIDE AIR DAMPER MOTOR
D-2	=	RETURN AIR DAMPER MOTOR
D-3	=	EXHAUST AIR DAMPER MOTOR
D-4	=	EXHAUST FAN INLET VANE DAMPER MOTOR
D-5	=	SUPPLY FAN INLET VANE DAMPER MOTOR
EP-1	=	SOLENOID AIR VALVE
R-1	=	HIGH SIGNAL SELECTOR
R-2	=	HIGH SIGNAL SELECTOR
RC-1	=	RECEIVER CONTROLLER
RC-2	=	RECEIVER CONTROLLER
RC-3	=	RECEIVER CONTROLLER
RC-4	=	RECEIVER CONTROLLER
S-1	=	MINIMUM POSITION SWITCH
SP-1	=	STATIC PRESSURE SENSOR

VAV Heating and Cooling Air-Handling Unit, Hot Water Heating, DX Cooling, Inlet Vanes on Supply and Exhaust Fans, O.A., R.A., and REL Dampers with Economizer Control Cycle, Variable Volume Control by Direct Building Pressure Control

Whenever the fan runs, the EP-1 is energized and O.A., R.A., and REL dampers are placed under automatic controls. When the fan stops, the dampers return to their normal positions.

Receiver controller RC-1 with sensor T-2 in the mix chamber and sensor T-1 in the O.A. control the O.A., R.A., and REL dampers. The system acts as an economizer control cycle. Minimum position switch S-1 allows through relays R-1 and R-2 a minimum setting for the O.A. damper regardless of the actions of the economizer thermostats. The system does not have a return air fan but has an exhaust fan. Receiver controller RC-3 with static pressure sensor SP-1 controls the supply fan inlet vanes. Sensor SP-1 resets the control point of receiver controller RC-4 controlling the exhaust air fan. Receiver controller RC-2 with sensor T-3 controls valve V-1 and PE switch PE-1 on the DX cooling coil. The damper for the exhaust fan is operated in conjunction with the other dampers.

The control of the supply and exhaust fans is done through building static pressure controls with SP-1 sensing the building pressure. Receiver controller RC-2 through sensor T-3 in the fan discharge controls valve V-1 and valve V-2 on the heating and cooling coils.

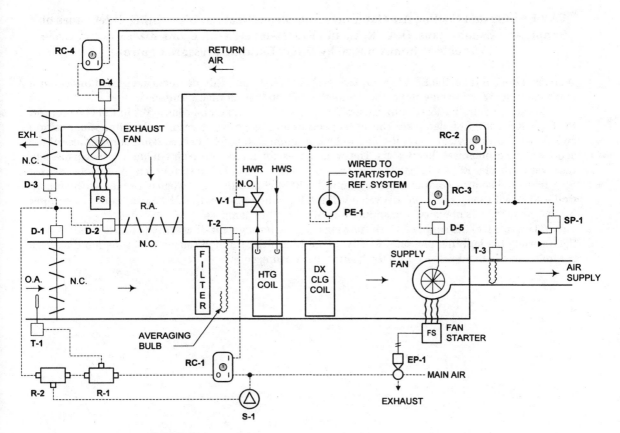

LEGEND:

T-1 = MODULATING CAPILLARY THERMOSTAT
T-2 = MODULATING CAPILLARY DUCT THERMOSTAT
T-3 = MODULATING CAPILLARY DUCT THERMOSTAT
V-1 = N.O. MODULATING 2-WAY HEATING COIL VALVE
V-2 = N.C. MODULATING 2-WAY COOLING COIL VALVE
D-1 = OUTSIDE AIR DAMPER MOTOR
D-2 = RETURN AIR DAMPER MOTOR
D-3 = EXHAUST AIR DAMPER MOTOR
D-4 = EXHAUST FAN INLET VANE DAMPER MOTOR
D-5 = SUPPLY FAN INLET VANE DAMPER MOTOR
EP-1 = SOLENOID AIR VALVE
R-1 = HIGH SIGNAL SELECTOR
R-2 = HIGH SIGNAL SELECTOR
RC-1 = RECEIVER CONTROLLER
RC-2 = RECEIVER CONTROLLER
RC-3 = RECEIVER CONTROLLER
RC-4 = RECEIVER CONTROLLER
S-1 = MINIMUM POSITION SWITCH
SP-1 = STATIC PRESSURE SENSOR

VAV Heating and Cooling Air-Handling Unit, Steam Heating, DX Cooling, Inlet vanes on Supply and Exhaust fans. O.A., R.A., and REL Dampers with Economizer Control Cycle, Variable Volume Control by Direct Building Pressure Control

Whenever the fan runs, the EP-1 is energized and O.A., R.A., and REL dampers are placed under automatic controls. When the fan stops, the dampers return to their normal positions.

Receiver controller RC-1 with sensor T-2 in the mix chamber and sensor T-1 in the O.A. control the O.A., R.A., and REL dampers. The system acts as an economizer control cycle. Minimum position switch S-1 allows through relays R-1 and R-2 a minimum setting for the O.A. damper regardless of the actions of the economizer thermostats. The system does not have a return air fan but has an exhaust fan. Receiver controller RC-3 with static pressure sensor SP-1 controls the supply fan inlet vanes. Sensor SP-1 resets the control point of receiver controller RC-4 controlling the exhaust air fan. Receiver controller RC-2 with sensor T-3 controls valve V-1 and PE switch PE-1 on the DX cooling coil. The damper for the exhaust fan is operated in conjunction with the other dampers.

The control of the supply and exhaust fans is done through building static pressure controls with SP-1 sensing the building pressure. Receiver controller RC-2 through sensor T-3 in the fan discharge controls valve V-1 and valve V-2 on the heating and cooling coils.

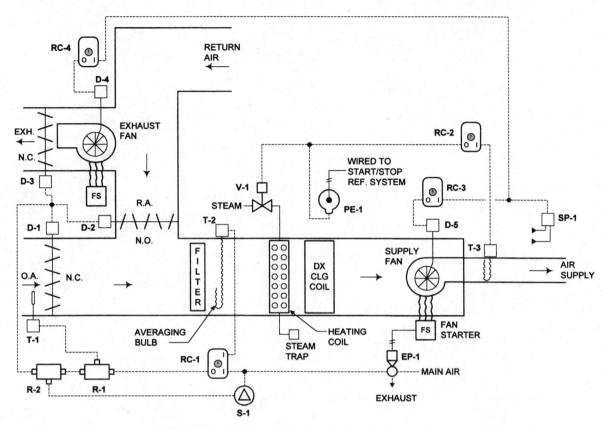

LEGEND:

T-1 = MODULATING CAPILLARY THERMOSTAT
T-2 = MODULATING CAPILLARY DUCT SENSOR
T-3 = MODULATING CAPILLARY DUCT SENSOR
V-1 = N.O. MODULATING STEAM COIL VALVE
D-1 = OUTSIDE AIR DAMPER MOTOR
D-2 = RETURN AIR DAMPER MOTOR
D-3 = EXHAUST AIR DAMPER MOTOR
D-4 = EXHAUST FAN INLET VANE DAMPER MOTOR
D-5 = SUPPLY FAN INLET VANE DAMPER MOTOR
EP-1 = SOLENOID AIR VALVE
PE-1 = PRESSURE/ELECTRIC SWITCH
R-1 = HIGH SIGNAL SELECTOR
R-2 = HIGH SIGNAL SELECTOR
RC-1 = RECEIVER CONTROLLER
RC-2 = RECEIVER CONTROLLER
RC-3 = RECEIVER CONTROLLER
RC-4 = RECEIVER CONTROLLER
S-1 = MINIMUM POSITION SWITCH
SP-1 = STATIC PRESSURE SENSOR

CHAPTER 19

Dual-Duct VAV Systems

Dual-Duct VAV Heating and Cooling Air-Handling Unit, Hot Water Heating Coil, DX Cooling Inlet Vanes on Supply and Exhaust Fans, O.A., R.A., and EX Dampers Using Economizer Control Cycle, Flow Control of Both Fans   406

Dual-Duct VAV Heating and Cooling Air-Handling Unit, Hot Water Heating Coil, Chilled Water Cooling Coil, Inlet Vanes on Supply and Return Fans, O.A., R.A., and REL Fans Using Economizer Control Cycle, Static Pressure Control on Both Fans

Whenever the fan runs, the EP-1 is energized and O.A., R.A., and REL dampers are placed under automatic controls. When the fan stops, the dampers return to their normal positions.

Duct thermostats T-1 and T-2 in the mix chamber control the O.A., R.A., and REL dampers and act as economizer thermostats. They can allow for up to 100% O.A. when conditions are favorable. Receiver controller RC-1 with sensor T-3 in the hot air duct controls the valve V-1 on the heating coil. Receiver controller RC-2 with sensor in the cold duct controls valve V-2 on the cooling coil. Receiver controllers RC-3 and RC-4 with flow sensors F-1 and F-2, through averaging relay R-1, control the inlet vanes on the supply fan. Flow sensor F-3 in the fan discharge resets the control point of receiver controller RC-5, which is also sensing the return fan sensor F-4 in the return air fan discharge to control the return air fan inlet vane dampers.

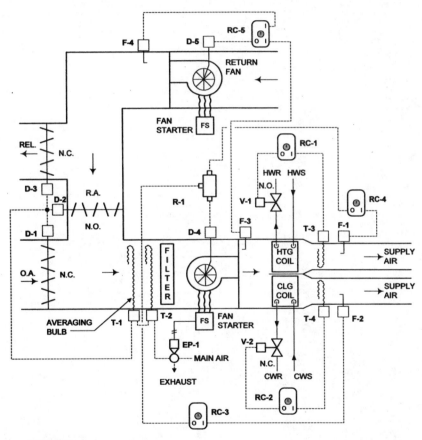

LEGEND:

T-1	= MODULATING CAPILLARY DUCT THERMOSTAT	EP-1	= SOLENOID AIR VALVE
T-2	= MODULATING CAPILLARY DUCT THERMOSTAT	F-1	= HOT DECK FLOW SENSOR
T-3	= MODULATING CAPILLARY DUCT THERMOSTAT	F-2	= COLD DECK FLOW SENSOR
T-4	= MODULATING CAPILLARY DUCT THERMOSTAT	F-3	= SUPPLY AIR FLOW SENSOR
V-1	= N.O. MODULATING 2-WAY HEATING COIL VALVE	F-4	= RETURN AIR FLOW SENSOR
V-2	= N.C. MODULATING 2-WAY COOLING COIL VALVE	RC-1	= RECEIVER CONTROLLER
D-1	= OUTSIDE AIR DAMPER MOTOR	RC-2	= RECEIVER CONTROLLER
D-2	= RETURN AIR DAMPER MOTOR	RC-3	= RECEIVER CONTROLLER
D-3	= RELIEF AIR DAMPER MOTOR	RC-4	= RECEIVER CONTROLLER
D-4	= SUPPLY FAN INLET VANE DAMPER MOTOR	RC-5	= RECEIVER CONTROLLER
D-5	= RETURN FAN INLET VANE DAMPER MOTOR	R-1	= HIGH SIGNAL SELECTOR

Dual-Duct VAV Heating and Cooling Air-Handling Unit, Steam Heating Coil, Chilled Water Cooling Coil, Inlet Vanes on Supply and Return Fans, O.A., R.A., and REL Fans Using Economizer Control Cycle, Static Pressure Control on Both Fans

Whenever the fan runs, the EP-1 is energized and O.A., R.A., and REL dampers are placed under automatic controls. When the fan stops, the dampers return to their normal positions.

Duct thermostats T-1 and T-2 in the mix chamber control the O.A., R.A., and REL dampers and act as economizer thermostats. They can allow for up to 100% O.A. when conditions are favorable. Receiver controller RC-1 with sensor T-3 in the hot air duct controls the valve V-1 on the heating coil. Receiver controller RC-2 with sensor in the cold duct controls valve V-2 on the cooling coil. Receiver controllers RC-3 and RC-4 with flow sensors F-1 and F-2, through averaging relay R-1, control the inlet vanes on the supply fan. Flow sensor F-3 in the fan discharge resets the control point of receiver controller RC-5, which is also sensing the return fan sensor F-4 in the return air fan discharge to control the return air fan inlet vane dampers.

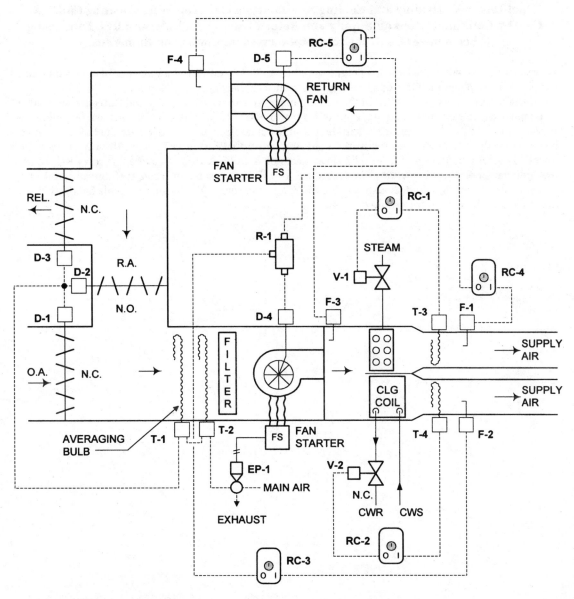

LEGEND:

T-1 = MODULATING CAPILLARY DUCT THERMOSTAT
T-2 = MODULATING CAPILLARY DUCT THERMOSTAT
T-3 = MODULATING CAPILLARY DUCT THERMOSTAT
T-4 = MODULATING CAPILLARY DUCT THERMOSTAT
V-1 = N.O. MODULATING STEAM COIL VALVE
V-2 = N.C. MODULATING 2-WAY COOLING COIL VALVE
D-1 = OUTSIDE AIR DAMPER MOTOR
D-2 = RETURN AIR DAMPER MOTOR
D-3 = RELIEF AIR DAMPER MOTOR
D-4 = SUPPLY FAN INLET VANE DAMPER MOTOR
D-5 = RETURN FAN INLET VANE DAMPER MOTOR

EP-1 = SOLENOID AIR VALVE
F-1 = HOT DECK FLOW SENSOR
F-2 = COLD DECK FLOW SENSOR
F-3 = SUPPLY AIR FLOW SENSOR
F-4 = RETURN AIR FLOW SENSOR
RC-1 = RECEIVER CONTROLLER
RC-2 = RECEIVER CONTROLLER
RC-3 = RECEIVER CONTROLLER
RC-4 = RECEIVER CONTROLLER
RC-5 = RECEIVER CONTROLLER
R-1 = HIGH SIGNAL SELECTOR

Dual-Duct VAV Heating and Cooling Air-Handling Unit, Hot Water Heating Coil, DX Cooling Coil, Inlet Vanes on Supply and Return Fans, O.A., R.A., and REL Fans Using Economizer Control Cycle, Static Pressure Control on Both Fans

Whenever the fan runs, the EP-1 is energized and O.A., R.A., and REL dampers are placed under automatic controls. When the fan stops, the dampers return to their normal positions.

Duct thermostats T-1 and T-2 in the mix chamber control the O.A., R.A., and REL dampers and act as economizer thermostats. They can allow for up to 100% O.A. when conditions are favorable. Receiver controller RC-1 with sensor T-3 in the hot air duct controls the valve V-1 on the heating coil. Receiver controller RC-2 with sensor in the cold duct controls PE switch PE-1 on the DX cooling coil. Receiver controllers RC-3 and RC-4 with flow sensors F-1 and F-2, through averaging relay R-1, control the inlet vanes on the supply fan. Flow sensor F-3 in the fan discharge resets the control point of receiver controller RC-5, which is also sensing the return fan sensor F-4 in the return air fan discharge to control the return air fan inlet vane dampers.

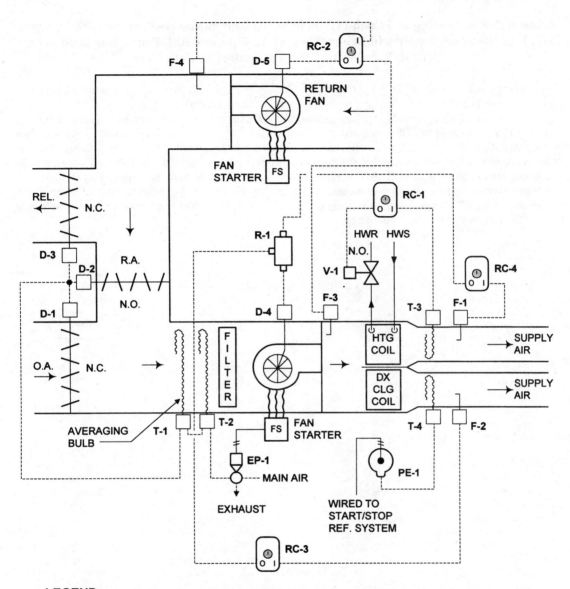

LEGEND:

T-1	=	MODULATING CAPILLARY DUCT THERMOSTAT
T-2	=	MODULATING CAPILLARY DUCT THERMOSTAT
T-3	=	MODULATING CAPILLARY DUCT THERMOSTAT
T-4	=	MODULATING CAPILLARY DUCT THERMOSTAT
V-1	=	N.O. MODULATING 2-WAY HEATING COIL VALVE
D-1	=	OUTSIDE AIR DAMPER MOTOR
D-2	=	RETURN AIR DAMPER MOTOR
D-3	=	RELIEF AIR DAMPER MOTOR
D-4	=	SUPPLY FAN INLET VANE DAMPER MOTOR
D-5	=	RETURN FAN INLET VANE DAMPER MOTOR
EP-1	=	SOLENOID AIR VALVE

PE-1	=	PRESSURE/ELECTRIC SWITCH
F-1	=	HOT DECK FLOW SENSOR
F-2	=	COLD DECK FLOW SENSOR
F-3	=	SUPPLY AIR FLOW SENSOR
F-4	=	RETURN AIR FLOW SENSOR
RC-1	=	RECEIVER CONTROLLER
RC-2	=	RECEIVER CONTROLLER
RC-3	=	RECEIVER CONTROLLER
RC-4	=	RECEIVER CONTROLLER
R-1	=	HIGH SIGNAL SELECTOR

Dual-Duct VAV Heating and Cooling Air-Handling Unit, Steam Heating Coil, DX Cooling Coil, Inlet Vanes on Supply and Return Fans, O.A., R.A., and REL Fans Using Economizer Control Cycle, Static Pressure Control on Both Fans

Whenever the fan runs, the EP-1 is energized and O.A., R.A., and REL dampers are placed under automatic controls. When the fan stops, the dampers return to their normal positions.

Duct thermostats T-1 and T-2 in the mix chamber control the O.A., R.A., and REL dampers and act as economizer thermostats. They can allow for up to 100% O.A. when conditions are favorable. Receiver controller RC-1 with sensor T-3 in the hot air duct controls the valve V-1 on the heating coil. Receiver controller RC-2 with sensor in the cold duct controls PE switch PE-1 on the DX cooling coil. Receiver controllers RC-3 and RC-4 with flow sensors F-1 and F-2, through averaging relay R-1, control the inlet vanes on the supply fan. Flow sensor F-3 in the fan discharge resets the control point of receiver controller RC-5, which is also sensing the return fan sensor F-4 in the return air fan discharge to control the return air fan inlet vane dampers.

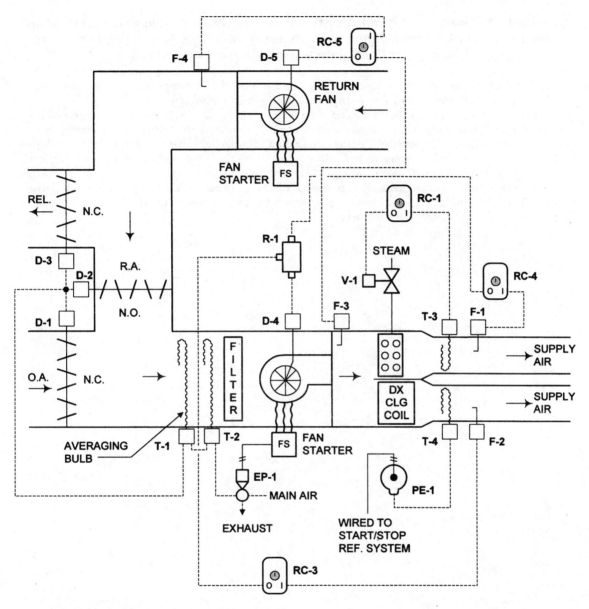

LEGEND:

T-1	= MODULATING CAPILLARY DUCT THERMOSTAT
T-2	= MODULATING CAPILLARY DUCT THERMOSTAT
T-3	= MODULATING CAPILLARY DUCT THERMOSTAT
T-4	= MODULATING CAPILLARY DUCT THERMOSTAT
V-1	= N.O. MODULATING STEAM COIL VALVE
D-1	= OUTSIDE AIR DAMPER MOTOR
D-2	= RETURN AIR DAMPER MOTOR
D-3	= RELIEF AIR DAMPER MOTOR
D-4	= SUPPLY FAN INLET VANE DAMPER MOTOR
D-5	= RETURN FAN INLET VANE DAMPER MOTOR
EP-1	= SOLENOID AIR VALVE

PE-1	= PRESSURE/ELECTRIC SWITCH
F-1	= HOT DECK FLOW SENSOR
F-2	= COLD DECK FLOW SENSOR
F-3	= SUPPLY AIR FLOW SENSOR
F-4	= RETURN AIR FLOW SENSOR
RC-1	= RECEIVER CONTROLLER
RC-2	= RECEIVER CONTROLLER
RC-3	= RECEIVER CONTROLLER
RC-4	= RECEIVER CONTROLLER
R-1	= HIGH SIGNAL SELECTOR

Dual-Duct VAV Heating and Cooling Air-Handling Unit, Steam Heating Coil, Chilled Water Cooling Coil, Inlet Vanes on Supply and Exhaust Fans, O.A., R.A., and REL Fans Using Economizer Control Cycle, Flow Control on Both Fans

Whenever the fan runs, the EP-1 is energized and O.A., R.A., and REL dampers are placed under automatic controls. When the fan stops, the dampers return to their normal positions.

Duct thermostats T-1 and T-2 in the mix chamber control the O.A., R.A., and REL dampers and act as economizer thermostats. They can allow for up to 100% O.A. when conditions are favorable. Receiver controller RC-1 with sensor T-3 in the hot air duct controls the valve V-1 on the heating coil. Receiver controller RC-2 with sensor in the cold duct controls valve V-2 on the cooling coil. Receiver controllers RC-3 and RC-4 with flow sensors F-1 and F-2, through averaging relay R-1, control the inlet vanes on the supply fan. Flow sensor F-3 in the fan discharge resets the control point of receiver controller RC-5, which controls the inlet vane dampers on the exhaust fan. The exhaust fan dampers are controlled with the O.A. and R.A. dampers.

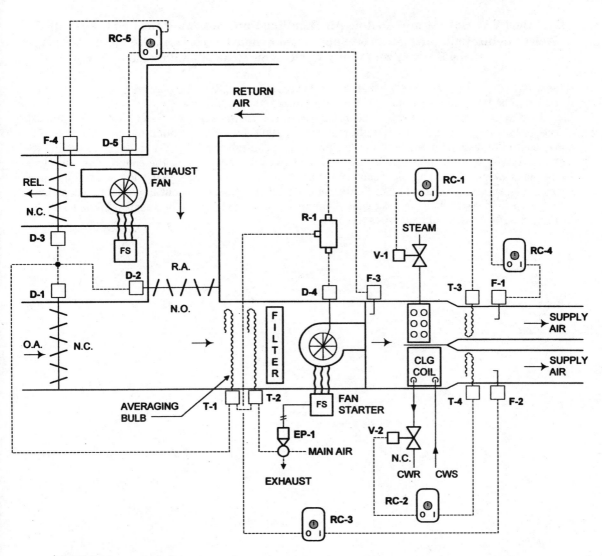

LEGEND:

T-1	=	MODULATING CAPILLARY DUCT THERMOSTAT
T-2	=	MODULATING CAPILLARY DUCT THERMOSTAT
T-3	=	MODULATING CAPILLARY DUCT SENSOR
T-4	=	MODULATING CAPILLARY DUCT SENSOR
V-1	=	N.O. MODULATING STEAM COIL VALVE
V-2	=	N.C. MODULATING 2-WAY COOLING COIL VALVE
D-1	=	OUTSIDE AIR DAMPER MOTOR
D-2	=	RETURN AIR DAMPER MOTOR
D-3	=	RELIEF AIR DAMPER MOTOR
D-4	=	SUPPLY FAN INLET VANE DAMPER MOTOR
D-5	=	EXHAUST FAN INLET VANE DAMPER MOTOR

EP-1	=	SOLENOID AIR VALVE
F-1	=	HOT DECK FLOW SENSOR
F-2	=	COLD DECK FLOW SENSOR
F-3	=	SUPPLY AIR FLOW SENSOR
F-4	=	RETURN AIR FLOW SENSOR
RC-1	=	RECEIVER CONTROLLER
RC-2	=	RECEIVER CONTROLLER
RC-3	=	RECEIVER CONTROLLER
RC-4	=	RECEIVER CONTROLLER
RC-5	=	RECEIVER CONTROLLER
R-1	=	HIGH SIGNAL SELECTOR

Dual-Duct VAV Heating and Cooling Air-Handling Unit, Hot Water Heating Coil, Chilled Water Cooling Coil, Inlet Vanes on Supply and Exhaust Fans, O.A., R.A., and EX Fans Using Economizer Control Cycle, Flow Control on Both Fans

Whenever the fan runs, the EP-1 is energized and O.A., R.A., and REL dampers are placed under automatic controls. When the fan stops, the dampers return to their normal positions.

Duct thermostats T-1 and T-2 in the mix chamber control the O.A., R.A., and EX dampers and act as economizer thermostats. They can allow for up to 100% O.A. when conditions are favorable. Receiver controller RC-1 with sensor T-3 in the hot air duct controls the valve V-1 on the heating coil. Receiver controller RC-2 with sensor in the cold duct controls valve V-2 on the cooling coil. Receiver controllers RC-3 and RC-4 with flow sensors F-1 and F-2, through averaging relay R-1, control the inlet vanes on the supply fan. Flow sensor F-3 in the fan discharge resets the control point of receiver controller RC-5, which controls the inlet vane dampers on the exhaust fan. The exhaust fan dampers are controlled with the O.A. and R.A. dampers.

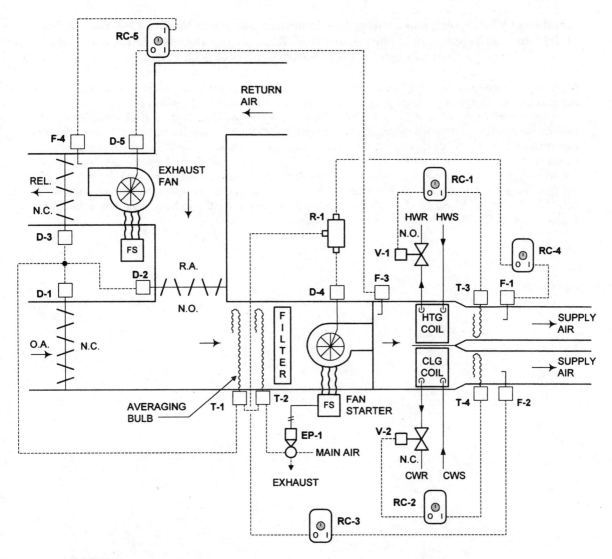

LEGEND:

T-1	=	MODULATING CAPILLARY DUCT THERMOSTAT
T-2	=	MODULATING CAPILLARY DUCT THERMOSTAT
T-3	=	MODULATING CAPILLARY DUCT SENSOR
T-4	=	MODULATING CAPILLARY DUCT SENSOR
V-1	=	N.O. MODULATING 2-WAY HEATING COIL VALVE
V-2	=	N.C. MODULATING 2-WAY COOLING COIL VALVE
D-1	=	OUTSIDE AIR DAMPER MOTOR
D-2	=	RETURN AIR DAMPER MOTOR
D-3	=	RELIEF AIR DAMPER MOTOR
D-4	=	SUPPLY FAN INLET VANE DAMPER MOTOR
D-5	=	EXHAUST FAN INLET VANE DAMPER MOTOR

EP-1	=	SOLENOID AIR VALVE
F-1	=	HOT DECK FLOW SENSOR
F-2	=	COLD DECK FLOW SENSOR
F-3	=	SUPPLY AIR FLOW SENSOR
F-4	=	RETURN AIR FLOW SENSOR
RC-1	=	RECEIVER CONTROLLER
RC-2	=	RECEIVER CONTROLLER
RC-3	=	RECEIVER CONTROLLER
RC-4	=	RECEIVER CONTROLLER
RC-5	=	RECEIVER CONTROLLER
R-1	=	HIGH SIGNAL SELECTOR

Dual-Duct VAV Heating and Cooling Air-Handling Unit, Steam Heating Coil, DX Cooling, Inlet Vanes on Supply and Exhaust Fans, O.A., R.A., and EX Dampers Using Economizer Control Cycle, Static Pressure Control of Both Fans

Whenever the fan runs, the EP-1 is energized and O.A., R.A., and REL dampers are placed under automatic controls. When the fan stops, the dampers return to their normal positions.

Duct thermostats T-1 and T-2 in the mix chamber control the O.A., R.A., and EX dampers and act as economizer thermostats. They can allow for up to 100% O.A. when conditions are favorable. Receiver controller RC-1 with sensor T-3 in the hot air duct controls the valve V-1 on the heating coil. Receiver controller RC-2 with sensor in the cold duct controls PE switch PE-1 on the DX cooling coil. Receiver controllers RC-3 and RC-4 with flow sensors F-1 and F-2, through averaging relay R-1, control the inlet vanes on the supply fan. Flow sensor F-3 in the fan discharge resets the control point of receiver controller RC-5, which controls the inlet vane dampers on the exhaust fan. The exhaust fan dampers are controlled with the O.A. and R.A. dampers.

LEGEND:

T-1	= MODULATING CAPILLARY DUCT THERMOSTAT
T-2	= MODULATING CAPILLARY DUCT THERMOSTAT
T-3	= MODULATING CAPILLARY DUCT SENSOR
T-4	= MODULATING CAPILLARY DUCT SENSOR
V-1	= N.O. MODULATING STEAM COIL VALVE
D-1	= OUTSIDE AIR DAMPER MOTOR
D-2	= RETURN AIR DAMPER MOTOR
D-3	= RELIEF AIR DAMPER MOTOR
D-4	= SUPPLY FAN INLET VANE DAMPER MOTOR
D-5	= EXHAUST FAN INLET VANE DAMPER MOTOR
EP-1	= SOLENOID AIR VALVE

PE-1	= PRESSURE/ELECTRIC SWITCH
F-1	= HOT DECK FLOW SENSOR
F-2	= COLD DECK FLOW SENSOR
F-3	= SUPPLY AIR FLOW SENSOR
F-4	= RETURN AIR FLOW SENSOR
RC-1	= RECEIVER CONTROLLER
RC-2	= RECEIVER CONTROLLER
RC-3	= RECEIVER CONTROLLER
RC-4	= RECEIVER CONTROLLER
R-1	= HIGH SIGNAL SELECTOR

Dual-Duct VAV Heating and Cooling Air-Handling Unit, Hot Water Heating Coil, DX Cooling Inlet Vanes on Supply and Exhaust Fans, O.A., R.A., and EX Dampers Using Economizer Control Cycle, Flow Control of Both Fans

Whenever the fan runs, the EP-1 is energized and O.A., R.A., and REL dampers are placed under automatic controls. When the fan stops, the dampers return to their normal positions.

Duct thermostats T-1 and T-2 in the mix chamber control the O.A., R.A., and EX dampers and act as economizer thermostats. They can allow for up to 100% O.A. when conditions are favorable. Receiver controller RC-1 with sensor T-3 in the hot air duct controls the valve V-1 on the heating coil. Receiver controller RC-2 with sensor in the cold duct controls PE switch PE-1 on the DX cooling coil. Receiver controllers RC-3 and RC-4 with flow sensors F-1 and F-2, through averaging relay R-1, control the inlet vanes on the supply fan. Flow sensor F-3 in the fan discharge resets the control point of receiver controller RC-5, which controls the inlet vane dampers on the exhaust fan. The exhaust fan dampers are controlled with the O.A. and R.A. dampers.

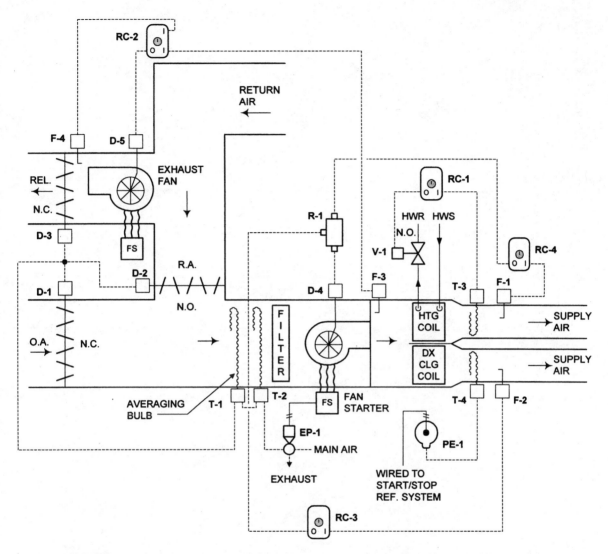

LEGEND:

T-1	= MODULATING CAPILLARY DUCT THERMOSTAT
T-2	= MODULATING CAPILLARY DUCT THERMOSTAT
T-3	= MODULATING CAPILLARY DUCT SENSOR
T-4	= MODULATING CAPILLARY DUCT SENSOR
V-1	= N.O. MODULATING 2-WAY HEATING COIL VALVE
D-1	= OUTSIDE AIR DAMPER MOTOR
D-2	= RETURN AIR DAMPER MOTOR
D-3	= RELIEF AIR DAMPER MOTOR
D-4	= SUPPLY FAN INLET VANE DAMPER MOTOR
D-5	= EXHAUST FAN INLET VANE DAMPER MOTOR
EP-1	= SOLENOID AIR VALVE

PE-1	= PRESSURE/ELECTRIC SWITCH
F-1	= HOT DECK FLOW SENSOR
F-2	= COLD DECK FLOW SENSOR
F-3	= SUPPLY AIR FLOW SENSOR
F-4	= RETURN AIR FLOW SENSOR
RC-1	= RECEIVER CONTROLLER
RC-2	= RECEIVER CONTROLLER
RC-3	= RECEIVER CONTROLLER
RC-4	= RECEIVER CONTROLLER
R-1	= HIGH SIGNAL SELECTOR

CHAPTER 20

Special Systems

Constant Temperature, Constant Humidity Air-Handling Unit, Chilled Water Cooling (Dehumidification) Hot Water Reheat, Steam Humidifier, O.A., R.A., and REL Dampers Using Mixed Air Control Cycle from Room Thermostat

Whenever the fan runs, the EP-1 is energized and O.A., R.A., and REL dampers are placed under automatic controls. When the fan stops, the dampers return to their normal positions.

Room thermostat T-1 controls the chilled water cooling coil valve V-1 and the reheat coil valve V-2 through relay R-1. Humidistat H-1 also controls the cooling coil through relay R-1, so that either the thermostat or the humidistat can call for cooling. The thermostat can call for reheat through reheat coil valve V-2 if the humidistat calls for too much cooling trying to dehumidify. The humidistat also controls the humidifier valve V-3 through high-limit humidistat H-2 to add humidity when required. Thermostat T-1 also controls in sequence the O.A., R.A., and REL dampers to maintain a fixed mix chamber temperature. Minimum position switch S-1 sets the minimum percentage of O.A., regardless of the actions of other controllers as long as the fan is running. The heating coil is a "reheat" coil and must be after the cooling coil.

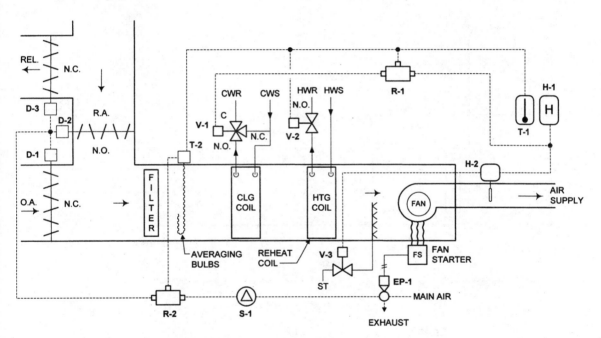

LEGEND:

T-1	=	MODULATING ROOM THERMOSTAT
T-2	=	MODULATING CAPILLARY DUCT THERMOSTAT
H-1	=	MODULATING ROOM HUMIDISTAT
H-2	=	HIGH LIMIT DUCT HUMIDISTAT
V-1	=	3-WAY CHILLED WATER COIL MIXING VALVE
V-2	=	N.O. REHEAT COIL HOT WATER VALVE
V-3	=	N.C. MODULATING HUMIDIFIER STEAM VALVE
D-1	=	OUTSIDE AIR DAMPER MOTOR
D-2	=	RETURN AIR DAMPER MOTOR
D-3	=	RELIEF AIR DAMPER MOTOR
EP-1	=	SOLENOID AIR VALVE
R-1	=	HIGH SIGNAL SELECTOR
R-2	=	HIGH SIGNAL SELECTOR
S-1	=	MINIMUM POSITION SWITCH

Constant Temperature, Constant Humidity Air-Handling Unit, Chilled Water Cooling (Dehumidification) Steam Reheat, Steam Humidifier, O.A., R.A., and REL Dampers Using Mixed Air Control Cycle from Room Thermostat

Whenever the fan runs, the EP-1 is energized and O.A., R.A., and REL dampers are placed under automatic controls. When the fan stops, the dampers return to their normal positions.

Room thermostat T-1 controls the chilled water cooling coil and the reheat coil through relay R-1. Humidistat H-1 also controls the cooling coil through relay R-1, so that either the thermostat or the humidistat can call for cooling. The thermostat can call for reheat through reheat coil valve V2 if the humidistat calls for too much cooling trying to dehumidify. The humidistat also controls the humidifier valve V-3 through high-limit humidistat H-2 to add humidity when required. Thermostat T-1 also controls in sequence the O.A., R.A., and REL dampers to maintain a fixed mix chamber temperature. Minimum position switch S-1 sets the minimum percentage of O.A., regardless of the actions of other controllers as long as the fan is running. The heating coil is a "reheat" coil and must be after the cooling coil.

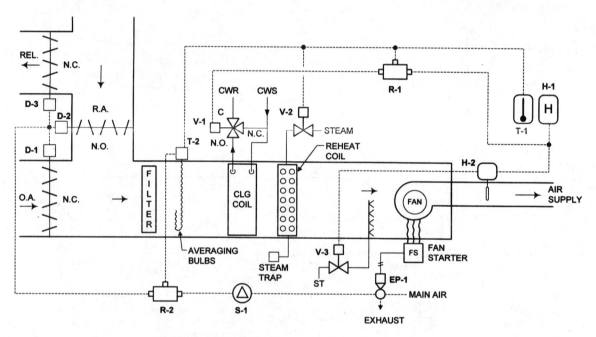

LEGEND:

T-1	= MODULATING ROOM THERMOSTAT
T-2	= MODULATING CAPILLARY DUCT THERMOSTAT
H-1	= MODULATING ROOM HUMIDISTAT
H-2	= HIGH LIMIT DUCT HUMIDISTAT
V-1	= 3-WAY CHILLED WATER COIL MIXING VALVE
V-2	= N.O. REHEAT COIL STEAM VALVE
V-3	= N.C. MODULATING HUMIDIFIER STEAM VALVE
D-1	= OUTSIDE AIR DAMPER MOTOR
D-2	= RETURN AIR DAMPER MOTOR
D-3	= RELIEF AIR DAMPER MOTOR
EP-1	= SOLENOID AIR VALVE
R-1	= HIGH SIGNAL SELECTOR
R-2	= HIGH SIGNAL SELECTOR
S-1	= MINIMUM POSITION SWITCH

Constant Temperature, Constant Humidity Air-Handling Unit, DX Cooling (Dehumidification) Hot Water Reheat, Steam Humidifier, O.A., R.A., and REL Dampers Using Mixed Air Control Cycle from Room Thermostat

Whenever the fan runs, the EP-1 is energized and O.A., R.A., and REL dampers are placed under automatic controls. When the fan stops, the dampers return to their normal positions.

Room thermostat T-1 controls the DX cooling coil through PE switch PE-1 and the reheat coil valve V-2 through relay R-1. Humidistat H-1 also controls the cooling coil through relay R-1, so that either the thermostat or the humidistat can call for cooling. The thermostat can call for reheat through reheat coil valve V-2 if the humidistat calls for too much cooling trying to dehumidify. The humidistat also controls the humidifier valve V-2 through high-limit humidistat H-2 to add humidity when required. Thermostat T-1 also controls in sequence the O.A., R.A., and REL dampers to maintain a fixed mix chamber temperature. Minimum position switch S-1 sets the minimum percentage of O.A., regardless of the actions of other controllers as long as the fan is running. The heating coil is a "reheat" coil and must be after the cooling coil.

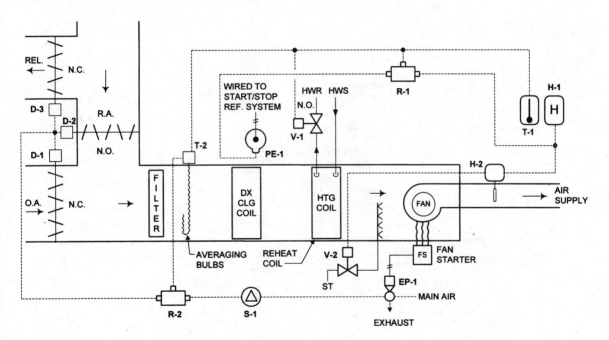

LEGEND:

T-1	=	MODULATING ROOM THERMOSTAT
T-2	=	MODULATING CAPILLARY DUCT THERMOSTAT
H-1	=	MODULATING ROOM HUMIDISTAT
H-2	=	HIGH LIMIT DUCT HUMIDISTAT
V-1	=	N.O. REHEAT COIL HOT WATER VALVE
V-2	=	N.C. MODULATING HUMIDIFIER STEAM VALVE
D-1	=	OUTSIDE AIR DAMPER MOTOR
D-2	=	RETURN AIR DAMPER MOTOR
D-3	=	RELIEF AIR DAMPER MOTOR
EP-1	=	SOLENOID AIR VALVE
PE-1	=	PRESSURE/ELECTRIC SWITCH
R-1	=	HIGH SIGNAL SELECTOR
R-2	=	HIGH SIGNAL SELECTOR
S-1	=	MINIMUM POSITION SWITCH

Constant Temperature, Constant Humidity Air-Handling Unit, DX Cooling (Dehumidification) Steam Reheat, Steam Humidifier, O.A., R.A., and REL Dampers Using Mixed Air Control Cycle from Room Thermostat

Whenever the fan runs, the EP-1 is energized and O.A., R.A., and REL dampers are placed under automatic controls. When the fan stops, the dampers return to their normal positions.

Room thermostat T-1 controls the DX cooling coil through PE switch PE-1 and the reheat coil valve V-2 through relay R-1. Humidistat H-1 also controls the cooling coil through relay R-1, so that either the thermostat or the humidistat can call for cooling. The thermostat can call for reheat through reheat coil valve V-2 if the humidistat calls for too much cooling trying to dehumidify. The humidistat also controls the humidifier valve V-2 through high-limit humidistat H-2 to add humidity when required. Thermostat T-1 also controls in sequence the O.A., R.A., and REL dampers to maintain a fixed mix chamber temperature. Minimum position switch S-1 sets the minimum percentage of O.A., regardless of the actions of other controllers as long as the fan is running. The heating coil is a "reheat" coil and must be after the cooling coil.

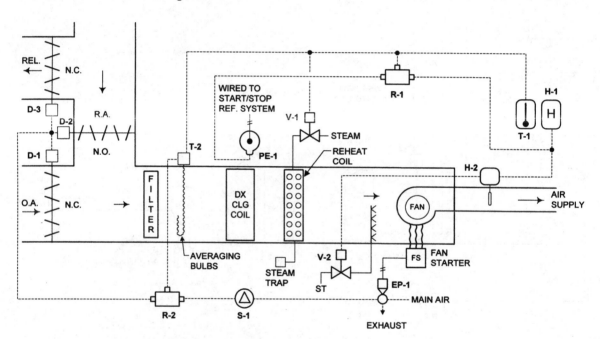

LEGEND:

T-1	= MODULATING ROOM THERMOSTAT
T-2	= MODULATING CAPILLARY DUCT THERMOSTAT
H-1	= MODULATING ROOM HUMIDISTAT
H-2	= HIGH LIMIT DUCT HUMIDISTAT
V-1	= N.O. REHEAT COIL STEAM VALVE
V-2	= N.C. MODULATING HUMIDIFIER STEAM VALVE
D-1	= OUTSIDE AIR DAMPER MOTOR
D-2	= RETURN AIR DAMPER MOTOR
D-3	= RELIEF AIR DAMPER MOTOR
EP-1	= SOLENOID AIR VALVE
PE-1	= PRESSURE/ELECTRIC SWITCH
R-1	= HIGH SIGNAL SELECTOR
R-2	= HIGH SIGNAL SELECTOR
S-1	= MINIMUM POSITION SWITCH

Constant Temperature, Constant Humidity Air-Handling Unit, Chilled Water Cooling, Hot Water Reheat, Steam Humidifier, O.A., R.A., and REL Dampers Using Economizer Control Cycle

Whenever the fan runs, the EP-1 is energized and O.A., R.A., and REL dampers are placed under automatic controls. When the fan stops, the dampers return to their normal positions.

Room thermostat T-1 controls the cooling coil valve V-1 and reheat coil valve V-2 through relay R-1. Humidistat H-1 also controls the cooling coil through relay R-1, so that either the thermostat or the humidistat can call for cooling. The thermostat can call for reheat through reheat coil valve V-2 if the humidistat calls for too much cooling trying to dehumidify. The humidistat also controls the humidifier valve V-3 through high-limit humidistat H-2 to add humidity when required. Thermostats T-2 and T-3 control the O.A., R.A., and REL dampers. They act as economizer thermostats that can allow up to 100% O.A. when conditions are favorable. Minimum position switch S-1 sets the minimum percentage of O.A., regardless of the actions of other controllers as long as the fan is running. The heating coil is a "reheat" coil and must be after the cooling coil.

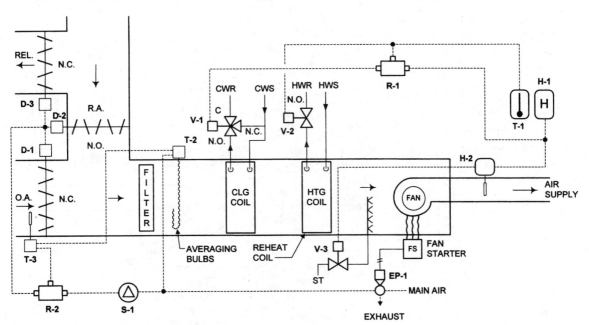

LEGEND:

T-1	=	MODULATING ROOM THERMOSTAT
T-2	=	MODULATING CAPILLARY DUCT THERMOSTAT
T-3	=	MODULATING CAPILLARY THERMOSTAT
H-1	=	MODULATING ROOM HUMIDISTAT
H-2	=	HIGH LIMIT DUCT HUMIDISTAT
V-1	=	3-WAY CHILLED WATER COIL MIXING VALVE
V-2	=	N.O. REHEAT COIL HOT WATER VALVE
V-3	=	N.C. MODULATING HUMIDIFIER STEAM VALVE
D-1	=	OUTSIDE AIR DAMPER MOTOR
D-2	=	RETURN AIR DAMPER MOTOR
D-3	=	RELIEF AIR DAMPER MOTOR
EP-1	=	SOLENOID AIR VALVE
R-1	=	HIGH SIGNAL SELECTOR
R-2	=	HIGH SIGNAL SELECTOR
S-1	=	MINIMUM POSITION SWITCH

Constant Temperature, Constant Humidity Air-Handling Unit, Chilled Water Cooling, Steam Reheat, Steam Humidifier, O.A., R.A., and REL Dampers Using Economizer Control Cycle

Whenever the fan runs, the EP-1 is energized and O.A., R.A., and REL dampers are placed under automatic controls. When the fan stops, the dampers return to their normal positions.

Room thermostat T-1 controls the cooling coil valve V-1 and reheat coil valve V-2 through relay R-1. Humidistat H-1 also controls the cooling coil through relay R-1, so that either the thermostat or the humidistat can call for cooling. The thermostat can call for reheat through reheat coil valve V-2 if the humidistat calls for too much cooling trying to dehumidify. The humidistat also controls the humidifier valve V-3 through high-limit humidistat H-2 to add humidity when required. Thermostats T-2 and T-3 control the O.A., R.A., and REL dampers. They act as economizer thermostats that can allow up to 100% O.A. when conditions are favorable. Minimum position switch S-1 sets the minimum percentage of O.A., regardless of the actions of other controllers as long as the fan is running. The heating coil is a "reheat" coil and must be after the cooling coil.

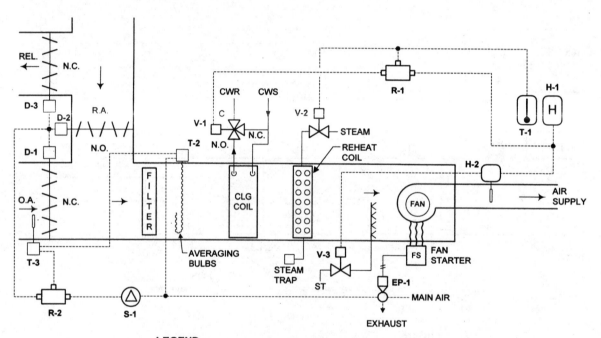

LEGEND:

T-1 = MODULATING ROOM THERMOSTAT
T-2 = MODULATING CAPILLARY DUCT THERMOSTAT
T-3 = MODULATING CAPILLARY THERMOSTAT
H-1 = MODULATING ROOM HUMIDISTAT
H-2 = HIGH LIMIT DUCT HUMIDISTAT
V-1 = 3-WAY CHILLED WATER COIL MIXING VALVE
V-2 = N.O. REHEAT COIL STEAM VALVE
V-3 = N.C. MODULATING HUMIDIFIER STEAM VALVE
D-1 = OUTSIDE AIR DAMPER MOTOR
D-2 = RETURN AIR DAMPER MOTOR
D-3 = RELIEF AIR DAMPER MOTOR
EP-1 = SOLENOID AIR VALVE
R-1 = HIGH SIGNAL SELECTOR
R-2 = HIGH SIGNAL SELECTOR
S-1 = MINIMUM POSITION SWITCH

Constant Temperature, Constant Humidity Air-Handling Unit, DX Cooling, Hot Water Reheat, Steam Humidifier, O.A., R.A., and REL Dampers Using Economizer Control Cycle

Whenever the fan runs, the EP-1 is energized and O.A., R.A., and REL dampers are placed under automatic controls. When the fan stops, the dampers return to their normal positions.

Room thermostat T-1 controls the cooling coil through PE switch PE-1 and reheat coil valve V-2 through relay R-1. Humidistat H-1 also controls the cooling coil through relay R-1, so that either the thermostat or the humidistat can call for cooling. The thermostat can call for reheat through reheat coil valve V-1 if the humidistat calls for too much cooling trying to dehumidify. The humidistat also controls the humidifier valve V-2 through high-limit humidistat H-2 to add humidity when required. Thermostats T-2 and T-3 control the O.A., R.A., and REL dampers. They act as economizer thermostats that can allow up to 100% O.A. when conditions are favorable. Minimum position switch S-1 sets the minimum percentage of O.A., regardless of the actions of other controllers as long as the fan is running. The heating coil is a "reheat" coil and must be after the cooling coil.

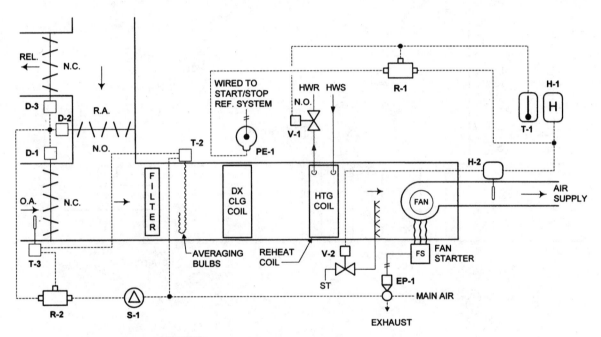

LEGEND:

T-1	= MODULATING ROOM THERMOSTAT
T-2	= MODULATING CAPILLARY DUCT THERMOSTAT
T-3	= MODULATING CAPILLARY THERMOSTAT
H-1	= MODULATING ROOM HUMIDISTAT
H-2	= HIGH LIMIT DUCT HUMIDISTAT
V-1	= N.O. REHEAT COIL HOT WATER VALVE
V-2	= N.C. MODULATING HUMIDIFIER STEAM VALVE
D-1	= OUTSIDE AIR DAMPER MOTOR
D-2	= RETURN AIR DAMPER MOTOR
D-3	= RELIEF AIR DAMPER MOTOR
EP-1	= SOLENOID AIR VALVE
PE-1	= PRESSURE/ELECTRIC SWITCH
R-1	= HIGH SIGNAL SELECTOR
R-2	= HIGH SIGNAL SELECTOR
S-1	= MINIMUM POSITION SWITCH

Constant Temperature, Constant Humidity Air-Handling Unit, DX Cooling, Steam Reheat, Steam Humidifier, O.A., R.A., and REL Dampers Using Economizer Control Cycle

Whenever the fan runs, the EP-1 is energized and O.A., R.A., and REL dampers are placed under automatic controls. When the fan stops, the dampers return to their normal positions.

Room thermostat T-1 controls the cooling coil through PE switch PE-1 and reheat coil valve V-1 through relay R-1. Humidistat H-1 also controls the cooling coil through relay R-1, so that either the thermostat or the humidistat can call for cooling. The thermostat can call for reheat through reheat coil valve V-1 if the humidistat calls for too much cooling trying to dehumidify. The humidistat also controls the humidifier valve V-2 through high-limit humidistat H-2 to add humidity when required. Thermostats T-2 and T-3 control the O.A., R.A., and REL dampers. They act as economizer thermostats that can allow up to 100% O.A. when conditions are favorable. Minimum position switch S-1 sets the minimum percentage of O.A., regardless of the actions of other controllers as long as the fan is running. The heating coil is a "reheat" coil and must be after the cooling coil.

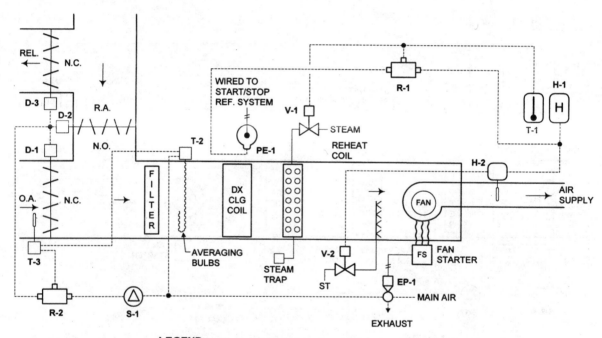

LEGEND:

T-1	= MODULATING ROOM THERMOSTAT
T-2	= MODULATING CAPILLARY DUCT THERMOSTAT
T-3	= MODULATING CAPILLARY THERMOSTAT
H-1	= MODULATING ROOM HUMIDISTAT
H-2	= HIGH LIMIT DUCT HUMIDISTAT
V-1	= N.O. REHEAT COIL STEAM VALVE
V-2	= N.C. MODULATING HUMIDIFIER STEAM VALVE
D-1	= OUTSIDE AIR DAMPER MOTOR
D-2	= RETURN AIR DAMPER MOTOR
D-3	= RELIEF AIR DAMPER MOTOR
EP-1	= SOLENOID AIR VALVE
PE-1	= PRESSURE/ELECTRIC SWITCH
R-1	= HIGH SIGNAL SELECTOR
R-2	= HIGH SIGNAL SELECTOR
S-1	= MINIMUM POSITION SWITCH

Control of O.A., R.A., and REL Dampers in Any Air-Handling Unit, through the Use of Minimum and Maximum Outdoor Dampers, Rest of the System (Coils, etc. Not Shown for Clarity)

Whenever the fan runs, an EP is energized and O.A., R.A., and REL dampers are placed under automatic controls. When the fan stops, the dampers return to their normal positions.

The system shown uses, instead of one O.A. damper, a set of two dampers. One is a minimum set for the ventilation air requirements and the other the maximum O.A. dampers used with, as an example, a thermostat T-1 in the mix plenum to control the maximum return air and relief air dampers.

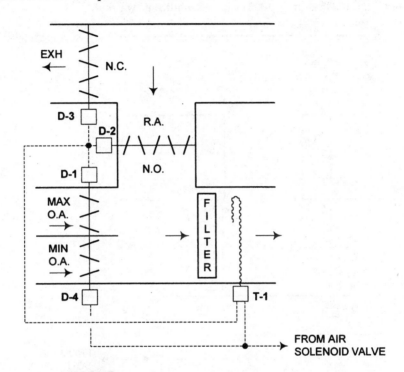

LEGEND:

T-1	=	MODULATING CAPILLARY THERMOSTAT
D-1	=	MAXIMUM OUTSIDE AIR DAMPER MOTOR
D-2	=	RETURN AIR DAMPER MOTOR
D-3	=	EXHAUST AIR DAMPER MOTOR
D-4	=	MINIMUM OUTSIDE AIR DAMPER MOTOR

Control of O.A., R.A., and REL Dampers in Any Air-Handling Unit, through the Use of Minimum and Maximum Outdoor Dampers, Rest of the System (Coils, etc., Not Shown for Clarity), Economizer Control Cycle Shown

Whenever the fan runs, an EP is energized and O.A., R.A., and REL dampers are placed under automatic controls. When the fan stops, the dampers return to their normal positions.

The system shown uses, instead of one O.A. damper, a set of two dampers. One is a minimum set for the ventilation air requirements and the other the maximum O.A. dampers used with, as an example, thermostat T-1 and thermostat T-2 that sense mixed air temperature and act as economizer thermostats. T-1 and T-2 can admit up to 100% O.A. as conditions warrant.

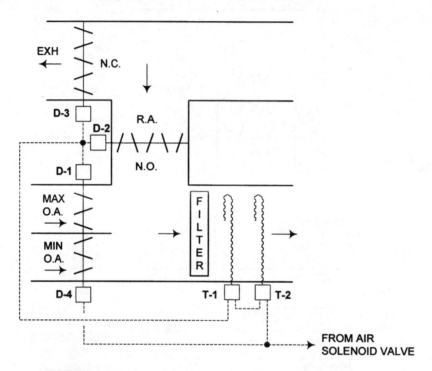

LEGEND:

T-1 = MODULATING CAPILLARY THERMOSTAT
T-2 = MODULATING CAPILLARY THERMOSTAT
D-1 = MAXIMUM OUTSIDE AIR DAMPER MOTOR
D-2 = RETURN AIR DAMPER MOTOR
D-3 = EXHAUST AIR DAMPER MOTOR
D-4 = MINIMUM OUTSIDE AIR DAMPER MOTOR

Capacity Control of DX Cooling Coil (Evaporator) Using Hot Gas Bypass to Reduce Coil Capacity

Room thermostat T-1 controls the solenoid valves V-1 and V-2 on the cooling coil. EV-1 is the expansion valve on the DX coil, and when reduced capacity in the space is required, T-1 opens the solenoid valve V-2 and closes valve V-1 to pass hot gas back to the DX cooling coil, thereby reducing its capacity.

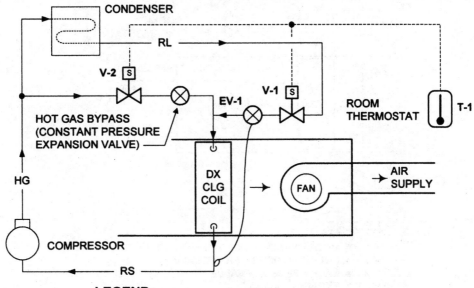

LEGEND:

T-1 = MODULATING ROOM THERMOSTAT
V-1 = SOLENOID REFRIGERANT VALVE
V-2 = SOLENOID REFRIGERANT VALVE
EV-1 = EXPANSION VALVE
RL   = REFRIGERANT LIQUID LINE
RS   = REFRIGERANT SUCTION LINE
HG   = HOT GAS LINE

Capacity Control of DX Cooling Coil (Evaporator) Using Back Pressure Control Valve to Reduce Coil Capacity

Room thermostat T-1 controls the solenoid valve V-4 and, through relay R-1 back pressure regulator V-2, maintains coil capacity. Expansion valve EV-1 operates the DX coil as normal.

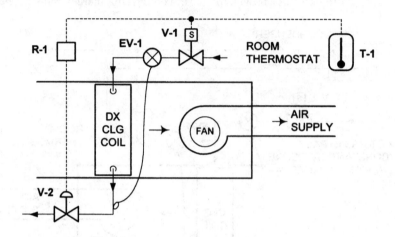

LEGEND:

T-1 = MODULATING ROOM THERMOSTAT
V-1 = SOLENOID REFRIGERANT VALVE
V-2 = BACK PRESSURE VALVE
EV-1 = EXPANSION VALVE
R-1 = RELAY

Control of Space Humidity with Preheat Coil and Spray System

Room humidistat H-1 controls the valve V-1 on the preheat coil before the 100% O.A. reaches the spray coil. The humidistat also controls the spray pump to keep the air stream saturated. The air from the spray coil, which is 100% saturated, is sent to a reheat coil, which if controlled by the humidistat can now control the humidity.

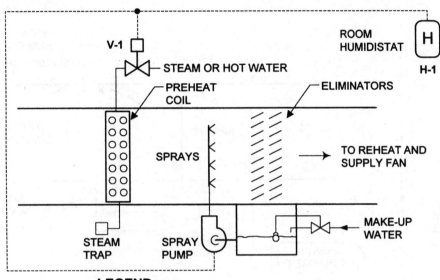

LEGEND:

H-1 = MODULATING ROOM HUMIDISTAT
V-1 = PREHEAT COIL VALVE

Control of Space Humidity with Hot Water and Chilled Water Exchangers Supplying Water to Evaporative Sprays

Room humidistat H-1 controls the valves V-1 and V-2 on the heat exchangers (heating and cooling water exchangers) to maintain proper water temperature to the sprays and allow for the reheat system to control the space humidity.

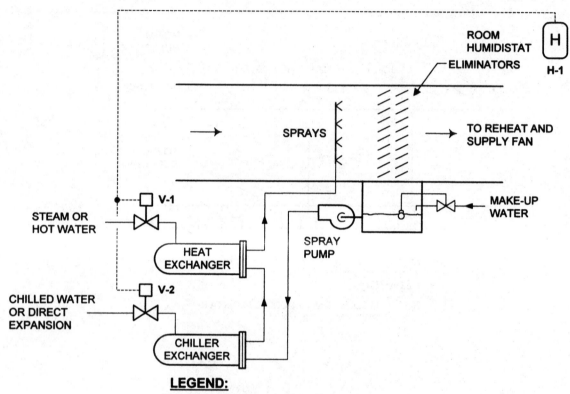

LEGEND:

H-1 = MODULATING ROOM HUMIDISTAT
V-1 = HEAT EXCHANGER STEAM OR HW VALVE
V-2 = CHILLER EXCHANGER VALVE

Control of Chemical Dehumidifier with Steam or Hot Water Regenerator and Chilled Water Coil

Room thermostat T-1 controls the valve V-1 on the chilled water coil that is cooling the air after it leaves the chemical dehumidifier. The (Kathabar) dehumidifier dehumidifies the air but also heats it up. The room humidistat H-1 controls the valve V-2 on the heating coil, which dries out the chemicals in the dehumidifier, supplying the space with dry air.

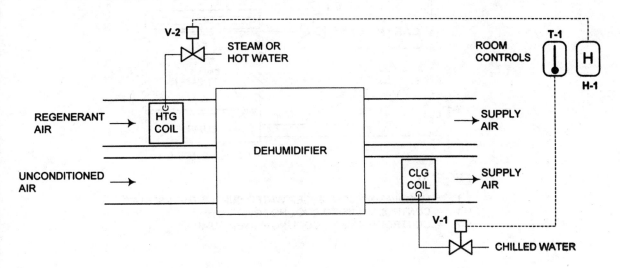

LEGEND:

T-1 = MODULATING ROOM THERMOSTAT
H-1 = MODULATING ROOM HUMIDISTAT
V-1 = CHILLED WATER COIL VALVE
V-2 = STEAM OR HOT WATER COIL VALVE

Cooling Tower Controls for Maintaining Condenser Water Temperature

Condenser water supply temperature from the cooling tower is sensed by T-1, which controls valves V-1 and V-2 to send water over the tower or bypass the tower and send water to the sump. T-1 also can control the fans through a VFD, as an example.

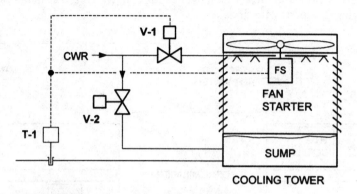

LEGEND:

T-1 = MODULATING CONDENSER WATER SUPPLY THERMOSTAT
V-1 = CONTROL VALVE TO COOLING TOWER
V-2 = CONTROL VALVE TO COOLING TOWER SUMP

Cooling Tower Controls for Maintaining Condenser Water Temperature Using Three-Way Bypass Valve

Condenser water supply temperature from the cooling tower is sensed by T-1, which controls bypass valve V-1 to send water over the tower or bypass the tower and send water back to the condenser. T-1 also can control the fans through a VFD, as an example.

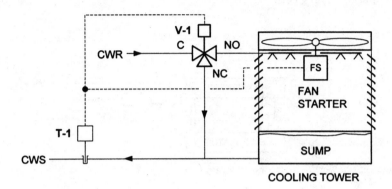

LEGEND:

T-1   = MODULATING CONDENSER WATER SUPPLY THERMOSTAT
V-1   = 3-WAY DIVERTING VALVE

Control of Heat Pump Supplying Chilled Water for Cooling and Hot Water for Heating

Condenser water supply temperature from the chiller condenser in this heat pump application is controlled by thermostat T-1, controlling the inlet vanes on the chiller. Room thermostat T-3 controls valve V-1 on the heating coil in the exterior spaces. Chilled water supply lines to the cooling coil are controlled only by low-limit thermostat T-2 in the chilled water supply line. Room thermostat T-4 controls valve V-2 on the interior cooling coil.

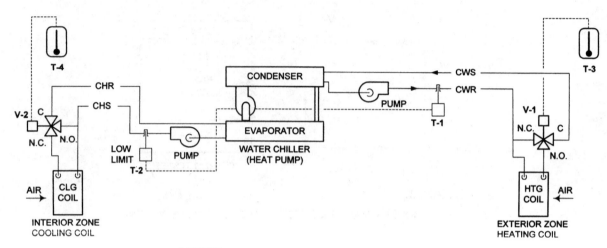

LEGEND:

T-1 = MODULATING CONDENSER WATER RETURN THERMOSTAT
T-2 = MODULATING CHILLED WATER THERMOSTAT
T-3 = MODULATING ROOM THERMOSTAT
T-4 = MODULATING ROOM THERMOSTAT
V-1 = MODULATING 3-WAY HEATING COIL VALVE
V-2 = MODULATING 3-WAY COOLING COIL VALVE

Control of Heat Recovery System Supplying Hot Water for Heating and Chilled Water from Cooling Using Double-Bundled Condenser

This split condenser system allows part of the hot condenser water to do heating when there is excess water from the system. The condenser has two separate bundles in it. One of them goes to the cooling tower as usual, and the other goes to a heating coil in the space. T-3 controls valve V-1 on the heating coil. T-4 controls the valve V-2 on the conventional cooling coil. T-2 controls the chiller as usual by operating the inlet vanes. Thermostat T-1 controls the supplemental heater and cooling tower pump when the system requires that all the condenser water go to the cooling tower.

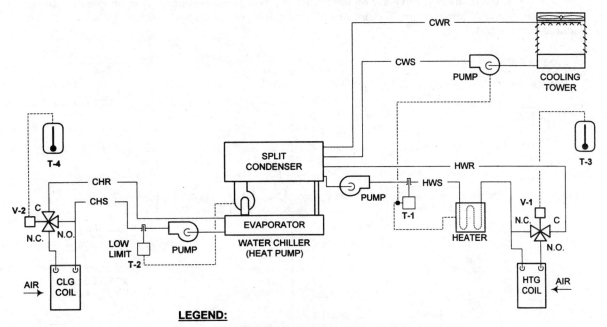

LEGEND:

T-1	= MODULATING CONDENSER WATER RETURN THERMOSTAT
T-2	= MODULATING CHILLED WATER THERMOSTAT
T-3	= MODULATING ROOM THERMOSTAT
T-4	= MODULATING ROOM THERMOSTAT
V-1	= MODULATING 3-WAY HEATING COIL VALVE
V-2	= MODULATING 3-WAY COOLING COIL VALVE

Typical Two-Pipe Control System with Boiler, Chiller, and Heat Exchangers for Hot Water/Chilled Water Switch-Over Valves; Heat Exchangers Isolate Boiler and Chiller from Hot Water and Chilled Water Lines

This two-pipe HW and CW system uses a common pipe to supply units with hot water in the heating season and chilled water in the cooling season. There is a boiler and a chiller. An O.A. thermostat T-1 resets the control point of submaster thermostat T-2, which controls valve V-1 on a on the boiler. T-2 also controls valve V-2 on the chilled water system. The purpose here is to give a variable hot water temperature in accordance with O.A. temperature by controlling T-3 in the boiler. In effect, T-2 controls the valves V-1 on the boiler and valve V-2 on the chiller to give a variable supply water temperature in accordance with O.A. temperature. T-1 also does the switching of the valves to isolate the hot water and chilled water lines.

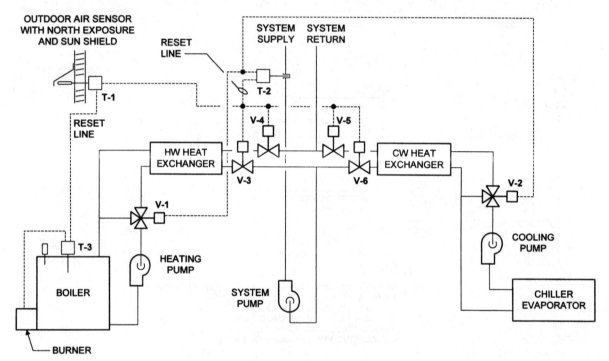

LEGEND:

T-1 = OUTDOOR AIR SWITCHOVER THERMOSTAT
T-2 = HW/CW SUPPLY LINE THERMOSTAT
T-3 = BOILER CONTROL SUBMASTER THERMOSTAT
T-4 = MODULATING ROOM THERMOSTAT
V-1 = HEATING SUPPLY LINE 3-WAY VALVE
V-2 = CHILLED WATER SUPPLY LINE 3-WAY VALVE
V-3 = HW SWITCHOVER VALVE
V-4 = HW SWITCHOVER VALVE
V-5 = CW SWITCHOVER VALVE
V-6 = CW SWITCHOVER VALVE

Three-Pipe Control System with Boiler, Chiller, and One Heat Exchanger for Hot Water Supply

This three-pipe HW and CW system uses a common return line and separate chilled water and hot water supply lines. O.A. thermostat T-2 resets the control point of T-3, which controls valve V-1 on the hot water supply to the heat exchanger. This gives a variable hot water temperature in accordance with O.A. temperature.

The chiller supplies a constant chilled water temperature to the system. This system allows chilled water and hot water to be supplied at all times.

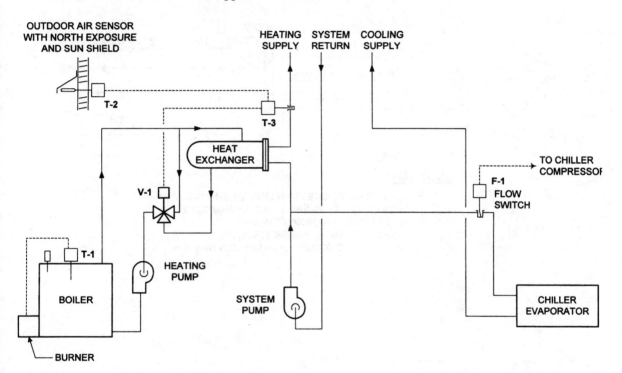

LEGEND:

T-1   = BOILER CONTROL THERMOSTAT
T-2   = OUTDOOR MASTER THERMOSTAT
T-3   = HOT WATER SUBMASTER THERMOSTAT
V-1   = 3-WAY HOT WATER VALVE
F-1   = CHILLED WATER FLOW SWITCH

Typical Control of Water Chillers

This shows the typical control of a chilled water chiller exchanger with thermostat T-1 in the return line to the chiller, controlling through a low-limit thermostat and a flow switch the solenoid valve V-1 on the DX chiller vessel.

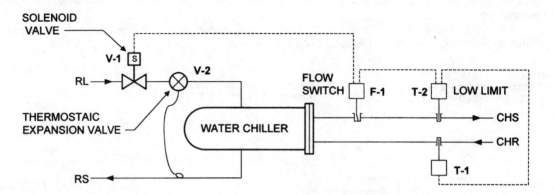

LEGEND:

T-1 = CHILLED WATER RETURN LINE THERMOSTAT
T-2 = CHILLED WATER SUPPLY LINE THERMOSTAT
V-1 = SOLENOID REFRIGERANT VALVE
V-2 = THERMOSTATIC EXPANSION VALVE
F-1 = CHILLED WATER SUPPLY LINE FLOW SWITCH

Typical Constant Flow Chilled Water Systems with Multiple Chillers

This shows the typical control of a chilled water system with multiple chillers and primary pumps only. T-1 controls the 3-way valve on a typical chilled water coil. There may be multiple coils. The valve in this case must be a 3-way valve.

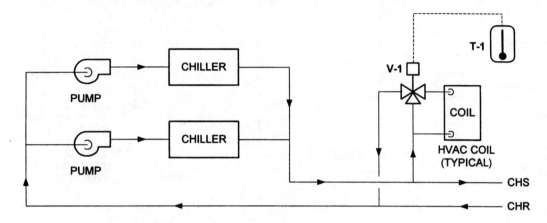

LEGEND:

T-1   = MODULATING ROOM THERMOSTAT
V-1   = MODULATING 3-WAY COIL VALVE

Typical Variable Volume Chilled Water Piping with Multiple Chillers and Primary Pumps

This shows the typical control of a chilled water system with multiple chillers and variable-speed primary pumps. T-1 controls 2-way valve V-1 on a typical chilled water coil. There may be multiple coils. The valve in this case must be a 2-way valve. Differential pressure controller DP-1 controls valve V-2 to maintain a correct differential pressure across the pumps as the system valves close off.

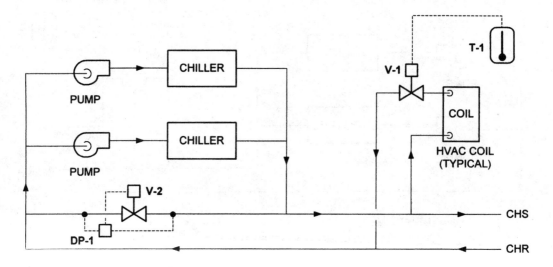

LEGEND:

T-1   = MODULATING ROOM THERMOSTAT
V-1   = 2-WAY COIL VALVE
V-2   = CHILLED WATER SUPPLY BYPASS VALVE
DP-1 = DIFFERENTIAL PRESSURE CONTROLLER

Glossary

ASHRAE American Society of Heating Refrigerating and Air Conditioning Engineers.

AX Automatic expansion.

BLDG Building.

CLG Cooling.

CW Chilled water.

CWR Chilled water return.

CWS Chilled water supply.

DP Differential pressure.

DPDT Double pole double throw.

DPST Double pole single throw.

DX Direct expansion.

EXH Exhaust.

FS Fan starter.

HG Hot gas.

HTG Heating.

HW Hot water.

HWR Hot water return.

HWS Hot water supply.

MA Main air.

NC Normally closed.

NO Normally open.

OA Outdoor air.

RA Return air.

REF Refrigeration.

REL Relief.

SPDT Single pole double throw.

SPST Single pole single throw.

VAV Variable air volume.

About the Author

John I. Levenhagen, P.E., is the author of several respected books on HVAC controls, including McGraw-Hill's *HVAC Controls and Systems* (1992), coauthored with Don Spethmann. A widely known lecturer and consultant in HVAC control design engineering, Mr. Levenhagen is a former member of the board of directors of the American Society of Heating, Refrigerating, and Air Conditioning Engineers. He has worked for a number of control firms, including Johnson Controls, where he was a market manager for a decade.